APPLIED MATHEMATICS
AND SCIENTIFIC
COMPUTING

APPLIED MATHEMATICS AND SCIENTIFIC COMPUTING

Edited by

Zlatko Drmač
Vjeran Hari
University of Zagreb
Zagreb, Croatia

Luka Sopta
University of Rijeka
Rijeka, Croatia

Zvonimir Tutek
University of Zagreb
Zagreb, Croatia

and

Krešimir Veselić
University of Hagen
Hagen, Germany

Kluwer Academic / Plenum Publishers
New York, Boston, Dordrecht, London, Moscow

Library of Congress Cataloging-in-Publication Data

Conference on Applied Mathematics and Scientific Computing (2nd: 2001: Dubrovnik, Croatia)
Applied mathematics and scientific computing/edited by Zlatko Drmac … [et al.].
p. cm.
Includes bibliographical references and index.
ISBN 0-306-47426-3
1. Mathematics—Congresses. 2. Science—Data processing—Congresses. I. Drmac, Zlatko. II. Title.

QA1 .C743 2001
510—dc21

2002034140

Proceedings of the Second Conference on Applied Mathematics and Scientific Computing, held June 4–9, 2001 in Dubrovnik, Croatia

ISBN 0-306-47426-3

233 Spring Street, New York, New York 10013

http://www.wkap.nl/

10 9 8 7 6 5 4 3 2 1

A C.I.P. record for this book is available from the Library of Congress

Printed in the United States of America

Preface

The Second Conference on Applied Mathematics and Scientific Computing was held June 4–9, 2001, at the Inter–University Center Dubrovnik, Croatia. The main idea of the conference was to bring together applied mathematicians both from and outside academia, as well as experts from other areas (engineering, applied sciences) whose work involves advanced mathematical techniques.

During the meeting there were one complete mini–course, invited presentations, contributed talks, and software presentations. A mini–course *Schwarz Methods for Partial Differential Equations* was given by Prof. Marcus Sarkis (Worcester Polytechnic Institute, USA), and invited presentations were given by active researchers from the fields of numerical linear algebra, computational fluid dynamics, matrix theory, and mathematical physics (fluid mechanics and elasticity).

This volume contains the mini–course and review papers by invited speakers (Part I), as well as selected contributed presentations from the fields of analysis, numerical mathematics, and engineering applications (Part II).

The generous support of the Croatian Ministry of Science and Technology is gratefully acknowledged. We are also indebted to the main organizer, Department of Mathematics, University of Zagreb, and the cosponsors, Inter–University Center Dubrovnik, and CV systems, Ltd. Special thanks go to Dr. Josip Tambača for his immeasurable help in organizing the Conference and during the preparation of this volume.

ZLATKO DRMAČ

VJERAN HARI

LUKA SOPTA

ZVONIMIR TUTEK

KREŠIMIR VESELIĆ

Contents

I

INVITED LECTURES

DOMAIN DECOMPOSITION METHODS

Schwarz Methods

Marcus Sarkis

Mathematical Sciences Department, Worcester Polytechnic Institute,
Worcester, MA 01609, USA

Abstract Domain decomposition is a major focus of contemporary research in numerical analysis of partial differential equations. Among the reasons for considering domain decomposition are: parallel computing, modeling of different physical phenomena in different subregions and complicated geometries, and its solid and elegant theoretical foundation. In this text, we provide an introduction to domain decomposition methods. We describe a general variational framework to construct and analyze domain decomposition methods in terms of subspaces and projectionlike operators. This allows a unified analysis of both Schwarz methods (where there is overlap of subregions) and substructuring methods (where there is no overlap). We also discuss important ingredients commonly used in this research area such as inexact solvers, coarse spaces and nonnested spaces. We provide the basics of the abstract Schwarz theory, giving several proofs, in order to demonstrate the requirements needed for designing good preconditioners.

Keywords: Schwarz preconditioner, domain decomposition, coarse spaces, partition of unity, elliptic equations, finite elements, eigenvalues.

1. Introduction

The widespread availability of parallel computers and their potential for the numerical solution of many difficult-to-solve problems, in linear and nonlinear continuum mechanics problems have led to a large body of research in domain decomposition methods; see the proceedings of the international conferences held every year since 1987; see *www.ddm.org*. Domain decomposition methods refer to a collection of techniques which revolve around the principle of divide-and-conquer. Let us assume we want to model an engineering problem on a partitioned domain $\Omega = \cup_{i=1}^{N} \Omega_i$, where the Ω_i may overlap or not. Domain decomposition methods attempt to solve the problem on the entire domain Ω by using solutions on the subdomains Ω_i. There are several reasons why such techniques can be advantageous. One obvious reason is that the subproblems can be small enough to be allocated to individual processors of a parallel or dis-

Edited by Drmač *et al.*, Kluwer Academic/Plenum Publishers, 2002

tributed system. Due to the good localization of the associated data, well-known sequential and fast algorithms can be used locally for each subproblem. The proper treatment of interfacing the local solutions is the key to these methods; see [17, 40, 47]. A second reason is that the domain of the physical problem can sometimes be split naturally into a small number of subregions where the modeling equations are different (e.g. Navier-Stokes's equations on one region and full potential in another) [4, 11, 18, 40, 39, 43]. Solving the full potential equation in regions of irrotational flow is not only more efficient but should also improve the accuracy. A third reason is that the entire domain Ω can sometimes be split naturally into a small number of subregions where the geometry of the Ω_i' is much simpler than that of the Ω. Discretization based on highly structured grids can be applied and corresponding fast solvers (and software) can be used easily in these parts of the domain [1, 2, 3, 9, 12, 16, 29, 30, 32, 50, 48, 49]. Those three techniques can be combined to generate a more general and powerful approach. The underlying structure of domain decomposition methods is therefore very rich; these methods combine geometrical, analytical, and algebraic ingredients. The rich structure can be used both to derive efficient algorithmic procedures as well as to analyze the properties of proposed algorithms in detail.

The main goal of this paper is to describe a general framework, the Schwarz convergence abstract theory (*abstract theory*) for constructing and analyzing the convergence of solvers using the domain decomposition techniques described above. The convergence of such solvers involves two phases: 1) the Schwarz convergence abstract theory, and 2) the verification of the assumptions of the abstract theory. The first phase is common to most of the algorithms. It has a unifying mathematical structure. The second phase depends strongly on the particular problem in consideration. This text is intended to describe the first phase for the symmetric positive definite case problems. In the text, on a step by step basis, we motivate and define the main ingredients of the abstract theory in order to introduce properly its general assumptions. We then prove the fundamental theorems of this theory. The second phase is a problem dependent and we refer to the international BibTex reference database of Domain Decomposition in *www.ddm.org*.

The text is organized as follows. In Section 2, we discuss general issues on iterative methods and preconditioners and motivate the reader for the use of domain decomposition methods. In Section 3, we describe the domain decomposition method proposed by Hermann Amandus Schwarz in 1869 and then introduce its variations and generalizations such as variational formulations, inexact solvers, nonnested meshes and preconditioning. In Section 4 we consider the case of multidomains and discuss the necessity of the so called coarse spaces. In Section 5, we remove the multiplicative aspect of the Schwarz algorithms and introduce the additive Schwarz methods. In Section 6, we consider some Schwarz hybrid algorithms; i.e. algorithms that combine the additive and the multiplicative Schwarz algorithms. Section 7 is devoted to the mathematical analysis of the algorithms; i.e. the abstract convergency theory for the additive, the symmetrized multiplicative, and the hybrid Schwarz algorithms. We conclude the paper by discussing the applicability of the abstract theory to other classes of domain decomposition methods.

2. Iterative Methods and Preconditioners

Assume we have a discrete nonlinear system of equation written generally as

$$\text{Given } F : \Re^n \to \Re^n, \quad \text{solve} \quad F(x) = 0,$$

where n is very large. A standard method for numerically solving the nonlinear problem is the classical Newton's method, which is widely used because of its ultimate quadratic convergence. At the jth step of Newton's method, the linearized equation

$$F'(x_j)\delta x_j = -F(x_j) \tag{1}$$

is solved and we then set $x_{j+1} = x_j + \delta x_j$. We note that in practice nonlinear adaptive strategies including pseudo-transient continuation and damping are used to enlarge the domain of convergence of Newton's method.

We next address efficient iterative methods for solving (1) so that they can be implemented efficiently on high-performance computers and do not require that $F'(x_j)$ be formed, stored and factored. The use of iterative methods is sometimes the only choice since it is a seriously challenge to the most efficient direct solvers available today to handle the memory and the computational requirements for solving large three-dimensional problems. In addition, using appropriate iterative methods, we can explore the freedom of solving for the Newton correction only approximately while controlling the degradation of the asymptotic rate of convergence.

A Newton iterative method is very naturally treated as an inexact Newton method in which the Newton equation (1) is relaxed to an inexact Newton condition

$$||F(x_j) + F'(x_j)\delta x_j|| \le \tau_j ||F(x_j)||, \qquad \tau_j \in [0, 1). \tag{2}$$

The local convergence of an inexact Newton method is governed by the choice of the tolerance τ_j; in particular, desirable fast local convergence can be obtained by choosing them to be sufficiently small. However, chosing τ_j small, especially away from a solution, may result in little or no progress toward a solution, and it may entail needless expense [26]. To solve (2), we will make use of Krylov methods; see [41].

Krylov methods are Galerkin or minimal residual methods for finding best approximations to the solutions of linear system in subspaces generated from an initial vector, by means of matrix-vector products. The spaces are expanded incrementally to meet the convergence requirements. Although these methods are well founded theoretically, they are all likely to suffer from slow convergence for n large or for problems which arise from typical applications such as fluid dynamics. Preconditioning is the key to the success of Krylov subspace methods. We can view preconditioning as simply a means of transforming the original linear system into one which has same solution, but which is likely to be easier to solve with an Krylov method. Preconditioning may easily be incorporated into matrix-free Newton-Krylov methods assuming that the action of a preconditioner C_j, an approximation to the $(F'(x_j))^{-1}$ is available from some source in terms of a subroutine call. If a left-preconditioning is considered, the condition (2) is transformed into

$$||C_jF(x_j) + C_jF'(x_j)\delta x_j)|| \leq \tau_j||C_jF(x_j)||, \quad \tau_j \in [0, 1). \tag{3}$$

The action of $C_jF'(x_j)$ on a vector v can be computed by an automatic differentiation routine or by an approximation like $C_jF'(x_j)v \approx \frac{1}{h}[C_jF(x_j + hv) - C_jF(x_j)]$. The important point is that the convergence of preconditioned Krylov subspace methods depends on the distribution of the eigenvalues or the pseudospectrum of the preconditioned operator $C_jF'(x_j)$. For symmetric positive definite matrices $F'(x_j)$ and C_j, the convergence, in the first approximation, is measured by the condition number of $C_jF'(x_j)$, which is simply the ratio of the largest eigenvalue to the smallest. Most proofs for domain decomposition algorithms involve bounds on the extreme eigenvalues or singular values; see [41, 42, 24]. In this paper, we concentrate our discussion to the symmetric positive definite case. We note however that several of the techniques introduced here can be extended to some indefinite and nonsymmetric; see case [14, 15].

The conjugate gradient algorithm (CG) [31] is often an effective iterative algorithm to solve symmetric positive definite systems $Ax = b$. For simplicity of notation, we let A and C to denote $F'(x_j)$ and B_k, respectively. When A is not well conditioned, which is generally the case for discretizations of elliptic problems, we can introduce a preconditioner C and solve the preconditioned linear system

$$CAx = Cb. \tag{4}$$

by the conjugate gradient algorithm; also denoted by preconditioned conjugate gradient algorithm (PCG); see [27]. An essential feature of PCG, and also for PGMRES, is that an explicit representation of A and C are not needed. We only need to know how to apply them to a given vector. The preconditioner C should be chosen with the following properties:

i) the computation of the Cr should be easily and efficiently attainable on scalar as well as parallel machines;

ii) the condition number $\kappa(CA)$ should be small to guarantee that the PCG converges in a small number of iterations.

The extreme eigenvalues of CA provide a standard estimate of the rate of convergence of the preconditioned conjugate gradient method: The iteration error, measured in the $A-$norm, of the conjugate gradient method will decrease by at least the factor

$$2(\frac{\sqrt{\kappa}-1}{\sqrt{\kappa}+1})^k \text{ where } \kappa = \frac{\lambda_{max}(CA)}{\lambda_{min}(CA)}$$

in k steps; cf., e.g. [41].

2.1 Domain Decomposition Preconditioners

An effective way of constructing such a preconditioner is based on domain decomposition techniques. Domain decomposition (DD) methods provide a very natural way of deriving parallel algorithms for the numerical solution of partial differential equations and have recently attracted much theoretical and practical interest; see *www.ddm.org*. DD methods usually are viewed as preconditioners for iterative methods like the conjugate gradient method or GMRES but also, as described on the introduction of this text, can be viewed on more general settings. Much has recently been learned about those general settings, but stronger contribution has been on how to design these special preconditioned conjugate gradient type methods to obtain very fast convergence.

Such a successful preconditioner method is built mainly from two major components: many local problems and one global problem. The local problems typically correspond to subproblems that correspond to the original (or a similar) problem restricted to subregions into which the given a domain has been divided and possibly extended. In a multiprocessor machine, with distributed memory, one approach is to assign a subregion to each of the processors, and then to form the distributed discretized system. The domain can be decomposed into *overlapping* or *nonoverlapping* subregions. In the former case the algorithms are often called as *overlapping Schwarz methods* or *overlapping domain decomposition methods*, in the latter they are called *iterative substructuring methods* or *Schur complement method*, or *nonoverlapping domain decomposition methods*. The distinction between overlapping and nonoverlapping based methods however is not always clear in the theoretical point of view because many iterative substructuring methods can be analyzed as Schwarz methods. In this overview paper, we describe domain decomposition preconditioners based on a general theoretical framework which we will be able to analyze and establish upper bounds for the condition number of a general class of preconditioner systems.

Before we start the theoretical foundation of Schwarz methods, we complete this section by pointing out the several important reasons why domain decomposition methods have became popular in recent years: They

i) are well suited for parallel computers;

ii) are of great intrinsic mathematical interest;

iii) apply to regions with complex geometries;

iv) have a solid theoretical foundation;

v) make it possible to use different numerical schemes for the different subproblems, such as the finite element method, h, p and hp versions, fast Poisson solvers, finite differences, mortar elements, spectral and collocation methods;

vi) allow the use of different kinds of equations in different subregions whenever the underlying physics is of a different nature;

vii) can be combined with multiple scale solutions and local refinement;

viii) apply to several important physical problems of nonlinear nature arising from mechanics, elasticity, fluid dynamics, aerodynamics and thermomechanics.

3. Classical Alternating Schwarz Algorithms

It is believed that the first domain decomposition method known was proposed by Hermann Amandus Schwarz in 1869 [46]. It was originally used to show the existence of the solution of the Poisson problem on domains that consist of a union of simple overlapping subregions. In each subregion it is a priori known how to solve the problem. For instance, assume that we have two overlapping rectangles Ω_1' and Ω_2'; see Fig. 1. Since we can use separation of variables techniques to solve in each Ω_i' separately, we can use Schwarz ideas to solve the elliptic problem in $\Omega_1' \cap \Omega_2'$. (If Ω_1' were a disk, we could use the Poisson formula for the corresponding subproblem.) For more than two overlapping subregions, subregions were constructed recursively by forming unions of pairs of regions starting with regions for which existence could be established by some more elementary means. We next describe the *classical alternating Schwarz algorithm*.

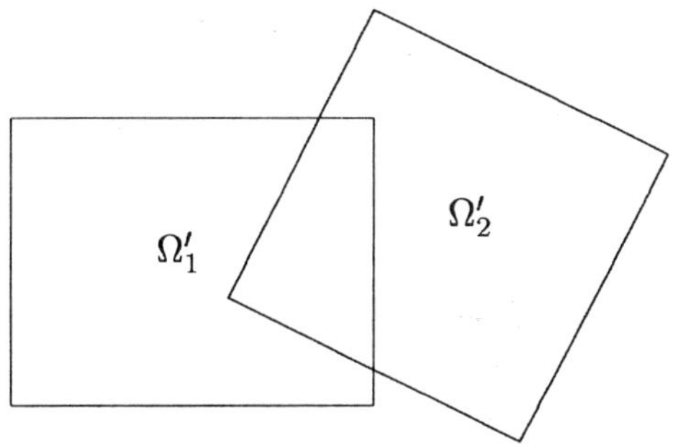

Fig. 1

Let $\Omega = \Omega_1' \cup \Omega_2'$ be the domain of interest; see Fig 1. Let us assume that we want to solve

$$\begin{cases} -\Delta u = f, & \text{in } \Omega, \\ u = 0, & \text{on } \partial\Omega. \end{cases} \tag{5}$$

Let Γ_i, $i = 1, 2$, denote the part of the boundary of Ω_i' which is in the interior of Ω; i.e. $\Gamma_i = \partial\Omega_i' \cap \Omega$. The classical alternating Schwarz algorithm consists of constructing a sequence of functions $\{u^k\}$ converging to the solution u of (5). The sequence is formally constructed in the following way:

Step i) Choose a suitable initial guess u^0 defined on Ω, and let $k = 0$.

Step ii) Solve

$$\begin{cases} -\Delta u_1^{k+1/2} = f & \text{in } \Omega_1', \\ u_1^{k+1/2} = u^k & \text{in } \overline{\Omega}_2' \backslash \Omega_1' \\ u_1^{k+1/2} = 0 & \text{on } \partial\Omega_1' \backslash \Gamma_1. \end{cases}$$

Step iii) Solve

$$\begin{cases} -\Delta u_2^{k+1} = f & \text{in } \Omega_2', \\ u_2^{k+1} = u_1^{k+1/2} & \text{in } \overline{\Omega}_1' \backslash \Omega_2' \\ u_2^{k+1} = 0 & \text{on } \partial\Omega_2' \backslash \Gamma_2. \end{cases}$$

Step iv) Let $k = k + 1$ and go to Step ii).

The convergence of the sequence $\{u^k\}$ was first established by Schwarz using maximum principle techniques. At the core of this work is a proof that this iterative method converges in the maximum norm at a geometric rate.

3.1 A Variational Formulation

The classical alternating Schwarz algorithm can be formulated in a variational framework. The formulation and related analysis were first introduced in a fundamental work due to Pierre Lions [34]. Consider the problem. Find $u \in H_0^1(\Omega)$ such that

$$a(u, v) = f(v) \;\; \forall v \in H_0^1(\Omega), \tag{6}$$

where

$$a(u, v) = \int_\Omega \nabla u \cdot \nabla v \, dx, \quad \text{and } f(v) = \int_\Omega f \, v \, dx.$$

In all that follows, we consider $H_0^1(\Omega_1')$ and $H_0^1(\Omega_2')$ as closed subspaces of $H_0^1(\Omega)$ by extending their elements to $\Omega \backslash \Omega_i$ by zero. We can use the alternating Schwarz algorithm for numerical computations by considering finite element subspaces $V^h \subset H_0^1(\Omega)$ and $V_i^h = V^h \cap H_0^1(\Omega_i')$. In order to consider the continuous and the finite element cases simultaneously, we let $V = H_0^1(\Omega)$ and $V_i = H_0^1(\Omega_i')$, or let $V = V^h$ and $V_i = V_i^h$.

The classical alternating Schwarz algorithm can be written as

Step i) choose an initial guess $u^0 \in V$. Let $k = 0$.

Step ii) find $u^{k+1/2} - u^k \in V_1$ such that

$$a(u^{k+1/2} - u^k, v) = f(v) - a(u^k, v) \quad \forall v \in V_1.$$

Step iii) find $u^{k+1} - u^{k+1/2} \in V_2$ such that

$$a(u^{k+1} - u^{k+1/2}, v) = f(v) - a(u^{k+1/2}, v) \quad \forall v \in V_2.$$

Step iv) let $k = k + 1$, and go to Step ii).

Let us define the orthogonal projection $P_i : V \to V_i$, by

$$a(P_i u, v) = a(u, v) \quad \forall v \in V_i, \;\; i = 1, 2,$$

and let us introduce the error e^k, for integer $k \geq 0$, by

$$e^k = u^k - u,$$
$$e^{k+1/2} = u^{k+1/2} - u.$$

Here, u is the solution of (6). It is easy to check that

$$e^{k+1/2} = (I - P_1)e^k,$$
$$e^{k+1} = (I - P_2)e^{k+1/2},$$

and that the error propagation operator E for a complete step of the classical alternating Schwarz algorithm is given by

$$E = (I - P_2)(I - P_1).$$

3.2 Inexact Solvers

The variational formalism also allows us to introduce inexact solvers for the subproblems. We note that the solution of the subproblems generates only an approximation correction of the global error. Therefore it is not required to solve the local problems exactly. Experience shows that solving the local problems by an appropriate incomplete LU or by another inexact iterative procedure (inner iterations) may increase considerably the performance (total CPU time and storage) of the whole algorithm. The recursive use of Schwarz preconditioners as a solver for the subproblems forms a natural and important class of inexact local solver.

For each subspace V_i, we introduce a symmetric positive definite bilinear form $b_i(\cdot, \cdot)$ defined on $V_i \times V_i$, and an operator $T_i : V \to V_i$ defined by

$$b_i(T_i u, v) = a(u, v) \quad \forall\, v \in V_i.$$

The $b_i(\cdot, \cdot)$ can be regarded as an approximation of $a(\cdot, \cdot)$. When we use exact solvers for the subproblems, $b_i(\cdot, \cdot) = a(\cdot, \cdot)$ and $T_i = P_i$ is a orthogonal with respect to the inner product $a(\cdot, \cdot)$. If we do not use exact solvers, the T_i are symmetric but not necessarily projections.

Next, we show how to view T_i in a matrix form. Let us define the linear operator $A : V \to V'$ by

$$(Au, v) = a(u, v) \quad \forall v \in V,$$

and the linear operators $B_i : V_i \to V_i'$ by

$$(B_i u, v) = b_i(u, v) \quad \forall v \in V_i.$$

Here, the V_i' and V' are the dual spaces of the V_i and V, respectively, with respect to the inner product $(\cdot, \cdot)$; i.e. ℓ_2 or L_2 inner product. It follows from the properties of the bilinear forms $b_i(\cdot, \cdot)$ that the B_i are symmetric positive definite operators with respect to $(\cdot, \cdot)$. It is easy to see that the T_i are given by

$$T_i = B_i^{-1} A.$$

The product $B_i^{-1} A$ is well defined since $V' \subset V_i'$. It is also easy to see that the T_i are symmetric with respect to the inner product $a(\cdot, \cdot)$. If we use exact solvers, then $b_i(\cdot, \cdot) = a(\cdot, \cdot)$, $B_i^{-1} = A_i^{-1}$. Here, the operator A_i is given by

$$(A_i u, v) = a(u, v) \quad \forall u, v \in V_i.$$

We can regard the classical alternating Schwarz algorithm as a simple iterative method for solving the equation

$$Tu = (I - E)u = g_T, (T_1 + T_2 - T_2 T_1)u = g_T,$$

where

$$g_T = (I - E)u = T_1 u + T_2 u - T_2 T_1 u. \tag{7}$$

We note that g_T can be computed without the knowledge of the solution of (6), since we can find the $g_i = T_i u$ and $g_{21} = T_2 T_1 u$ by solving

$$a(g_i, v) = f(v), \quad v \in V_i, \; i = 1, 2,$$

and

$$a(g_{21} u, v) = a(g_1, v), \quad v \in V_2.$$

3.3 Nonnested Spaces

The variational formalism above also allows us to consider more general cases of local solvers by dropping the assumption that $V_i \subset V$. We refer to the case $V_i \subset V$ as the nested case, and the case $V_i \not\subset V$ as the nonnested case. The use of nonnested spaces V_i appears in several applications. We can consider local solvers (multigrid and fast Poisson solvers) with structured grids even if the original problem is formulated on unstructured grids. We can consider lower order degree discrete solvers or preconditioners as local solvers even if the original problem uses higher-order discretizations; or the opposite; i.e. the use of higher-order discrete preconditioners (on a coarser grid) for lower-order discretizations. It also allows the development of preconditioners for discretizations on overlapping nonmatching grids domains and the use of preconditioners based on simpler equations than the equation in which the problem is formulated.

Let the V_i be a set of auxiliary spaces. We assume that there exists a set of interpolation operators $I_i : V_i \rightarrow V$. The nested case the I_i are simply embedding operators. We also assume that on each space V_i there is a symmetric positive definite bilinear form $b_i(\cdot, \cdot)$ defined on $V_i \times V_i$, and an operator $\tilde{T}_i : V \rightarrow V_i$ defined by

$$b_i(\tilde{T}_i u, v) = a(u, I_i v) \quad \forall\, v \in V_i.$$

Now we define the operator $T_i : V \rightarrow I_i V_i$ as

$$T_i = I_i \tilde{T}_i. \tag{8}$$

It is easy to see that the T_i are given by

$$T_i = I_i B_i^{-1} I_i^T A.$$

It is possible to avoid the entire issue of nonnested spaces by simply redefining $V_i = I_i V_i$. We note however that the stiffness matrices from the subproblems are different in the two approaches. We adopt the first approach since in general is not practical to build the stiffness matrices of the subproblems for the second approach.

4. Multidomain Schwarz Methods and Coarse Spaces

We now consider algorithms based on many overlapping open subdomains Ω_i', $i = 1, \cdots N$, which cover Ω. For each subdomain Ω_i', $i = 1, \cdots N$, we introduce a space $I_i V_i \subset V$, a symmetric positive definite bilinear form $b_i(\cdot, \cdot)$, and a local operators T_i given as in (8). The *multiplicative Schwarz algorithm* is a straightforwardly extension of the classical alternating Schwarz algorithm to N subdomains once has been an order selected for the subproblems. We note, however, if a large number subdomains is used then the convergence rates of this method will typically deteriorate rapidly with the number of subdomains. This happens because such a method does not provide global communication of information in each iteration; information is passed only between neighboring subdomains. The most commonly used mechanism of transmitting global information is the use of a *coarse space* $I_0 V_0$. The coarse space could be the finite element problem on a coarser mesh or a space with one or just a few degrees of freedom for each each subdomain Ω_i. We also introduce a bilinear form $b_0(\cdot, \cdot)$, and an operator T_0 for the subspace $I_0 V_0$.

Coarse spaces play at least two critical roles on domain decomposition methods. They may be used to weaken or remove the dependence of the convergence rate on the number (or size) of the subdomains and on the jumps in the coefficients of the differential equation. The space $I_0 V_0$ represents the coarse subspace and it is added to the algorithm to provide global and balanced communication of information in each iteration. We note that Widlund [53] has shown that, without a global coarse space problem, the condition number of the relevant iteration operator must grow at least like $1/H^2$, where H is the diameter of the subregions. For the *discontinuous coefficient* case, we note that Oswald [38] shows that, for the natural choice (basis functions on a coarser mesh) of a coarse space on three-dimensional case, the conditioning number of the preconditioning system grow linearly with $1/h$. See [5, 6, 19, 20, 23, 44] for examples of appropriate choices of coarse spaces for elliptic problems with discontinuous coefficients on the three-dimensional space case.

Each cycle of the multidomain multiplicative Schwarz algorithm can be described as follows:

Multiplicative Schwarz Algorithm

i) *compute* $g_i = T_i u$, *for* $i = 0, 1, \cdots, N$;

ii) *given* u^k, *compute* u^{k+1} *in* $N+1$ *fractional steps*:

$$u^{k+\frac{i+1}{N+1}} = u^{k+\frac{i}{N+1}} + (g_i - T_i u^{k+\frac{i}{N+1}}), \quad i = 1, \cdots, N.$$

This algorithm can also be viewed as a simple iterative method for solving

$$T_{ms} u = (I - (I - T_N) \cdots (I - T_0))u = g_{ms},$$

with

$$g_{ms} = T_{ms}u.$$

Like in (7), g_{ms} also can be computed without knowledge of the solution of (6).

It is easy to see that the error $e^k = u^k - u$ satisfies

$$e^{k+1} = E_N e^k,$$

where E_N is the error propagation operator

$$E_N = (I - T_N) \cdots (I - T_0).$$

We notice that the operator T_{ms} is generally a nonsymmetric operator. It can be accelerated by the GMRES method (or by another conjugate gradient type method designed for nonsymmetric operators):

Accelerated multiplicative Schwarz algorithm

i) *compute* $g = (I - E_N)u$.

ii) *solve the nonsymmetric operator equation*

$$(I - E_N)v = g$$

by the GMRES algorithm.

We can also solve the problem (6 by PCG using the $a(\cdot,\cdot)$ inner product, by considering the following *symmetrized multiplicative Schwarz* operator:

$$T_{sms} = I - (I - T_0) \cdots (I - T_N)(I - T_N) \cdots (I - T_0). \tag{9}$$

Other symmetric variations are to replace T_{sms} by

$$I - (I - T_0) \cdots (I - T_{N-1})(I - T_N)(I - T_{N-1}) \cdots (I - T_0),$$

or to change the order of the multiplications.

We note that the T_i are symmetric operators with respect to the inner product $a(\cdot,\cdot)$. Hence $T_i^T = T_i$, and we guarantee that the T_{sms} and the variations are symmetric operators with respect to $a(\cdot,\cdot)$. By construction, the solution u of (6) is also a solution of (9). Under appropriate assumptions on the b_i, I_i, and V_i, the operator T_{msm} is symmetric with respect to the inner product $a(\cdot,\cdot)$ and invertible. Hence, the solution u of (9) is the only solution of (6). The reason for replacing the problem (6) by (11) is that, by a suitable choice of the b_i, I_i, and V_i, we can transform a large ill-conditioned system into a very well conditioned system problem at the expense of solving many small independent linear problems. The equation (9) is typically solved by a conjugate gradient method, without further preconditioning, using $a(\cdot,\cdot)$ as the inner product. In order to see that the problem (9) has u as the unique solution, and also to estimate the rate of convergence of the preconditioned conjugate gradient method, we need to obtain upper and lower bounds for the spectrum of T_{sms}. The bounds are obtained later through the Theorem 3.

5. Additive Schwarz Methods

To remove completely the inherent sequential behavior of the T_{ms}, Matsokin and Nepomnyaschikh [37], and independently also by Dryja and Widlund [21], introduced a variant of the Schwarz method: the *additive Schwarz methods*, namely

$$T_{as} = T_0 + T_1 + \cdots + T_N.$$

An additive Schwarz method is defined by decomposing V into a sum of $N+1$ subspaces:

$$V = I_0 V_0 + I_1 V_1 + \cdots + I_N V_N. \tag{10}$$

We note that this decomposition is not necessarily a direct sum of subspaces; in many applications of interest, the representation of an element of V in terms of components of the $I_i V_i$ is not unique. Often, these subspaces are related to a decomposition of the domain Ω into overlapping subregions Ω'_i.

We now replace (6) by the problem:

Find $u \in V$ such that

$$T_{as} u = g_{as}, \quad g_{as} = \sum_{i=0}^{N} g_i, \quad g_i = I_i \tilde{g}_i = T_i u \tag{11}$$

We note that g_{as} can also be computed, without knowledge of the solution of (6), since we can find the $\tilde{g}_i$ by solving

$$b_i(\tilde{g}_i, v_i) = a(u, I_i v_i) = f(I_i v_i) \quad \forall v_i \in V_i.$$

or using operator notation

$$g_{as} = \sum_{i=0}^{N} I_i B_i^{-1} I_i^T A f.$$

Under appropriate assumptions on the b_i and I_i and V_i, the operator T_{as} is symmetric with respect to the inner product $a(\cdot,\cdot)$ and invertible, and the lower and upper bounds for the spectrum of T_{as} are obtained through the Theorems 1 and 2.

6. Hybrid Schwarz Methods

We note that all preconditioned system that we have introduced here in this paper can be written as

$$T_{poly} u = poly(T_0, T_1, \cdots, T_N) u = g_{poly},$$

where T is a polynomial of the T_i's such that $poly(0, 0, \cdots, 0) = 0$ so that g_{poly} can be computed without knowing the solution u itself. It is easy to see that if T_{poly} is invertible, then the solution u of (6) is the only solution of $T_{poly}u = g_{poly}$. The additive Schwarz method is equivalent the idea of choosing the simplest (nontrivial lowest degree) polynomial of operators T_i's.

In the multiplicative Schwarz algorithm, each iteration involves $N+1$ sequential fractional steps and this is not ideal for parallel computing if N is large; on an abstract level, T_{poly} is not ideal for parallel computing if the degree of the polynomial *poly* is large. We can decrease the degree of the polynomial by grouping the subregions using the following coloring strategy. Associate with the decomposition $\{\Omega_i'\}$, an undirected graph in which the nodes represent the subdomains $\{\Omega_i'\}$ and the edges intersections of subdomains. This graph can be colored, using colors $1, \cdots, J$, such that no connected nodes have the same color. We now group the T_i in term of the color that the subdomain has been assigned. We obtain

$$T_{smcolor} = I - (I - T_J) \cdots (I - T_0).$$

We note that different T_i of the same color correspond to domains that are mutually disconnected; these subproblems can be solved in parallel.

In practice, despite less parallelism, the multiplicative Schwarz methods are often faster than the additive Schwarz methods [28]. As we have noted, the parallelism of $T_{smcolor}$ results mainly from the fact that, for each $j = 1, \cdots, J$, T_j is a sum of a number of local, independent subproblems that can be handled in parallel.

A variation of Schwarz method that combines both additive and multiplicative Schwarz components was introduced in Mandel [35, 36]

$$T_{hyb} = I - (I - P_0)(I - \sum_{i=1}^{N} T_i)(I - P_0), \tag{12}$$

which gives

$$T_{hyb} = P_0 + (I - P_0)(\sum_{i=1}^{N} T_i)(I - P_0).$$

This important algorithm is used much in practice due to the potential for parallelism of the local problems $T_i, i = 1, \cdots, N$ and the fast convergence rates due to the multiplicative aspect of the algorithm between the local problems and the coarse problem. This is a symmetric operator with respect to $a(\cdot, \cdot)$ and which we can work essentially with the same cost as the additive Schwarz method. When forming powers of the operator (12), which is the case of PCG, we can use the fact that $I - P_0$ is a projection, and therefore $(I - P_0)^2 = I - P_0$; this is one reason why solve the coarse problem exactly. We also will see later in the analysis of T_{hyb} that the use of P_0 will imply that the T_{hyb} has a smaller conditioning number than of the T_{as} as we will see in Theorem 4.

In addition, if the coarse problem is not too large, it is natural to solve coarse problems exactly since sometimes inexact coarse problems do not handle global communications properly or do not remove singularities of the discrete solution due to discontinuities of coefficients.

We also have considered the use of polynomials of the form

$$P_0 + \sum T_{ij} \text{ where } T_{ij} = I - (I - T_i)(I - T_j).$$

T_0 corresponds to the coarse problem, and sum corresponds the T_i and T_j which are adjacent subregions, where Ω_i is downstream to Ω_j [13]. The problems considered are convection-diffusion equations.

7. Abstract Schwarz Theory

We will now develop a quite general theory that has proven very useful in the development and analysis of many Schwarz methods. The abstract convergence Schwarz theory centers around three parameters C_0^2, ω and $\rho(\mathcal{E})$ which measure the iterations of the spaces V_i, the interpolators I_i, and the bilinear forms $b_i(\cdot,\cdot)$, and their suitability in the construction of the preconditioners.

7.1 Additive Schwarz Methods

The convergence properties of the additive Schwarz algorithm depends of the spectrum of T_{as}. The next theorem provides a bound on the condition number for the abstract additive Schwarz method in terms of the parameters C_0^2, ω and $\rho(\mathcal{E})$.

Theorem 1. *Suppose the following three assumptions hold:*

i) There exists a constant C_0 such that for all $u \in V$ there exists a decomposition $u = \sum_{i=0}^N I_i u_i$, $u_i \in V_i$, such that

$$\sum_{i=0}^{N} b_i(u_i, u_i) \leq C_0^2 a(u, u);$$

ii) There exits a constant $\omega > 0$ such that

$$a(I_i u, I_i u) \leq \omega\, b_i(u, u) \;\; \forall u \in V_i, \;\; i = 0, \ldots, N;$$

iii) There exist constants ϵ_{ij}, $i, j = 1, \ldots, N$, such that

$$a(I_i u_i, I_j u_j) \leq \epsilon_{ij} a(I_i u_i, I_i u_i)^{1/2} a(I_j u_j, I_j u_j)^{1/2} \; \forall u_i \in V_i \; \forall u_j \in V_j.$$

Then, T_{as} is invertible and

$$C_0^{-2}\, a(u, u) \leq a(T_{as} u, u) \leq (\rho(\mathcal{E}) + 1)\, \omega\, a(u, u) \;\forall u \in V. \tag{13}$$

Here $\rho(\mathcal{E})$ is the spectral radius of the matrix $\mathcal{E} = \{\epsilon_{ij}\}_{i,j=1}^N$.

Proof. Upper bound: We first note that $\|T_i\|_a \leq \omega$. Indeed, using the *assumption*, the definition of the T_i given in (8), and the Cauchy-Schwarz inequality, we obtain

$$a(T_iu, T_iu) \leq \omega b_i(\tilde{T_i}u, \tilde{T_i}u) = \omega\, a(u, T_iu) \qquad (14)$$
$$\leq \omega a(u,u)^{1/2}\, a(T_iu, T_iu)^{1/2}.$$

Therefore, $a(T_iu, T_iu) \leq \omega^2 a(u,u)$, which implies $\|T_i\|_a \leq \omega$.

Using the *assumptions* (iii) and (ii), the positive semidefiniteness of T_i, we obtain

$$a(\sum_{i=1}^N T_iu, \sum_{j=1}^N T_ju) = \sum_{i,j=1}^N a(T_iu, T_ju)$$

$$\leq \sum_{i,j=1}^N \epsilon_{ij} a(T_iu, T_iu)^{1/2} a(T_ju, T_ju)^{1/2}$$

$$\leq \sum_{i,j=1}^N \epsilon_{ij} (\|T_i\|_a \|T_j\|_a)^{1/2} a(T_iu, u)^{1/2} a(T_ju, u)^{1/2}$$

$$\leq \rho(\mathcal{E})\omega \sum_{i=1}^N a(T_iu, u) \leq \rho(\mathcal{E})\omega a(u,u)^{1/2} a(\sum_{i=1}^N T_iu, \sum_{i=1}^N T_iu)^{1/2}.$$

Hence,

$$a(\sum_{i=1}^N T_iu, \sum_{j=1}^N T_ju) \leq \rho(\mathcal{E})^2 \omega^2 a(u,u),$$

and therefore,

$$a(\sum_{i=1}^N T_iu, u) \leq \rho(\mathcal{E})\omega a(u,u). \qquad (15)$$

This last inequality added to

$$a(T_0u, u) \leq \omega a(u,u),$$

completes the proof of the upper bound.

Lower bound: Using the definition of T_i, the Cauchy-Schwarz inequality, and the *assumption* (i), we obtain

$$a(u,u) = \sum_{i=0}^N a(u, I_iu_i) = \sum_{i=0}^N b_i(\tilde{T_i}u, u_i)$$

$$\leq \Big(\sum_{i=0}^{N} b_i(\tilde{T}_i u, \tilde{T}_i u)\Big)^{1/2} \Big(\sum_{i=0}^{N} b_i(u_i, u_i)\Big)^{1/2}$$

$$\leq \Big(\sum_{i=0}^{N} a(u, T_i u)\Big)^{1/2} C_0 (a(u, u))^{1/2}.$$

Hence, T_{as} is invertible and $a(u, u) \leq C_0^2 a(T_{as} u, u)$. □

Remark 1. *We have seen that assumption i) implies that $\lambda_{min}(T_{as}) \leq C_0^2$. This result is often known as Lions's Lemma [34] and it is the most fundamental tool in estimating the minimum eigenvalue of T_{as}. Usually assumption i) is, among the three assumptions, the most difficult to be verified. If the C_0 can be bounded from above independently of grid parameters and on the jumps in the coefficients of the differential equations, then the $I_i V_i$ are said to provide a stable decomposition of V.*

Remark 2. *The parameter $\rho(\mathcal{E})$ [25] measures the orthogonality of the local subspaces $I_i V_i$, $i = 1, \cdots, N$. A bound on $\rho(\mathcal{E})$ can often be obtained quite easily for many domain decomposition methods. For multilevel and multigrid methods, it often involves strengthened Cauchy-Schwarz inequalities, of the form $\epsilon_{ij} \leq q^{-|i-j|}$ with $q < 1$; see [52].*

Remark 3. *The upper bound of T_{as} can be replaced by $(\rho(\hat{\mathcal{E}}) + \|T_0\|_a)$, where the elements of $\hat{\mathcal{E}}$ are given by $\|T_i\|_a^{1/2} \epsilon_{ij} \|T_j\|_a^{1/2}$. We note that the rate of convergence of an additive Schwarz method is affected by the scaling of bilinear forms $b_i(\cdot,\cdot)$, relative to each other, i.e. by the scaling of the T_i; both the upper and lower bounds of T_{as} are affected. It is easy to show that if the norm of the T_i all are made equal, then the bound for the condition number of additive algorithm at least as good as before.*

We remark that the lower bound in (15) is sharp for the smallest C_0^2 for which *assumption* i) is satisfied. This smallest value for C_0^2 is easily characterized [52] by the next theorem.

Theorem 2. *Define $T = \sum_{i=0}^{N} T_i$. Then*

$$a(T_{as}^{-1} u, u) = \min_{u = \sum_{i=0}^{N} I_i u_i} \sum_{i=0}^{N} b_i(u_i, u_i). \tag{16}$$

Hence,

$$\frac{1}{\lambda_{min}(T_{as})} = \max_{u \in V \setminus \{0\}} \min_{u = \sum_{i=0}^{N} I_i u_i} \frac{\sum_{i=0}^{N} b_i(u_i, u_i)}{a(u, u)}.$$

Proof. Let the $u_i \in V_i$ be any decomposition of u such that $\sum_{i=0}^N I_i u_i = u$. Then,

$$a(T_{as}^{-1}u, u) = a(T_{as}^{-1}u, \sum_{i=0}^N I_i u_i) = \sum_{i=0}^N a(T_{as}^{-1}u, I_i u_i) = \sum_{i=0}^N b_i(\tilde{T}_i T_{as}^{-1}u, u)$$

$$\leq \Big(\sum_{i=0}^N b_i(\tilde{T}_i T_{as}^{-1}u, \tilde{T}_i T_{as}^{-1}u)\Big)^{1/2} \Big(\sum_{i=0}^N b_i(u_i, u_i)\Big)^{1/2}\Big)$$

$$\leq \Big(\sum_{i=0}^N a(T_{as}^{-1}u, T_i T_{as}^{-1}u)\Big)^{1/2} \Big(\sum_{i=0}^N b_i(u_i, u_i)\Big)^{1/2}$$

$$= a(T_{as}^{-1}u, u)^{1/2} \Big(\sum_{i=0}^N b_i(u_i, u_i)\Big)^{1/2}\Big)^{1/2}.$$

Therefore,

$$a(T_{as}^{-1}u, u) \leq \sum_{i=0}^N b_i(u_i, u_i) \quad \forall u_i \in V_i, \ \sum_{i=0}^N I_i u_i = u.$$

To conclude the proof, we construct a particular decomposition for u which the equality in (16) holds. Let us choose $u_i = \tilde{T}_i T_{as}^{-1}u$. Then,

$$\sum_{i=0}^N b_i(u_i, u_i) = \sum_{i=0}^N b_i(\tilde{T}_i T_{as}^{-1}u, \tilde{T}_i T_{as}^{-1}u)$$

$$= \sum_{i=0}^N a(T_{as}^{-1}u, T_i T_{as}^{-1}u) = a(T_{as}^{-1}u, u).$$

□

7.2 Multiplicative Schwarz Methods

We next develop the abstract convergence theory for multiplicative Schwarz methods. This theory was introduced by Bramble, Pasciak, Wang and Xu [8] and further developed by [23, 28, 51]. We consider only the symmetrized version T_{sms}. We note that several of the nonsymmetric version of the multiplicative can be analyzed similarly; see [8, 28].

Theorem 3. *Assume that* $\omega < 2$ *and* $\hat{\omega} = \max\{1, \omega\}$. *Then the symmetric multiplicative Schwarz method satisfies*

$$\kappa(T_{sms}) \leq \frac{[1 + 2\hat{\omega}^2 \rho^2(\mathcal{E})] C_0^2}{2 - \omega}.$$

Proof. Let us define

$$E_i = (I - T_i)E_{i-1} = (I - T_i)\dots(I - T_0), \quad E_{-1} = I. \tag{17}$$

Then, $T_{msm} = I - E_N^T E_N$, where the transpose is in the $a(\cdot,\cdot)$ inner product; i.e, $a(E_N^T u, v) = a(u, E_N v)$.

Upper Bound: Using that $a(\cdot,\cdot)$ is positive definite, we have

$$\lambda_{max}(T_{sms}) = \max_{u\in V\setminus\{0\}} \frac{a(T_{sms}u, u)}{a(u,u)}$$

$$= \frac{a(u - E_N^T E_N u, u)}{a(u,u)} = 1 - \min_{u\in V\setminus\{0\}} a(E_N^T u, E_N^T u) \le 1.$$

Lower Bound: Using (17) we have

$$a(T_{sms}u, u) = a(u,u) - a(E_N u, E_N u)$$

$$= \sum_{i=0}^{N} \{a(E_{i-1}u, E_{i-1}u) - a(E_i u, E_i u)\}$$

$$= \sum_{i=0}^{N} \{a(E_{i-1}u, E_{i-1}u) - a((E_{i-1} - T_i E_{i-1})u, (E_{i-1} - T_i E_{i-1})u)\}$$

$$= \sum_{i=0}^{N} \{2a(E_{i-1}u, T_i E_{i-1}u) - a(T_i E_i u, T_i E_i u)\}$$

$$\ge (2 - \omega) \sum_{i=0}^{N} a(E_{i-1}u, T_i E_{i-1}u). \tag{18}$$

In the last inequality we have used assumption (ii), the definition of ω and the positive semidefiniteness of T_i to obtain

$$a(T_i E_{i-1}u, T_i E_{i-1}u) \le \omega b_i(\tilde{T}_i E_{i-1}u, \tilde{T}_i u) = \omega a(E_{i-1}u, T_i E_{i-1}u).$$

By using the definition of E_i we have

$$I = T_0 + E_{i-1} + \sum_{j=1}^{i-1} T_j E_{j-1},$$

and therefore,

$$a(T_i u, u) = a(T_i u, T_0 u) + a(T_i u, E_{i-1}u) + \sum_{j=1}^{i-1} a(T_i u, T_j E_{j-1}u).$$

By using a Schwarz's inequality, the upper bound $\|T_i\|_a \leq \omega$, the positive semidefiniteness of T_i, and the definition of ϵ_{ij} we obtain

$$a(T_iu,u) \leq a(T_iu,u)^{1/2}\{a(T_iT_0u,T_0u)^{1/2} + a(T_iE_{i-1}u,E_{i-1}u)^{1/2}$$

$$+\omega\sum_{j=1}^{i-1}\epsilon_{ij}a(T_jE_{j-1}u,E_{j-1}u)^{1/2}\}$$

$$\leq a(T_iu,u)^{1/2}\{a(T_iT_0u,T_0u)^{1/2} + \hat{\omega}\sum_{j=1}^{i}\epsilon_{ij}a(T_jE_{j-1}u,E_{j-1}u)^{1/2}\}.$$

In the last inequality we have used $\epsilon_{ii} = 1$, the definition of $\hat{\omega}$, and the combination of the second and third terms together.

Denote by x the vector with components $x_j = a(T_jE_{j-1},E_{j-1})^{1/2}$. We next cancel the common factor and square both sides and use the relation $(a+b)^2 \leq 2a^2 + 2b^2$ to obtain

$$a(T_iu,u) \leq 2a(T_iT_0u,T_0u) + 2\hat{\omega}^2\Big(\sum_{j=1}^{i}\epsilon_{ij}x_j\Big)^2.$$

Summing from $i = 1$ to N and extending the previous sum to N will only increase the right hand side,

$$\sum_{i=1}^{N}a(T_iu,u) \leq 2\sum_{i=1}^{N}a(T_iT_0u,T_0u) + 2\hat{\omega}^2\Big(\sum_{i=1}^{N}\sum_{j=1}^{N}\epsilon_{ij}x_j\Big)^2.$$

For the first term we use the bound (15) combined with inequality (14) to obtain

$$2a(\sum_{i=1}^{N}a(T_iT_0u,T_0u) \leq 2\rho(\mathcal{E})\omega a(T_0u,T_0u) \leq 2\rho(\mathcal{E})\omega^2 a(T_0u,u)$$

For the second term, we use that $(\mathcal{E}x)^T(\mathcal{E}x) \leq \rho^2(\mathcal{E})x^Tx$. We now add the term $a(T_0u,u)$ to both sides, and $a(T_0u,u) = a(T_0E_{-1}u,E_{-1}u)$ to obtain

$$a(T_{as}u,u) \leq (1+\hat{\omega}^2\rho^2(\mathcal{E}))\sum_{i=0}^{N}a(T_iE_{i-1}u,E_{i-1}u).$$

The proof can now be completed by using (18) and the lower bound of Theorem 1. □

7.3 Hybrid Schwarz Methods

We next consider the abstract convergence theory for the hybrid preconditioner T_{hyb} introduced by Mandel [36],

$$T_{hyb}\, u = g_{hyb}.$$

We note that g_{hyb} can be calculated without knowledge of the solution u; in matrix notation g_{hyb} is given by

$$g_{hyb} = [A_0^{-1} + (I - A_0^{-1}A)(\sum_{i=1}^{N} I_i B_i^{-1} I_i^T)(I - AA_0^{-1})]f.$$

Note that,

$$I - T_{hyb} = (I - P_0)(I - \sum_{i=1}^{N} T_i)(I - P_0)$$

and therefore, $a(g_{hyb} - T_{hyb}\, g_{hyb}, v) = 0, \forall v \in V_0$; i.e. $g_{hyb} - T_{hyb}\, g_{hyb} \in V_0^{\perp}$. In addition, $T_{hyb}\, w \in V_0^{\perp}$ if $w \in V_0^{\perp}$. Hence, instead of solving $T_{hyb}\, u = g_{hyb}$, we solve, by PCG, the reduced problem

$$T_{hyb} e = g_{hyb} - T_{hyb}\, g_{hyb} = g'_{hyb}, \tag{19}$$

and recover u by $u = e + g_{hyb}$. The PCG algorithm searchs for the best approximation to the solution in subspaces generated by powers of $T_{hyb}^k g'_{hyb}$, and therefore, on subspaces belonging to $V_0^{\perp}$. Hence, the extreme eigenvalues of T_{hyb} restricted to $V_0^{\perp}$ will provide the estimate of the rate of convergence of the PCG. Taking the advantage that we need to estimate extreme eigenvalues of T_{hyb} on a smaller space than V, we show $\kappa(T_{hyb})$ restricted to $V_0^{\perp}$, denoted by $\kappa(T_{hyb|V_0^{\perp}})$, is smaller or equal than $\kappa(T_{as})$. In addition, it is easy to see that $\kappa(T_{hyb|V_0^{\perp}})$ is strictly smaller than $\kappa(T_{as})$ when $(\cup_{i=1}^{N} I_i V_i) \cap V_0 \neq \emptyset$.

Theorem 4. *We have:*

$$\lambda_{\max}(T_{hyb|V_0^{\perp}}) \leq \max_{\gamma \in \Re} \lambda_{\max}(\gamma P_0 + \sum_{i=1}^{N} T_i) \leq \lambda_{\max}(T_{as}),$$

and

$$\lambda_{\min}(T_{hyb|V_0^{\perp}}) \geq \max_{\gamma \in \Re} \lambda_{\min}(\gamma P_0 + \sum_{i=1}^{N} T_i) \geq \lambda_{\min}(T_{as}).$$

Proof. Upper Bound: Using Rayleigh quotient arguments and properties of the orthogonal projection P_0; i.e. that $(I - P_0)P_0 = 0$, we obtain

$$\lambda_{\max}(T_{hyb|V_0^{\perp}}) = \max_{u \in V_0^{\perp} \setminus \{0\}} \frac{a(T_{hyb} u, u)}{a(u, u)}$$

$$= \max_{u \in V_0^{\perp} \setminus \{0\}} \frac{a(\sum_{i=1}^{N} T_i u, u)}{a(u,u)} = \max_{u \in V_0^{\perp} \setminus \{0\}} \frac{a([\gamma P_0 + \sum_{i=1}^{N} T_i]u, u)}{a(u,u)}$$

$$= \min_{\gamma \in \Re} \max_{u \in V_0^{\perp} \setminus \{0\}} \frac{a([\gamma P_0 + \sum_{i=1}^{N} T_i]u, u)}{a(u,u)}$$

$$\leq \min_{\gamma \in \Re} \max_{u \in V \setminus \{0\}} \frac{a([\gamma P_0 + \sum_{i=1}^{N} T_i]u, u)}{a(u,u)}$$

$$\leq \max_{u \in V \setminus \{0\}} \frac{a([P_0 + \sum_{i=1}^{N} T_i]u, u)}{a(u,u)} = \lambda_{\max}(T_{as}).$$

The proof for the lower bound is similar to the upper bound.

□

8. Final Remarks

The overlapping Schwarz methods just described are special cases of a more general class of methods. Special cases include: 1) the traditional Jacobi (additive) and Gauss-Seidel, (multiplicative) where the subspaces are unit vectors corresponding to the nodal basis functions, 2) the multigrid, in which the subspaces are hierarchically nested by wavenumber, 3) the block diagonal preconditioners, which are subdomains block Jacobi or Gauss-Seidel methods (with possibly overlapping blocks), and 4) the two-level methods where a global coarse problem with a few degrees of freedom per subdomain is added to the one-level Schwarz preconditioner. The design of coarse problems for unstructured meshes and corresponding parallel implementations are still a very active area of research. The improved performance resulting from introducing this second level problem has been demonstrated repeatedly and very convincingly in a number of application areas. Efficient overlapping additive Schwarz methods are: the *additive Schwarz method with small overlap* [22], the *BPX method* [7], and the *restricted additive Schwarz method* [13, 45, 33, 10]. We also note that there are several variants of Schwarz algorithms, additive, multiplicative, and several hybrid types; a number of them are discussed in detail in [47].

Another important family of domain decomposition methods is the ones that are based on *Schur complement methods*, also known as nonoverlapping domain decomposition methods. In general, these methods are simpler to implement: the coarse problem is usually easier to construct, and certain difficulties related to obtaining overlapping subregions are avoided. We note however, that these methods typically require the use of exact local solvers to compute residuals, well-shaped and sometimes connected subregions; it is still a seriously challenge for partitioners codes available today to attain these requirements for three-dimensional problems on unstructured meshes. The use of such methods begins by partitioning the elements of the initial grid into nonoverlapping subset of elements $\Omega_1, ..., \Omega_N$ such that $\cup_{i=1}^N \overline{\Omega}_i = \overline{\Omega}$. This partition generates interface and interior nodes. The interface nodes are those that belong to $\Gamma = (\cup_{i=1}^N \partial\Omega_i) \backslash \partial\Omega$. The interior nodes are the ones that belong to the interior of a subdomain Ω_i. In the Shur complement methods all the variables placed on interior nodes are eliminated. The remaining system, the Schur complement system, is then solved by a preconditioned conjugate gradient algorithm. A key point is that the Shur complement system is equivalent to the original system when restricted to a subspace of the original space V. This subspace, which we denote by V_{har}, is the set of functions belonging to V and which are discrete harmonic on all the interior nodes; i.e the $v \in V$ such that $a(v, \phi_k) = 0$ for discrete basis functions $\phi_k \in V$ associated to all the interior nodes x_k. Because of this equivalence, all the abstract theory also can be applied to the Shur complement methods. We emphasize that the decomposition (10) does not require that the $I_i V_i$ should be generated from an overlapping subregion Ω_i'. The abstract theory holds for any decomposition of V and bilinear forms b_i satisfying the three assumptions of Theorem 1. The space V of the the abstract theory can be chosen more generally than the space of finite element functions defined on the whole original region Ω; the only requirement is the ellipticity and boundness of the operator generated by $a(\cdot, \cdot)$ with respect to a space V. Using the equivalence, we also have ellipticity and boundness of the Shur complement system since $V_{harm} \subset V$. Hence, we can consider V_{har} as the V of the abstract theory. The abstract theory asks basically that we should have a decomposition V_{har} as

$$V_{har} = I_0 V_0 + I_1 V_1 + \cdots + I_p V_p,$$

define the bilinear forms $b_i(\cdot, \cdot)$ on $V_i \times V_i$, and check the three assumptions. Note that we require that $I_i V_i \subset V_{har}$. We can choose, for instance, $V_i, i = 1, \cdots, N$ as the space of piecewise linear functions on Γ with degrees of freedom on certain pieces of Γ and zero value at the remaining nodes of Γ. The space $I_i V_i$ is defined as V_i on Γ, zero on $\partial\Omega$, and discrete harmonic extension elsewhere (interior nodes). The operator I_i stands for the lift operator (discrete harmonic extension) from Γ to $\overline{\Omega}$. There are several possible choices for selecting the pieces. The most common chosen pieces are the faces, the edges and the corners between subdomains Ω_i, or pieces which are the whole boundary of a $\Omega_i \backslash \partial\Omega$. It is easy to see that the support of a function is on $I_i V_i$ is the union of all the subdomains $\overline{\Omega}_i$ which share that piece. There are several choices of coarse spaces V_0. Most of them are based on selecting coarse basis functions on Γ with one or few degrees of freedom per piece and with support (nodewise) only on that piece. There are many choices for the bilinear forms $b_i(\cdot, \cdot)$ on $V_i \times V_i$. A simple choice is the use of exact solvers; it can be defined as $b_i(v_i, w_i) = a(I_i v_i, I_i w_i)$. We note that inexact bilinear forms also can be considered. A popular Schur complement based method is the *balancing domain decomposition* method [35]. This method is a hybrid one of the form of

T_{hyb}. The V_i are the degrees of freedom associated to the nodes on $\Gamma \cap \partial\Omega_i$. The local solvers are Neumann problems on Ω_i. The coarse problem, with one degree of freedom per Ω_i, is used multiplicatively with the local solvers to make all the residuals associated to the local problems perpendicular to the null space of the Neumann local problems. A detailed discussion on these and others nonoverlapping domain decompositions methods can be found in [47].

Acknowledgments

A special thanks goes to Zlatko Drmač and local organizers of the conference for the hospitality and support. The work was supported in part also by the NSF grant CCR-9984404.

References

[1] Abdoulaev G., Achdou Y., Hontand J., Kuznetsov Y., Pironneau O., and Prud'homme C. (1998). *Nonmatching grids for fluids*. Domain Decomposition Methods 10, Contemporary Mathematics, J. Mandel, C. Farhat and X.-C. Cai, eds, AMS, Vol. 218.

[2] Achdou Y., Hontand J., and Pironneau. (1997). *A mortar element method for fluids*. Domain Decomposition Methods in Science and Engineering, R. Glowinski, J. Periaux, Z.-C. Shi and O. Widlund, eds, Wiley & Sons.

[3] Bernardi C., Maday Y., Patera A. (1994). *A new nonconforming approach to domain decomposition: The mortar element method*. Collége de France Seminar, H. Brezis and J. Lions, eds., Pitman.

[4] Berkman M. E., Sankar L. N., Berezin C. R., and Torok M. S. (1997). *Navier–Stokes/full potential/free–wake method for rotor flows*. J. of Aircraft, Vol. 34(5), pp. 635–640.

[5] Mandel J., and Brezina M. (1993). *Balancing domain decomposition: Theory and computations in two and three Dimensions*. Center for Computational Mathematics, University of Colorado at Denver, UCD/CCM 2.

[6] Bramble J. H., Pasciak J. E., and Schatz A. H. (1989). *The construction of preconditioners for elliptic problems by substructuring IV*. Math. Comp., Vol. 53, pp. 1–24.

[7] Bramble J. H., Pasciak J. E., and Xu J. (1990). *Parallel multilevel preconditioners*. Math. Comp., Vol. 55, pp. 1–22.

[8] Bramble J. H., Pasciak J. E., Wang J., and Xu J. (1991). *Convergence estimates for product iterative methods with applications to domain decomposition*. Math. Comp., Vol. 57(195), pp. 1–21.

[9] Cai X.-C., Dryja M., Sarkis M. (1999). *Overlapping nonmatching grid mortar element methods for elliptic problems*. SIAM J. Numer. Anal., Vol. 36(2), pp. 581-606.

[10] Cai X.-C, Dryja M., and Sarkis M. (2002). *A restricted additive Schwarz preconditioner with harmonic overlap for symmetric positive definite linear systems*. SIAM J. Numer. Anal, (in review).

[11] Cai X.-C., Keyes D., Paraschivoiu M., Sarkis M., and Young D. (1999). *Multi-domain multi-model formulation for compressible flows: Conservative interface coupling and parallel implicit solvers for 3D unstructured meshes*. AIAA Paper 99-0784.

[12] Cai X.-C., Mathew T., and Sarkis M. (2000). *Maximum norm analysis of overlapping nonmatching grid discretizations of elliptic equations*. SIAM J. Numer. Anal., SIAM J. Numer. Anal., 37, pp. 1709–1728.

[13] Cai X.-C, and Sarkis M. (1998) *Local multiplicative Schwarz algorithms for steady and unsteady convection-diffusion equations*. East-West J. Numer. Math., Vol. 6.

[14] Cai X.-C., and Widlund O. (1992). *Domain decomposition algorithms for indefinite elliptic problems*. SIAM J. Sci. Statist. Comput., Vol. 13(1), pp. 243–258.

[15] Cai X.-C., and Widlund O. (1993), *Multiplicative Schwarz algorithms for some nonsymmetric and indefinite problems*. SIAM J. Numer. Anal., Vol. 30(4), pp. 936–952.

[16] Chesshire G., Henshaw W. (1990). *Composite overlapping meshes for the solution of partial differential equations*. J. of Comp. Phys., Vol. 90, pp. 1–64.

[17] Chan T., and Mathew T. (1994). *Domain decomposition algorithms*. Acta Numerica 1994. Cambridge University Press, pp. 61–143.

[18] Coclici C., and Wendland W. L. (1998). *Domain decomposition methods and far–field boundary conditions for 2D compressible viscous flows*. Proc. of ECCOMAS 98.

[19] Dryja M., Sarkis M., and Widlund O. (1996). *Multilevel Schwarz methods for elliptic problems with discontinuous coefficients in three dimensions*. Numer. Math., Vol 72, pp. 313–348.

[20] Dryja M., Smith B., and Widlund O. (1994). *Schwarz analysis of iterative substructuring algorithms for elliptic problems in three dimensions*. SIAM J. Numer. Anal., Vol. 31(6), 1662–1694.

[21] Dryja M., and Widlund O. (1987). *An additive variant of the Schwarz alternating method for the case of many subregions*. Department of Computer Science, Courant Institute, TR:339, also Ultracomputer Note 131.

[22] Dryja M., and Widlund O. (1994). *Domain Decomposition Algorithms with Small Overlap*. SIAM J. Sci.Comput., Vol. 15(3), pp. 604–620.

[23] Dryja M., and Widlund O. (1995). *Schwarz methods of Neumann-Neumann type for three-dimensional elliptic finite element problems*. Comm. Pure Appl. Math., Vol. 48(2), pp. 121–155.

[24] Eisenstat S. C., Elman H. C., and Schultz M. H. (1983). *Variational iterative methods for nonsymmetric systems of linear equations*. SIAM J. Numer. Anal., Vol. 20 (2), pp. 345–357.

[25] Eijkhout V., and Vassilevski P. (1991). *The role of the strengthened Cauchy Buniakowskii Schwarz inequality in multilevel methods*. SIAM Review, Vol. 33(3), pp. 405–419.

[26] Eisenstat S. C., and Walker H. F. (1994). *Choosing the forcing terms in an inexact Newton method*. SIAM J. Optimization, Vol. 4, pp. 393–422.

[27] Golub G. H., and Van Loan G. V. (1989). *Matrix computations*. Johns Hopkins Univ. Press, Second Edition.

[28] Griebel M, and Oswald P. (1995). *On the abstract theory of additive and multiplicative Schwarz algorithms*. Numer. Math., Vol. 70, pp. 163-170.

[29] Henshaw W. (1996). *Automatic grid generation*. Acta Numerica, pp. 121–148.

[30] Henshaw W., Brislawn K., Brown D., Chesshire G., Pao K., Quinlan D., and Saltzman J. (1996). *Overture: An object-oriented framework for solving PDEs on overlapping grids*. LA-UR-97-4033, in Third Symposium on Composite Overset Grid and Solution Technology, Los Alamos, New Mexico.

[31] Hestenes M. R. (1956). *The conjugate gradient method for solving linear systems*. Proc. Symp. Appl. Math VI, AMS, McGraw-Hill, New York, pp. 83–102.

[32] Kuznetsov Y. (1998). *Overlapping domain decomposition with nonmatching grids*. Proceedings of the Ninth International Conference on Domain Decomposition Methods, P. Bjørstad, M. Espedal and D. Keyes, eds, Wiley & Sons.

[33] M. Lesoinne, M. Sarkis, U. Hetmaniu, and C. Farhat. (2001). *A linearized method for the frequency analysis of three-dimensional fluid/structure interaction problem in all flow regimes*. Comp. Meth. Appl. Mech. Eng., Vol. 190, pp. 3121–3146.

[34] Lions P. (1988). *On the Schwarz alternating method I.*. First International Symposium on Domain Decomposition Methods for Partial Differential Equations, R. Glowinski, G. Golub, G. Meurant, and J. Périaux, eds., SIAM, Philadelphia, PA.

[35] Mandel J. (1993). *Balancing domain decomposition*. Comm. Numer. Meth. Engrg., Vol. 9, pp. 233–241.

[36] Mandel J. (1994). *Hybrid domain decomposition with unstructured subdomains*. Contemporary Mathematics. Vol. 157, pp. 103–112.

[37] Matsokin A. M., and Nepomnyaschikh S. V. (1985). *A Schwarz alternating method in a subspace*. Soviet Mathematics, Vol. 29(10), pp. 78–84.

[38] Oswald P. (1999) *On the robustness of the BPX-preconditioner with respect to jumps in the coefficients*. Math. Comp., Vol. 68, pp. 633-650.

[39] Paraschivoiu M., Cai X.-C., and Sarkis M. (2000). *An explicit multi-model compressible flow formulation based on the full potential equation and the Euler Equations on 3D unstructured meshes*. Contemporary Mathematics.

[40] Quarteroni A., and Valli A. (1999). *Domain decomposition methods for partial differential equations*. Oxford University Press, Oxford.

[41] Saad Y. (1996). *Iterative methods for sparse linear systems*. PWS Publishing Company, ITP.

[42] Saad Y., and Schultz M. (1986). *GMRES: A generalized minimum residual algorithm for solving nonsymmetric linear systems*. SIAM J. Sci. Stat. Comput., Vol. 7, pp. 856–869.

[43] Sankar L. N., Bharadvaj B. K., and Tsung F. (1993). *Three–dimensional Navier–Stokes/full potential coupled analysis for transonic viscous flow*. AIAA J., Vol. 31, pp. 1857–1862.

[44] Sarkis M. (1997). *Nonstandard coarse spaces and Schwarz methods for elliptic problems with discontinuous coefficients using nonconforming elements*. Numer. Math., Vol 77, pp. 383–406.

[45] Sarkis M., and Koobus B. (2000). *A scaled and minimum overlap restricted additive Schwarz method with application on aerodynamics*. Comp. Meth. Appl. Mech. and Eng., 184, pp. 391–400.

[46] Schwarz A. H. (1890). *Gesammelte mathematische abhaundlungen*. Springer, Vol. 2, pp. 133–143.

[47] Smith B., Bjørstad P., Gropp W. (1996). *Domain ecomposition: Parallel multilevel methods for elliptic partial differential equations*. Cambridge University Press.

[48] Starius G. (1977). *Composite mesh difference methods for elliptic boundary value problems*. Numer. Math., Vol. 28, pp. 243–258.

[49] Steger J., and Benek J. (1987). *On the use of composite grid schemes in computational aerodynamics*. Comp. Methods Appl. Mech. Engr., Vol. 64, pp. 301–320.

[50] Suhs N., and Tramel R. (1991). *PEGSUS 4.0 User's manual*, Calspan Corp./AEDC Operation, June.

[51] Xu J. (1992). *Iterative methods by space decomposition and subspace correction.* SIAM Review, Vol 34, pp. 581–613.

[52] Zhang X. (1992). *Multilevel Schwarz methods.* Numer. Math., Vol. 63, Vol. 4, pp. 512–539.

[53] Widlund O. (1988). *Iterative substructuring methods: algorithms and theory for elliptic problems in the plane.* First International Symposium on Domain Decomposition Methods for Partial Differential Equations, Edited by Roland Glowinski and Gene H. Golub and Gérard A. Meurant and Jacques Périaux, SIAM, Philadelphia, PA.

MODIFICATION AND MAINTENANCE OF ULV DECOMPOSITIONS

Jesse L. Barlow
The Pennsylvania State University
Department of Computer Science and Engineering, USA
barlow@cse.psu.edu

Abstract The ULV decomposition (ULVD) is an important member of a class of rank-revealing two-sided orthogonal decompositions used to approximate the singular value decomposition (SVD). It is useful in the many applications of the SVD where we are only interested in separating the right singular subspace associated with "large" singular values from that associated with "small" ones.

The ULVD may be computed from scratch only a little bit faster than the SVD. However, for $m \times n$ matrices, there are simple algorithms to update and downdate a ULVD in $O(mn)$ flops whereas most software to perform this operation for the SVD requires $O(mn^2)$ flops.

In this article, we survey the advances in update and downdate algorithms. We also discuss algorithms for refinement of a ULVD, that is, for improvement of its approximation to the SVD. We give a reasonable refinement algorithm that requires only $O(mn)$ flops.

Keywords: Subspace approximation, orthogonal decompositions, modifying matrix factorizations, refinement.

1. Introduction

The singular value decomposition (SVD) of a matrix $X \in \Re^{m\times n}, m \geq n$, is written

$$X = Y \begin{pmatrix} \Sigma \\ 0 \end{pmatrix} W^T \tag{1}$$

where

$$Y = (\mathbf{y}_1, \ldots, \mathbf{y}_m) \in \Re^{m\times m}, \quad W = (\mathbf{w}_1, \ldots, \mathbf{w}_n) \in \Re^{n\times n}, \tag{2}$$

are orthogonal and

$$\Sigma = \mathrm{diag}(\sigma_1, \ldots, \sigma_n) \in \Re^{n\times n}, \quad \sigma_1 \geq \ldots \geq \sigma_n. \tag{3}$$

Applied Mathematics and Scientific Computing
Edited by Drmač *et al.*, Kluwer Academic/Plenum Publishers, 2002

It contains useful information about a linear operator including rank, important subspaces, and conditioning. It also gives us the ability to compute low rank approximations.

Some of the most important applications of the SVD include solving ill-conditioned least squares problems, total least squares problems, subspace tracking, and isolating principal components.

In these applications, our interest in the singular value decomposition is to write X in the form

$$X = Y_1\Sigma_1 W_1^T + Y_2\Sigma_2 W_2^T \tag{4}$$

where

$$\Sigma_1 = \text{diag}(\sigma_1, \ldots, \sigma_k), \quad \Sigma_2 = \text{diag}(\sigma_{k+1}, \ldots, \sigma_n)$$

and Y_1, Y_2, W_1 and W_2 are the corresponding matrices of left and right singular vectors. The other information in the SVD is not needed. Moreover, accurate representations of Range(W_1) and/or Range(W_2) are often satisfactory.

We use $\|\cdot\|$ to denotes the two norm (matrix and vector), and $\|\cdot\|_F$ to denote the Frobenius norm. The notation $\sigma_k(X)$ denotes the kth largest singular value of X.

If the SVD is computed only once, its computational cost is only a constant factor times that of computing an LU decomposition or a QR decomposition. Thus, its computation is a reasonable price to pay for obtaining the dichotomy in (4). However, if the decomposition must be modified by adding rows (updating) and/or deleting rows (downdating), then the SVD can be computationally expensive [Gu and Eisenstat, 1995]. Thus alternative two-sided orthogonal decompositions or complete orthogonal decompositions (COD) have been proposed. These provide approximations to Range(W_1) and Range(W_2), but can be updated and downdated in $O(mn)$ flops.

The CODs have the form

$$X = U \begin{pmatrix} C \\ 0 \end{pmatrix} V^T \tag{5}$$

where C is a triangular matrix partitioned according to

$$C = \begin{array}{c} \\ \left(\right. \end{array}\!\!\begin{array}{cc} k & n-k \\ C_1 & C_2 \end{array}\!\!\begin{array}{c} \\ \left. \right) \end{array}, \quad \|C_2\| \approx \|\Sigma_2\| = \sigma_{k+1}$$

and $U \in \Re^{m\times m}$, $V \in \Re^{n\times n}$ are orthogonal.

The first such decompositions were proposed independently in [Faddeev et al., 1968] and in [Hanson and Lawson, 1969]. These used maximal column pivoting [Businger and Golub, 1965]. More recent enhancements, such as the URV and ULV decompositions, discussed in [Stewart, 1992; Stewart, 1993], made use of condition estimators. In this overview, we will use the ULV decomposition (ULVD) which we will define in the form (5) where, for a fixed integer k and tolerance ϵ, we have

$$U = \begin{pmatrix} \overset{k}{U_1} & \overset{m-k}{U_2} \end{pmatrix}, \quad V = \begin{pmatrix} \overset{k}{V_1} & \overset{n-k}{V_2} \end{pmatrix}, \tag{6}$$

$$C = \begin{matrix} k \\ n-k \end{matrix} \overset{\begin{matrix} k & n-k \end{matrix}}{\begin{pmatrix} L & 0 \\ F & G \end{pmatrix}} \tag{7}$$

and

$$\|L^{-1}\|^{-1} \geq \epsilon, \quad \|(\ F \quad G \ )\| \leq \Phi(n)\epsilon. \tag{8}$$

Here $\Phi(n)$ is modestly growing function of n, say, $\sqrt{n}$ or $\sqrt{\lceil \log_2 n \rceil}$. This formulation is a variant of one that has been used in the work of this author with others in [Yoon and Barlow, 1998; Barlow et al., 2001a; Barlow et al., 1999; Barlow et al., 1996; Erbay and Barlow, 2000].

Since, in many applications of interest, X is "tall and thin", that is, $m \gg n$, we will use the "short form" of the decomposition (5). It is written

$$X = UCV^T \tag{9}$$

where C and V are as in (5), but $U \in \Re^{m \times n}$ is a rectangular matrix with orthonormal columns.

The matrix V yields good approximations of the two singular subspaces, Range(W_1) and Range(W_2). Using the partitioning of W in (4) and that of V in (6), from a variant of an argument in [Fierro and Bunch, 1995] given in [Barlow et al., 2001a], the sin θ measure [Davis and Kahan, 1970] of error in subspaces is bounded by

$$|\sin \theta| = \|W_1^T V_2\| \leq \frac{\|F^T G\|}{\mathbf{gap}} \tag{10}$$

where

$$\mathbf{gap} = \max\{\sigma_k^2 \left[\begin{pmatrix} L \\ F \end{pmatrix} \right] - \sigma_{k+1}^2(X), \sigma_k^2(X) - \|G\|^2\}.$$

In this overview, we discuss three kinds of procedures for the ULVD.

- **Computation from scratch.** (See §3). Here we give the truncated SVD procedure from [Van Huffel and Vandewalle, 1991, pp.106–107], a divide–and–conquer procedure from [Erbay and Barlow, 2000], and triangular matrix based procedure from [Fierro and Bunch, 1995].
- **Updating and Downdating.** (See §4) Updating was already given in [Stewart, 1993], though some improvements upon this have been made

[Barlow et al., 1999; Barlow and Yoon, 1997; Yoon and Barlow, 1998]. For downdating, we give a Gram–Schmidt based procedure from [Barlow et al., 1999].

- **Refinement.** (See §5) Often after an update or downdate, the conditions (8) are not satisfied. Thus procedures which make $\|(\; F \quad G \;)\|_F$ smaller are necessary. We discuss procedures from [Mathias and Stewart, 1993; Fierro and Bunch, 1995; Barlow et al., 2001a].

MATLAB codes for the ULVD and the URV decompositions have been made available in [Fierro et al., 1999], but that software does not incorporate the newer ideas on downdating, norm estimation, and refinement discussed in this report .

Throughout the paper, we use MATLAB notation for submatrices, thus $X(i\colon j, :)$ denotes rows i through j of X, $X(:, k\colon \ell)$ denotes columns k through ℓ, and $X(i\colon j, k\colon \ell)$ denotes the intersection of the two. In the operation counts given here, a flop is an addition or a multiplication.

In the next section, we introduce the necessary matrix computational tools to build the algorithms for the ULVD.

2. Matrix Computational Tools for the ULVD

2.1 Q–L Decomposition

Before reducing an $m \times n$ matrix X to the form (5)–(7), we need to perform the factorizaton

$$X = U_0 \begin{pmatrix} C_0 \\ 0 \end{pmatrix} \tag{11}$$

where U_0 is the product of the n Householder transformations

$$U_0 = H_1 \cdots H_n.$$

Here H_k is a Householder transformation that sets entries $1, \ldots, k-1, n+1, \ldots, m$ of column k to zero.

To obtain a "short" form,

$$X = \bar{U}_0 C_0, \quad \bar{U}_0 \in \Re^{m \times n} \tag{12}$$

we compute

$$\bar{U}_0 = H_1 \cdots H_n \begin{pmatrix} I_n \\ 0 \end{pmatrix}.$$

Our complexity estimates will assume that this is done. Computing the decomposition in the form (12) requires $4mn^2 - 4/3n^3$ flops.

2.2 Norm and Condition Estimates

Once we have produced a lower triangular matrix $C \in \Re^{n \times n}$, we need to estimate its smallest singular value and associated vectors given by the triple $(\sigma_n, \mathbf{u}_n, \mathbf{v}_n)$. Realistically, we need to compute $\bar{\sigma}_n$, $\bar{\mathbf{u}}_n$, and $\bar{\mathbf{v}}_n$ such that

$$\begin{aligned} C^{-1}\bar{\mathbf{u}}_n &= \bar{\sigma}_n^{-1}\bar{\mathbf{v}}_n, &(13)\\ C^{-T}\bar{\mathbf{v}}_n &= \bar{\sigma}_n^{-1}\bar{\mathbf{u}}_n + \mathbf{r}_{inv}, &(14)\\ \mathbf{r}_{inv}^T\bar{\mathbf{u}}_n &= 0, \quad \|\mathbf{r}_{inv}\| \leq \bar{\sigma}_n^{-1}\delta &(15) \end{aligned}$$

where δ is some prescribed tolerance which should probably be much smaller than ϵ in (6). In practice, we expect to use only a few Lanczos or power iterations and some back and forward solves with C to perform this computation.

Likewise, to estimate the largest singular value and associated vectors, given by the triple $(\sigma_1, \mathbf{u}_1, \mathbf{v}_1)$, we compute a triple $(\bar{\sigma}_1, \bar{\mathbf{u}}_1, \bar{\mathbf{v}}_1)$ such that

$$\begin{aligned} C\bar{\mathbf{v}}_1 &= \bar{\sigma}_1\bar{\mathbf{u}}_1, &(16)\\ C^T\bar{\mathbf{u}}_1 &= \bar{\sigma}_1\bar{\mathbf{v}}_1 + \mathbf{r}, &(17)\\ \mathbf{r}^T\bar{\mathbf{v}}_1 &= 0, \quad \|\mathbf{r}\| \leq \bar{\sigma}_1\delta. &(18) \end{aligned}$$

Again, this can be achieved with a few Lanczos or power iterations. Here we do not restrict C to be lower triangular or even square. For complexity estimates, we will assume a fixed number iterations, c_{iter}, for both of these procedures.

When these operations are counted in complexity estimates, we will count (13)–(15) as $c_{iter}n^2$ flops and for $C \in \Re^{m \times n}$ we will count (16)–(18) as $2c_{iter}mn$ flops.

When using these procedures in the analysis of other algorithms, we will ignore the errors $\mathbf{r}$ and $\mathbf{r}_{inv}$, though they will affect the accuracy with which the subspaces Range(V_1) and Range(V_2) approximate Range(W_1) and Range(W_2).

2.3 Chasing Algorithms

Once we have computed the approximate singular vectors in §2.2, we will need the two chasing procedures described in this section.

First, suppose that we have computed $(\bar{\sigma}_n, \bar{\mathbf{u}}_n, \bar{\mathbf{v}}_n)$ from (13)–(15). To "deflate" this smallest singular value, we find an orthogonal matrix $\bar{U}$ such that

$$\bar{U}^T\bar{\mathbf{u}}_n = \pm\mathbf{e}_n \qquad (19)$$

and an orthogonal matrix $\bar{V}$ such that

$$\bar{U}^T C\bar{V} = \hat{C} \qquad (20)$$

is lower triangular. The matrices $\bar{U}$ and $\bar{V}$ are each the products of $n-1$ Givens rotations.

A Givens rotation, denoted by $J(i, j, \theta)$, is a rotation through angle θ applied to rows (or columns) i and j where $(cn) = \cos\,\theta$, $(sn) = \sin\,\theta$ are chosen based on two scalars u_1 and u_2 so that

$$\begin{pmatrix} (cn) & (sn) \\ -(sn) & (cn) \end{pmatrix} \begin{pmatrix} u_1 \\ u_2 \end{pmatrix} = \begin{pmatrix} \rho \\ 0 \end{pmatrix}$$

where $\rho = \pm\|(u_1, u_2)^T\|$. For the chasing algorithm (19)–(20), i and j are always two neighboring rows or columns.

In Figure 1, we illustrate how to perform (19)–(20) on a 5×5 example. Two copies of the symbol $\rightarrow$ denote a Givens rotation applied to two rows from the left, and two copies of the symbol $\downarrow$ denotes a Givens rotation applied to two columns from the right. A hat over an element, say, $\hat{u}$ or $\hat{c}$ means that element will be set to zero.

The algorithm (19)–(20) has interesting consequences for $\hat{C}$. If

$$\hat{\mathbf{v}}_n = \bar{V}^T \bar{\mathbf{v}}_n, \quad \hat{\mathbf{r}}_{inv} = U^T \mathbf{r}_{inv}, \tag{21}$$

then (13)–(15) becomes

$$\begin{aligned} \hat{C}\hat{\mathbf{v}}_n &= \pm\bar{\sigma}_n \mathbf{e}_n, & (22) \\ \pm\hat{C}^T \mathbf{e}_n &= \bar{\sigma}_n \hat{\mathbf{v}}_n + \bar{\sigma}_n \hat{C}^T \hat{\mathbf{r}}_{inv}, \quad \|\hat{\mathbf{r}}_{inv}\| \le \delta\bar{\sigma}_n^{-1}. & (23) \end{aligned}$$

It is easily concluded from (22) that

$$\hat{\mathbf{v}}_n = \pm\mathbf{e}_n, \quad \hat{c}_{nn} = \bar{\sigma}_n,$$

$$\|\hat{C}^T \mathbf{e}_n\|^2 = \bar{\sigma}_n^2 + \bar{\sigma}_n^2 \|\hat{C}^T \hat{\mathbf{r}}_{inv}\|^2 \le \bar{\sigma}_n^2 + \|\hat{C}\|^2 \delta^2,$$

and that

$$\begin{pmatrix} 0 & \hat{C}(n, 2\colon n) \end{pmatrix} = \mp\bar{\sigma}_n \hat{\mathbf{r}}_{inv}^T \hat{C}.$$

Thus

$$\|\hat{C}(n, 2\colon n)\| \le \delta\|C\|.$$

Likewise, if C is lower trapezoidal with more columns than rows, we can develop a chasing procedure similar to that above that starts with $\bar{\mathbf{u}}_1$ in (16)–(18) and produces orthogonal matrices $\bar{U}$ and $\bar{V}$ such that

$$\bar{U}^T \bar{\mathbf{u}}_1 = \pm\mathbf{e}_1, \tag{24}$$

and $\bar{V}$ such that $\hat{C}$ in (20) remains lower trapezoidal.

If $\hat{\mathbf{v}}_1 = V^T \bar{\mathbf{v}}_1$ and $\hat{\mathbf{r}} = V^T \mathbf{r}$, then

$$\hat{C}\hat{\mathbf{v}}_1 = \pm\bar{\sigma}_1 \mathbf{e}_1,$$

$$\begin{array}{cccccccc} & & & & & & \\ \rightarrow & c & & & & & \hat{u} \\ \rightarrow & c & c & & & & u \\ & c & c & c & & & u \\ & c & c & c & c & & u \\ & c & c & c & c & c & u \end{array} \qquad \begin{array}{cccccc} \downarrow & \downarrow & & & & \\ c & \hat{c} & & & & 0 \\ c & c & & & & u \\ c & c & c & & & u \\ c & c & c & c & & u \\ c & c & c & c & c & u \end{array}$$

$$\begin{array}{cccccccc} & & & & & & \\ & c & & & & & 0 \\ \rightarrow & c & c & & & & \hat{u} \\ \rightarrow & c & c & c & & & u \\ & c & c & c & c & & u \\ & c & c & c & c & c & u \end{array} \qquad \begin{array}{cccccc} & \downarrow & \downarrow & & & \\ c & & & & & 0 \\ c & c & \hat{c} & & & 0 \\ c & c & c & & & u \\ c & c & c & c & & u \\ c & c & c & c & c & u \end{array}$$

$$\begin{array}{cccccccc} & & & & & & \\ & c & & & & & 0 \\ & c & c & & & & 0 \\ \rightarrow & c & c & c & & & \hat{u} \\ \rightarrow & c & c & c & c & & u \\ & c & c & c & c & c & u \end{array} \qquad \begin{array}{cccccc} & & \downarrow & \downarrow & & \\ c & & & & & 0 \\ c & c & & & & 0 \\ c & c & c & \hat{c} & & 0 \\ c & c & c & c & & u \\ c & c & c & c & c & u \end{array}$$

$$\begin{array}{cccccccc} & & & & & & \\ & c & & & & & 0 \\ & c & c & & & & 0 \\ & c & c & c & & & 0 \\ \rightarrow & c & c & c & c & & \hat{u} \\ \rightarrow & c & c & c & c & c & u \end{array} \qquad \begin{array}{cccccc} & & & \downarrow & \downarrow & \\ c & & & & & 0 \\ c & c & & & & 0 \\ c & c & c & & & 0 \\ c & c & c & c & \hat{c} & 0 \\ c & c & c & c & c & u \end{array}$$

$$\begin{array}{cccccc} c & & & & & 0 \\ c & c & & & & 0 \\ c & c & c & & & 0 \\ c & c & c & c & & 0 \\ c & c & c & c & c & u \end{array}$$

Figure 1. Diagram of Chasing Procedure (19)–(20) on a 5×5 matrix

$$\pm\hat{C}^T\mathbf{e}_1 = \bar{\sigma}_1\hat{\mathbf{v}}_1 + \hat{\mathbf{r}}, \quad ||\hat{\mathbf{r}}|| \leq \delta\bar{\sigma}_1.$$

Since $\hat{\mathbf{r}}^T\hat{\mathbf{v}}_1 = 0$, we have

$$||\hat{C}^T\mathbf{e}_1||^2 = \bar{\sigma}_1^2 + ||\hat{\mathbf{r}}||^2, \tag{25}$$

thus

$$\bar{\sigma}_1 \leq ||\hat{C}^T\mathbf{e}_1|| \leq \bar{\sigma}_1(1+\delta^2)^{1/2}. \tag{26}$$

One goal of this computation is to make the first row of $\hat{C}$ as close to collinear to the approximate singular vector $\hat{\mathbf{v}}_1$ as possible. In fact, (25) suggests that $\hat{C}^T\mathbf{e}_1$ is a slightly better approximate singular vector. To show how close they are, let

$$\cos\,\theta = (\hat{\mathbf{v}}_1^T\hat{C}^T\mathbf{e}_1)/||\hat{C}^T\mathbf{e}_1||.$$

A simple calculation shows that

$$\cos\,\theta = \mp\bar{\sigma}_1/||\hat{C}^T\mathbf{e}_1||$$

and thus

$$(1+\delta^2)^{-1/2} \leq |\cos\,\theta| \leq 1. \tag{27}$$

The role of these chasing routines is discussed in [Fierro and Bunch, 1995] and, to a lesser extent, in [Stewart, 1993].

3. Forming a ULVD

3.1 SVD and partial SVD algorithms

One way to obtain a ULVD from scratch is to just use the SVD. Good SVD software is readily available from MATLAB [Mathworks, Inc., 1996], LAPACK [Anderson et al., 1992], and many other places. In MATLAB, the following "two liner" gets you the ULVD of a matrix X.

```
[U,C,V]=svd(X);
k=sum(diag(C)> ϵ );
```

A ULVD of X is given if we partition U and V according to (6) and C according to (7).

If we need only compute the SVD once, the savings from other methods for computing the ULVD is unimportant. Moreover, recent advances in computing the SVD [Fernando and Parlett, 1994; Fernando, 1999; Parlett and Marques, 2000] have made this approach harder to beat.

If $n-k \ll n$, one method for saving operations is the partial SVD. The algorithm given below is from [Van Huffel and Vandewalle, 1991, pp.106–107] with some details changed.

Algorithm 1 (Partial SVD).

Step 1. *Using the Q–L algorithm in §2.1, factor X into*

$$X = UC$$

where $U \in \Re^{m \times n}$ has orthonormal columns and C is lower triangular.

Step 2. *Using the algorithm from [Golub and Kahan, 1965], find orthogonal matrices $U^{(1)} \in \Re^{n \times n}$ and $V \in \Re^{n \times n}$ such that*

$$[U^{(1)}]^T CV = \begin{matrix} n \\ m-n \end{matrix} \begin{pmatrix} B \\ 0 \end{pmatrix} \tag{28}$$

where

$$B = \begin{pmatrix} \gamma_1 & 0 & \cdot & \cdot & \cdot & \cdot \\ \phi_2 & \gamma_2 & \cdot & \cdot & \cdot & \cdot \\ \cdot & \cdot & \cdot & \cdot & \cdot & \cdot \\ 0 & \cdot & \cdot & \phi_{n-1} & \gamma_{n-1} & 0 \\ 0 & \cdot & \cdot & \cdot & \phi_n & \gamma_n \end{pmatrix}. \tag{29}$$

has bidiagonal form. Update U according to

$$U \leftarrow UU^{(1)}.$$

Step 3. *Use the SVD version [Golub and Reinsch, 1970] of the QR algorithm with Wilkinson shift [Wilkinson, 1968] to compute all singular values of B that are less than ϵ with their associated singular vectors.* (This is likely to result in computing some singular values that are greater than ϵ.) *That produces the reduction of B into*

$$B = U^{(2)} C [V^{(2)}]^T$$

where

$$C = \begin{matrix} k_1 \\ k_2 \\ k_3 \end{matrix} \overset{\begin{matrix} k_1 & k_2 & k_3 \end{matrix}}{\begin{pmatrix} C_1 & & \\ & C_2 & \\ & & C_3 \end{pmatrix}}, \tag{30}$$

C_1 is a bidiagonal matrix of the form

$$C_1 = \begin{pmatrix} \hat{\gamma}_1 & 0 & \cdot & \cdot & \cdot & \cdot \\ \hat{\phi}_2 & \hat{\gamma}_2 & \cdot & \cdot & \cdot & \cdot \\ \cdot & \cdot & \cdot & \cdot & \cdot & \cdot \\ 0 & \cdot & \cdot & \hat{\phi}_{k_1-1} & \hat{\gamma}_{k_1-1} & 0 \\ 0 & \cdot & \cdot & \cdot & \hat{\phi}_{k_1} & \hat{\gamma}_{k_1} \end{pmatrix}, \quad \|C_1^{-1}\|^{-1} > \epsilon, \tag{31}$$

C_2 is a diagonal matrix of singular values greater than ϵ and C_3 is diagonal matrix of singular values less than ϵ. Update U and V according to

$$U \leftarrow UU^{(2)}, \quad V \leftarrow VV^{(2)}$$

To get the form of (7) we let $k = k_1 + k_2$, $L = \mathrm{diag}(C_1, C_2)$, $G = C_3$ and $F = 0$. The resulting representation produces exactly the subspaces we need.

The complexity of this algorithm depends upon how Step 3 is done. The Q–L decomposition and the bidiagonal reduction together cost $6mn^2 + 5n^3 + O(n^2)$ flops. The QR algorithm adds a cost of about $3c_{QR}(m+n)(n^2 - k_1^2) + O(mn)$ flops where c_{QR} is the average number of iterations per singular value, typically 2 or 3. Assuming $c_{QR} = 2$, the total cost of this procedure is $12mn^2 + 13n^3 - 6(m+n)k_1^2 + O(mn)$ flops. The norm inequality in (31) is easily verified using the SVD version of Sturm sequences [Golub and Loan, 1996, §8.5.2].

This algorithm can be improved somewhat using more recent methods of computing the bidiagonal SVD. For instance, if $k_3 = n - k \ll n$, the zero-shift QR algorithm in [Demmel and Kahan, 1990] is likely to produce a smaller C_2 block than the Wilkinson shift version. Newer SVD algorithms promise to produce the singular values and vectors of a bidiagonal in $O(n^2)$ flops [Fernando and Parlett, 1994; Fernando, 1999; Parlett and Marques, 2000]. Thus the cost of this algorithm and the standard SVD algorithm would be about the cost of the Q–L decomposition, the bidiagonal reduction, and two matrix multiplies, amounting to $8mn^2 + 7n^3 + O(n^3)$ flops.

In the next section, we give an alternative approach that uses the bidiagonal reduction above.

3.2 A Divide–and–Conquer Procedure

This algorithm from [Erbay and Barlow, 2000] also makes use of a bidiagonal reduction, but it reduces the bidiagonal matrix to the ULVD in $O(mn^{3/2} + mn \log n)$ operations. A more complex version will be discussed in a paper in preparation [Barlow et al., 2001a].

Algorithm 2 (Divide–and–Conquer ULVD Algorithm).

Step 1. *Do steps 1 and 2 of Algorithm 1 to obtain the bidiagonal matrix B and the resulting matrices U and V. Set $n_0 = n$.*

Step 2. *Partition the matrix B as*

$$B = \begin{array}{c} \\ r \\ 1 \\ n-r-1 \end{array} \begin{array}{c} \begin{array}{cc} r & n-r \end{array} \\ \begin{pmatrix} B_1 & 0 \\ \phi_{r+1}\mathbf{e}_r^T & \alpha_{r+1}\mathbf{e}_1^T \\ 0 & B_2 \end{pmatrix} \end{array}.$$

Step 3. ***If*** $r > n_0^{1/2}$ ***then*** *take* $B = B_1$, $n = r$ *and go to step 2 and obtain the ULVD*

$$B_1 = U^{(1)} \begin{pmatrix} L_1 & 0 \\ F_1 & G_1 \end{pmatrix} V^{(1)T}$$

such that $\epsilon \| L_1^{-1} \| \leq 1$,
else *compute the SVD of* B_1 *to obtain*

$$B_1 = U^{(1)} \begin{pmatrix} \Sigma_{11} & 0 \\ 0 & \Sigma_{12} \end{pmatrix} V^{(1)T}$$

such that $\epsilon \| \Sigma_{11}^{-1} \| < 1$, $\| \Sigma_{12} \| < \epsilon$. ***end if***
Define $L_1 = \Sigma_{11}$, $G_1 = \Sigma_{12}$, $F_1 = 0$.
If $n - r - 1 > n_0^{1/2}$ ***then*** *take* $B = B_2$, $n = n - r - 1$ *and go to step 2 to obtain the ULVD*

$$B_2 = U^{(2)} \begin{pmatrix} L_2 & 0 \\ F_2 & G_2 \end{pmatrix} [V^{(2)}]^T$$

such $\epsilon \| L_2^{-1} \| \leq 1$,
else *compute the SVD of* B_2 *to obtain*

$$B_2 = U^{(2)} \begin{pmatrix} \Sigma_{21} & 0 \\ 0 & \Sigma_{22} \end{pmatrix} [V^{(2)}]^T$$

such that $\epsilon \| \Sigma_{21}^{-1} \| < 1$, $\| \Sigma_{22} \| < \epsilon$. ***end if***
Define $L_2 = \Sigma_{21}$, $G_2 = \Sigma_{22}$, $F_2 = 0$.

Step 4. *Merge the matrices* B_1 *and* B_2 *to get*

$$\begin{pmatrix} U^{(1)T} & 0 & \\ 0 & 1 & 0 \\ 0 & 0 & U^{(2)T} \end{pmatrix} B \begin{pmatrix} V^{(1)T} & 0 \\ 0 & V^{(2)T} \end{pmatrix} = \begin{pmatrix} L_1 & 0 & 0 & \\ F_1 & G_1 & 0 & 0 \\ \mathbf{z}_1^T & \mathbf{z}_2^T & \mathbf{z}_3^T & \mathbf{z}_4^T \\ 0 & 0 & L_2 & 0 \\ 0 & 0 & F_2 & G_2 \end{pmatrix} \tag{32}$$

Step 5. *Reorder the matrix on the right side of (32) to obtain*

$$\begin{pmatrix} U^{(3)T} & 0 & \\ 0 & 1 & 0 \\ 0 & 0 & U^{(4)T} \end{pmatrix} B \begin{pmatrix} V^{(3)T} & 0 \\ 0 & V^{(4)T} \end{pmatrix} = \begin{pmatrix} L_1 & 0 & 0 & \\ 0 & L_2 & 0 & 0 \\ \mathbf{z}_1^T & \mathbf{z}_3^T & \mathbf{z}_2^T & \mathbf{z}_4^T \\ F_1 & 0 & G_1 & 0 \\ 0 & F_2 & 0 & G_2 \end{pmatrix}$$

Step 6. *Construct orthogonal matrices $U^{(5)}, V^{(5)}$ of appropriate dimensions such that*

$$[V^{(5)}]^T \begin{pmatrix} \mathbf{z}_2 \\ \mathbf{z}_4 \end{pmatrix} = \gamma \mathbf{e}_1, \quad \gamma = \pm\| \begin{pmatrix} \mathbf{z}_2 \\ \mathbf{z}_4 \end{pmatrix} \|$$

such that

$$\hat{G} = [U^{(5)}]^T \begin{pmatrix} G_1 & 0 \\ 0 & G_2 \end{pmatrix} V^{(5)}$$

remains lower trapezoidal ($\hat{G}$ is not square in general). Let

$$\hat{F} = [U^{(5)}]^T \begin{pmatrix} F_1 & 0 \\ 0 & F_2 \end{pmatrix}.$$

Define

$$\hat{L} = \begin{pmatrix} L_1 & 0 \\ 0 & L_2 \end{pmatrix}, \tag{33}$$

$$\hat{\mathbf{z}} = \begin{pmatrix} \mathbf{z}_1 \\ \mathbf{z}_3 \end{pmatrix}. \tag{34}$$

Step 7. *Using techniques similar to those in §4.3, construct orthogonal matrices $U^{(6)}, V^{(6)} \in \Re^{n\times n}$ such that*

$$\begin{pmatrix} L & 0 \\ F & G \end{pmatrix} = [U^{(6)}]^T \begin{pmatrix} \hat{L} & 0 \\ \hat{\mathbf{z}}^T & \gamma \mathbf{e}_1^T \\ \hat{F} & \hat{G} \end{pmatrix} V^{(6)}$$

such that $\epsilon\|L^{-1}\| \leq 1$.

The complexity of Algorithm 2 is $6mn^2 + 5n^3 + O(mn^{3/2} + mn \log n)$ flops where the $6mn^2 + 5n^3$ term comes from the bidiagonal reduction step only.

Theorem 1 tells us about the quality of the decomposition obtained by Algorithm 2.

Theorem 1. *Let $X \in \Re^{m\times n}$, $m \geq n$. Then Algorithm 2 produces a decomposition of X such that*

$$X = U \begin{pmatrix} L & 0 \\ F & G \end{pmatrix} V^T \tag{35}$$

where

$$U \in \Re^{m\times n},\ V \in \Re^{n\times n} \ \textit{orthogonal}, \tag{36}$$

$$\epsilon\|L^{-1}\| < 1, \tag{37}$$

$$\|(\ F \quad G\ )\|_F \leq \Phi_F(n)\epsilon, \quad \Phi_F(n) = n^{1/2}, \tag{38}$$

$$\|(F \quad G)\| \leq \Phi(n)\epsilon, \quad \Phi(n) = \left(\frac{1}{2}\lceil \log_2 n \rceil + 1\right)^{1/2}. \tag{39}$$

Note that here the bound on $\|(F \quad G)\|$ is significantly smaller than that on $\|(F \quad G)\|_F$, and that bound grows slowly with n.

3.3 A Triangular Chasing Based Algorithm

The following algorithm is that from [Fierro and Bunch, 1995], but with two important differences:

- Instead of using a condition estimator in [Cline et al., 1982], we use the algorithm in §2.2 to obtain the triplet $(\bar{\sigma}_n, \bar{\mathbf{u}}_n, \bar{\mathbf{v}}_n)$ in (13)–(15).
- C is lower triangular instead of upper triangular.

This algorithm makes use of ideas that the authors of [Foster, 1986; Chan, 1987] used to develop rank revealing QR decompositions.

Algorithm 3 (Triangular Based ULVD Algorithm).

Step 1. *Perform a Q–L decomposition of X to determine $U \in \Re^{m\times n}$ with orthonormal columns and lower triangular $C \in \Re^{n\times n}$ such that*

$$X = UC.$$

Let $V = I_n$ and let $k = n$. Using the procedure discussed in §2.2, determine the triplet $(\bar{\sigma}_n, \bar{\mathbf{u}}_n, \bar{\mathbf{v}}_n)$ that satisfies (13)–(15).

Step 2. *While $\bar{\sigma}_k < \epsilon$ and $k > 0$, use the procedure outlined in §2.3 to find orthogonal $\bar{U}_k, \bar{V}_k \in \Re^{k\times k}$ such that*

$$\bar{U}_k^T \bar{\mathbf{u}}_k = \pm \mathbf{e}_k$$

and such that the operation

$$C(1\colon k, 1\colon k) \leftarrow \bar{U}_k^T C(1\colon k, 1\colon k)\bar{V}_k$$

produces a lower triangular matrix. Then do the computations

$$U(:,1\colon k) \leftarrow U(:,1\colon k)\bar{U}_k, \quad V(:,1\colon k) \leftarrow V(:,1\colon k)\bar{V}_k,$$
$$C(k+1\colon n, 1\colon k) \leftarrow C(k+1\colon n, 1\colon k)\bar{V}_k,$$
$$k \leftarrow k-1.$$

If $k > 0$, determine an approximate smallest singular triplet $(\bar{\sigma}_k, \bar{\mathbf{u}}_k, \bar{\mathbf{v}}_k)$ such that

$$C(1\colon k, 1\colon k)\bar{\mathbf{v}}_k = \bar{\sigma}_k \bar{\mathbf{u}}_k.$$

The complexity of this routine is $7mn^2 - 3mk^2 + 4/3n^3 - 3nk^2 + \frac{c_{iter}}{3}(n^3 - k^3)$ where c_{iter} is the constant from §2.2. If $n - k = o(n)$, the dominant cost is that of the Q–L decomposition, making it much faster than an SVD or a partial SVD or the divide–and–conquer algorithm above. If however, $n-k = O(n)$, its cost will approach that of the other methods and could be even more expensive.

The approach here is a "bottom up" approach, start with a triangular matrix, and working backward to isolate $\text{Range}(V_2)$. A "top down" approach, which works forward to isolate $\text{Range}(V_1)$ is given in [Fierro et al., 1997].

For this method, our best guarantee on $\|(\ F \quad G\ )\|$ is

$$\|(\ F \quad G\ )\| \leq \|(\ F \quad G\ )\|_F \leq (n-k)^{1/2}\,\epsilon$$

which is not a good as the guarantee for the divide–and–conquer method. However, our own tests of the Algorithms 2 and 3 detected little difference in the size of $\|(\ F \quad G\ )\|$ or $\|(\ F \quad G\ )\|_F$ between them.

4. Modifying a ULVD

Probably the most important motivation for the ULVD is ease of modification. The addition of a row to X (called *updating*) or deletion of a row from X (called *downdating*) requires $O(mn)$ flops.

Most schemes to update the SVD require $O(mn^2)$ flops. The exception is a scheme by Gu and Eisentstat [Gu and Eisenstat, 1995] that requires $c_\epsilon mn \log n$ flops, but this is an approximation scheme. Moreover, c_ϵ depends upon the tolerance ϵ desired in that approximation.

We begin our discussion with downdating, since it is more complex than updating. To set the stage for ULVD downdating, in §4.1, we give a brief outline of a Gram–Schmidt procedure for downdating the Q–R decomposition.

In §4.2 and §4.3, we give the ULVD downdating algorithm and show how important parameters in the ULVD may be maintained efficiently in the course of the downdate. In §4.4, we develop an updating algorithm that allows us to maintain these same important parameters.

4.1 Gram-Schmidt-type Downdating Algorithms

Much has been written about downdating algorithms. For Gram–Schmidt based algorithms, recommended reading includes [Daniel et al., 1976; Yoo and Park, 1996; Ruhe, 1983; Björck, 1994]. An analysis of a class of such procedures given in [Barlow et al., 2001b] is the basis of the discussion in this section. That analysis arose out of understanding the behavior of the method in [Yoo and Park, 1996]. Techniques discussed in [Ruhe, 1983; Björck, 1994] are important in formulating this analysis.

To describe downdating for the ULVD, we need to describe it for a Q–R decomposition of the form

$$X = UR \tag{40}$$

where $X \in \Re^{m\times n}, m \geq n$, $U \in \Re^{m\times n}$, and $R \in \Re^{n\times n}$. The matrix U is expected to have orthonormal columns and R is upper triangular.

Suppose that we want a Q–R decomposition of $\bar{X} \in \Re^{(m-1)\times n}$ such that

$$X = \begin{matrix} 1 \\ m-1 \end{matrix} \begin{pmatrix} \mathbf{y}^T \\ \bar{X} \end{pmatrix}.$$

To find that decomposition, we find a unit vector $\mathbf{u}_0 \in \Re^m$ and a vector $\mathbf{d} \in \Re^{n+1}$ such that

$$U^T\mathbf{u}_0 = 0, \quad \begin{pmatrix} \mathbf{u}_0 & U \end{pmatrix}\mathbf{d} = \mathbf{e}_1. \tag{41}$$

We then find $Q \in \Re^{(n+1)\times(n+1)}$, a product of the Givens rotations

$$Q = G_2G_3\cdots G_{n+1}$$

such that $G_k = J(1, k, \theta_k)$ and

$$Q^T\mathbf{d} = \pm\|\mathbf{d}\|_2\mathbf{e}_1. \tag{42}$$

We compute

$$\begin{pmatrix} \mathbf{z}^T \\ \bar{R} \end{pmatrix} = Q^T \begin{pmatrix} 0 \\ R \end{pmatrix}, \tag{43}$$

followed by the computation

$$\begin{matrix} \\ 1 \\ m-1 \end{matrix} \begin{matrix} \begin{matrix} 1 & n \end{matrix} \\ \begin{pmatrix} \gamma & \mathbf{s}^T \\ \mathbf{p} & \bar{U} \end{pmatrix} \end{matrix} = \begin{matrix} \begin{matrix} 1 & n \end{matrix} \\ \begin{pmatrix} \mathbf{u}_0 & U \end{pmatrix} Q. \end{matrix}$$

Finally, we accept

$$\bar{X} = \bar{U}\bar{R}$$

as the Q–R decomposition of $\bar{X}$. If $\mathbf{d}$ exactly satisfies (41) then $\mathbf{p} = 0$. If U is exactly orthogonal then $\mathbf{s} = 0$. In practice, we expect $\|\mathbf{s}\|$ and $\|\mathbf{p}\|$ to be small. Their relation to (41) and the orthogonality of $\bar{U}$ is discussed in [Barlow et al., 2001b].

The simplest procedure in [Barlow et al., 2001b] is a variant of the classical Gram–Schmidt procedure in [Daniel et al., 1976]. It assumes that

$$U^TU = I + E + E^T$$

where E is a strictly lower triangular matrix, and $\|E\| = \eta \leq 0.1$. The parameter η is a measure of the distance between U and an orthogonal matrix. At the very least, it is desirable that $\mathbf{u}_0$ satisfy

$$\|U^T\mathbf{u}_0\| \leq \eta. \tag{44}$$

Under these assumptions, the following steps are used to obtain $\mathbf{u}_0$ and $\mathbf{d}$. First, compute

$$\mathbf{f}_1 = U^T\mathbf{e}_1; \tag{45}$$

$$\mathbf{r}_1 = \mathbf{e}_1 - U\mathbf{f}_1; \quad \beta_1 = \|\mathbf{r}_1\|; \tag{46}$$

$$\mathbf{z}_1 = \mathbf{r}_1/\beta_1; \tag{47}$$

This is one classical Gram–Schmidt step. We then perform a second.

$$\mathbf{f}_2 = U^T\mathbf{z}_1; \tag{48}$$

$$\mathbf{r}_2 = \mathbf{z}_1 - U\mathbf{f}_2; \quad \beta_2 = \|\mathbf{r}_2\|; \tag{49}$$

$$\mathbf{z}_2 = \mathbf{r}_2/\beta_2; \tag{50}$$

Using a criterion from [Hoffman, 1989], we check the size of β_2 to determine our success in achieving orthogonality. If $\beta_2 \geq \sqrt{0.8}$ (Hoffman's threshold is 0.5), excluding the effects of rounding, it is shown in [Barlow et al., 2001b] that

$$\|U^T\mathbf{z}_2\| \leq \eta.$$

Thus, to obtain (44), we accept

$$\mathbf{u}_0 = \mathbf{z}_2, \quad \mathbf{d} = \begin{pmatrix} \beta_1\beta_2 \\ \mathbf{f}_1 + \beta_1\mathbf{f}_2 \end{pmatrix}. \tag{51}$$

The vector $\mathbf{d}$ solves the second equation in (41).

If $\beta_2 < \sqrt{0.8}$, we use a different method for generating a vector $\mathbf{u}_0$.

We let $k \in \{1, \ldots, m\}$ be an index such that

$$\|U^T\mathbf{e}_k\| = \min_{1\leq j\leq m} \|U^T\mathbf{e}_j\|. \tag{52}$$

We then use two classical Gram–Schmidt steps to compute $\mathbf{u}_0$ by orthogonalizing $\mathbf{e}_k$ against U.

$$\mathbf{h}_1 = U^T\mathbf{e}_k; \tag{53}$$

$$\mathbf{t}_1 = \mathbf{e}_k - U\mathbf{h}_1; \quad \xi_1 = \|\mathbf{t}_1\|; \tag{54}$$

$$\mathbf{s}_1 = \mathbf{t}_1/\xi_1; \tag{55}$$

$$\mathbf{h}_2 = U^T\mathbf{s}_1; \tag{56}$$

$$\mathbf{t}_2 = \mathbf{s}_1 - U\mathbf{h}_2; \quad \xi_2 = \|\mathbf{t}_2\|. \tag{57}$$

$$\mathbf{u}_0 = \mathbf{t}_2/\xi_2; \tag{58}$$

Finally, we do one additional step to produce a vector $\mathbf{d}$ that makes

$$\|\mathbf{e}_1 - (\ \mathbf{u}_0 \quad U\ )\,\mathbf{d}\|$$

as small as possible. That step is

$$\delta\mathbf{d} = (\ \mathbf{u}_0 \quad U\ )^T \mathbf{z}_2, \tag{59}$$

$$\mathbf{d} = \begin{pmatrix} 0 \\ \mathbf{f}_1 + \beta_2 \mathbf{f}_2 \end{pmatrix} + \beta_1\beta_2\delta\mathbf{d}. \tag{60}$$

These alternative values of $\mathbf{u}_0$ and $\mathbf{d}$ are shown in [Barlow et al., 2001b] to satisfy

$$\|U^T\mathbf{u}_0\| \leq 4\eta^2\sqrt{n/(m-n)}, \qquad \|\mathbf{e}_1 - (\ \mathbf{u}_0 \quad U\ )\mathbf{d}\| \leq 4\eta.$$

Qualitatively, there is little difference between using classical Gram–Schmidt and modified Gram–Schmidt in this procedure. If modified Gram–Schmidt is used, the coefficients in the above equation are 1 instead of 4 and the threshold value for β_2 is $\sqrt{0.5}$ instead of $\sqrt{0.8}$. The above results change little if rounding error is considered [Barlow et al., 2001b].

If $\beta_2 \geq \sqrt{0.8}$, then the above procedure requires $8mn + O(m+n)$ flops. If $\beta_2 < \sqrt{0.8}$, it requires $18mn + O(m+n)$ flops. In practice, $\beta_2 \geq \sqrt{0.8}$ unless $\mathbf{e}_1$ is very close to Range(U), so, on average, the cost will be close to $8mn$ flops.

4.2 Deleting a Row

We now use the routine described by equations (45)–(60) to produce a method to downdate the ULVD. The big difference for the ULVD is that we must

- Preserve lower triangular form instead of upper triangular form.
- Keep the blocks F and G small throughout the computation (for stability reasons analyzed in [Barlow et al., 1996]).
- Transform the problem so that the downdate affects only the kth row, thus allowing us to update $\|F\|_F$, $\|G\|_F$, $\|L^{-1}\|_F$, and an upper bound on $\|L^{-1}\|$ cheaply.

The algorithm is from [Barlow et al., 1999]. It produces the decomposition of $\bar{X}$ given by

$$\bar{X} = \bar{U}\bar{C}\bar{V}^T \tag{61}$$

where

$$\bar{C} = \begin{matrix} s \\ n-s \end{matrix}\overset{\begin{matrix} s & n-s \end{matrix}}{\begin{pmatrix} \bar{L} & 0 \\ \bar{F} & \bar{G} \end{pmatrix}}. \tag{62}$$

Here either $s = k - 1$ or $s = k$ as determined by step 5 of the algorithm.

Algorithm 4 (Procedure for ULVD downdating).

Step 1. *Using the procedure in (45)–(60), construct vectors* $\mathbf{d} \in \Re^{n+1}$, $\mathbf{u}_0 \in \Re^m$ *satisfying (41). Let* $\mathbf{d}$ *be partitioned*

$$\mathbf{d} = \begin{matrix} 1 \\ k \\ n-k \end{matrix} \begin{pmatrix} \alpha \\ \mathbf{d}_1 \\ \mathbf{d}_2 \end{pmatrix}$$

conformally with (7).

Step 2. *Using the chasing procedures from §2.3, construct orthogonal matrices* $U^{(1)}, V^{(1)} \in \Re^{k \times k}$ *and* $U^{(2)}, V^{(2)} \in \Re^{(n-k)\times(n-k)}$ *such that*

$$L^{(1)} = [U^{(1)}]^T L V^{(1)}, \quad [U^{(1)}]^T \mathbf{d}_1 = \xi_1 \mathbf{e}_k, \quad \xi_1 = \pm\|\mathbf{d}_1\|,$$
$$G^{(1)} = [U^{(2)}]^T G V^{(2)}, \quad [U^{(2)}]^T \mathbf{d}_2 = \xi_2 \mathbf{e}_1, \quad \xi_2 = \pm\|\mathbf{d}_2\|,$$

where $L^{(1)}$ *and* $G^{(1)}$ *are lower triangular. Also compute*

$$F^{(1)} = [U^{(2)}]^T F V^{(1)},$$

and

$$U \leftarrow U \begin{pmatrix} U^{(1)} & 0 \\ 0 & U^{(2)} \end{pmatrix}, \quad V \leftarrow V \begin{pmatrix} V^{(1)} & 0 \\ 0 & V^{(2)} \end{pmatrix}.$$

Step 3. *For* $i = 1, 2$, *let* $\mathbf{f}_1^{(i)} = [F^{(i)}]^T \mathbf{e}_1$, *let* $g_{11}^{(i)} = \mathbf{e}_1^T G^{(i)} \mathbf{e}_1$, *and* $\ell_k^{(i)} = L^T \mathbf{e}_k$. *Find Givens rotations* $\tilde{U}^{(3)} = J(1, k+2, \theta_1)$, $\tilde{U}_4 = J(1, k+1, \theta_2)$ *where* $(cn)_j = \cos\,\theta_j$, $(sn)_j = \sin\,\theta_j$, $j = 1, 2$ *are chosen so that*

$$\begin{pmatrix} (cn)_1 & (sn)_1 \\ -(sn)_1 & (cn)_1 \end{pmatrix} \begin{pmatrix} \alpha \\ \xi_2 \end{pmatrix} = \begin{pmatrix} \alpha_1 \\ 0 \end{pmatrix},$$

$$\begin{pmatrix} (cn)_2 & (sn)_2 \\ -(sn)_2 & (cn)_2 \end{pmatrix} \begin{pmatrix} \alpha_1 \\ \xi_1 \end{pmatrix} = \begin{pmatrix} 1 \\ 0 \end{pmatrix}.$$

Let

$$\mathbf{f}_1^{(2)} = (cn)_1 \mathbf{f}_1^{(1)}, \quad g_{11}^{(2)} = (cn)_1 g_{11}^{(1)},$$

$$\ell_k^{(2)} = (cn)_2 \ell_k^{(1)} - (sn)_1 (sn)_2 \mathbf{f}_1^{(1)}, \quad \ell_{k,k+1}^{(2)} = -(sn)_1 (sn)_2 g_{11}^{(1)}.$$

Then perform the operations

$$\begin{pmatrix} \mathbf{u}_0 & U(:,k+1) \end{pmatrix} \leftarrow \begin{pmatrix} \mathbf{u}_0 & U(:,k+1) \end{pmatrix} \begin{pmatrix} (cn)_1 & -(sn)_1 \\ (sn)_1 & (cn)_1 \end{pmatrix}$$

$$\begin{pmatrix} \mathbf{u}_0 & U(:,k) \end{pmatrix} \leftarrow \begin{pmatrix} \mathbf{u}_0 & U(:,k) \end{pmatrix} \begin{pmatrix} (cn)_2 & -(sn)_2 \\ (sn)_2 & (cn)_2 \end{pmatrix}.$$

Step 4. *To eliminate $\ell^{(2)}_{k,k+1}$ and restore triangular form, find a Givens rotation $V^{(3)} = J(k, k+1, \theta_3)$ such that*

$$\begin{pmatrix} (cn)_3 & (sn)_3 \\ -(sn)_3 & (cn)_3 \end{pmatrix} \begin{pmatrix} \ell^{(2)}_{kk} \\ \ell^{(2)}_{k,k+1} \end{pmatrix} = \begin{pmatrix} \ell^{(3)}_{kk} \\ 0 \end{pmatrix}.$$

$$\begin{pmatrix} F^{(3)} & G^{(3)} \end{pmatrix} = \begin{pmatrix} F^{(2)} & G^{(2)} \end{pmatrix} V^{(3)}.$$

$$\begin{pmatrix} V(:,k) & V(:,k+1) \end{pmatrix} \leftarrow$$
$$\leftarrow \begin{pmatrix} V(:,k) & V(:,k+1) \end{pmatrix} \begin{pmatrix} (cn)_3 & -(sn)_3 \\ (sn)_3 & (cn)_3 \end{pmatrix}.$$

Get rid of the first row of U.

$$U \leftarrow U(2\colon m, :).$$

Step 5. *Using the algorithms discussed in §4.3, compute or bound $\|[L^{(3)}]^{-1}\|$. If $\|[L^{(3)}]^{-1}\|^{-1} \le \epsilon$ set $\bar{L} = L^{(3)}$, $\bar{F} = F^{(3)}$, $\bar{G} = G^{(3)}$, $\bar{U} = U$, and $\bar{V} = V$. Otherwise, as discussed in §4.3, determine the ULVD of $L^{(3)}$ and if necessary modify $F^{(3)}$ and $G^{(3)}$ accordingly.*

Step 1 requires between $8mn$ and $18mn$ operations if the classical Gram–Schmidt procedure in §4.1 is used. Steps 2 and 3 require $6mn+12n^2+O(m+n)$ flops. In the worst case, steps 4 and 5 require an additional $6mk+(6+c_{iter})k^2+O(m+n)$ flops where c_{iter} is the parameter given in §2.2.

Remark 1. *From the terminology of Algorithm 4, if there is no change in the dimensions of the block in the last step, the following formulas relate $\|\bar{F}\|_F$ and $\|\bar{G}\|_F$ to $\|F\|_F$ and $\|G\|_F$,*

$$\begin{aligned} \delta_{FG} &= \|G^{(3)}\mathbf{e}_1\|^2 - \|G^{(2)}\mathbf{e}_1\|^2, \\ \|\bar{F}\|_F^2 &= \|F\|_F^2 - \delta_{FG} - (sn)_1^2\|\mathbf{f}_1^{(1)}\|^2, && (63) \\ \|\bar{G}\|_F^2 &= \|G\|_F^2 + \delta_{FG} - (sn)_1^2[g_{11}^{(1)}]^2. && (64) \end{aligned}$$

These expressions can be combined to yield

$$\| \begin{pmatrix} \bar{F} & \bar{G} \end{pmatrix} \|_F^2 = \| \begin{pmatrix} F & G \end{pmatrix} \|_F^2 - (sn)_1^2 \| \begin{pmatrix} F^{(1)} & G^{(1)} \end{pmatrix}^T \mathbf{e}_1 \|^2, \tag{65}$$

and

$$\| \begin{pmatrix} F^{(2)} & G^{(2)} \end{pmatrix} \mathbf{z} \|^2 = \| \begin{pmatrix} F^{(1)} & G^{(1)} \end{pmatrix} \mathbf{z} \|^2 - (sn)_1^2 \| \mathbf{e}_1^T \begin{pmatrix} F^{(1)} & G^{(1)} \end{pmatrix} \mathbf{z} \|^2, \quad \mathbf{z} \in \Re^n. \tag{66}$$

From (66), we may conclude that

$$\| \begin{pmatrix} \bar{F} & \bar{G} \end{pmatrix} \| \leq \| \begin{pmatrix} F & G \end{pmatrix} \|. \tag{67}$$

The expressions (63) and (64) require $4n$ flops to compute altogether. The expression (65) requires only $4(k+1)$ flops. Since $\|G\|_F$ is an the estimate of the error in the total least squares problem [Van Huffel and Vandewalle, 1991], quick updates to these values are important. If

$$\| \begin{pmatrix} \bar{F} & \bar{G} \end{pmatrix} \|_F \ll \| \begin{pmatrix} F & G \end{pmatrix} \|_F$$

it is desirable to recompute $\| \begin{pmatrix} \bar{F} & \bar{G} \end{pmatrix} \|_F$ from scratch. Since that is only $2n(n-k)$ flops, such an expense is not overwhelming.

If the dimensions of the blocks change by one in Step 5, it is still easy to update $\|\bar{F}\|_F, \|\bar{G}\|_F$, and $\| \begin{pmatrix} \bar{F} & \bar{G} \end{pmatrix} \|_F$ in $O(n)$ flops.

Remark 2. *We can also develop fast expressions for updating $\|L^{-1}\|_F$. Since Algorithm 4 downdates only the last row of $L^{(1)}$ we have*

$$\|\bar{L}^{-1}\|_F^2 = \|L^{-1}\|_F^2 + \|[L^{(3)}]^{-T}\mathbf{e}_k\|^2 - \|[L^{(1)}]^{-T}\mathbf{e}_k\|^2. \tag{68}$$

It would be nice to be able to use only these computations. Unfortunately, the use of (68) has three significant weaknesses: (1) It is too conservative an upper bound on $\|L^{-1}\|$; (2) Over many updates and downdates the computation (68) can become very inaccurate in floating point arithmetic. Recomputation from scratch is $O(k^3)$ and thus out of the question ; (3) The Frobenius norm $\|\cdot\|_F$ is not an operator norm. It does not give us a specific vector to deflate if $\|\bar{L}^{-1}\|_F$ is too large.

4.3 Tracking $\|L^{-1}\|$ and Related Issues

Let $\bar{L} = L^{(3)} \in \Re^{k \times k}$ from steps 1–4 of Algorithm 4. A classical theorem on rank-one modifications [Wilkinson, 1965, pp.94–97] tells us that if C has k singular values greater than ϵ, then the $\bar{C}$ resulting from Algorithm 4 has either $k-1$ or k singular values greater than ϵ. To determine which of k or $k-1$ is correct, we must estimate $\|\bar{L}^{-1}\|$.

We have that

$$\bar{L}^{-1} = \begin{matrix} k-1 \\ 1 \end{matrix} \overset{\begin{matrix} k-1 & \qquad\qquad\qquad\qquad\qquad 1 \end{matrix}}{\begin{pmatrix} \bar{L}^{-1}(1\colon k-1, 1\colon k-1) & 0 \\ -\bar{L}^{-1}(k,k)\bar{L}(k, 1\colon k-1)\bar{L}^{-1}(1\colon k-1, 1\colon k-1) & \bar{L}^{-1}(k,k) \end{pmatrix}}. \tag{69}$$

A well–known bound on norms [Golub and Loan, 1996, p.484] yields

$$\|\bar{L}^{-1}\| \leq \kappa \stackrel{def}{=} \| \begin{pmatrix} a & 0 \\ c & b \end{pmatrix} \| \tag{70}$$

where

$$a = \|\bar{L}^{-1}(1\colon k-1, 1\colon k-1)\|, \quad b = |\bar{L}^{-1}(k,k)|,$$
$$c = b\|\bar{L}(k, 1\colon k-1)\bar{L}^{-1}(1\colon k-1, 1\colon k-1)\|.$$

Thus κ in (70) is given by

$$\kappa = \left(a^2 + b^2 + c^2 + \sqrt{(a^4 + 2a^2(c^2 - b^2) + (b^2 + c^2)^2)}\right)^{1/2} / \sqrt{2}. \tag{71}$$

The exact value of a will not be available to us, but since only the last row of $\bar{L}$ is altered by the downdate, we keep an upper bound $\hat{a}$ such that

$$\hat{a} \geq \|L^{-1}\| \geq \|L^{-1}(1\colon k-1, 1\colon k-1)\| = a.$$

We compute $\hat{\kappa}$ from

$$\hat{\kappa} = \left(\hat{a}^2 + b^2 + c^2 + \sqrt{(\hat{a}^4 + 2\hat{a}^2(c^2 - b^2) + (b^2 + c^2)^2)}\right)^{1/2} / \sqrt{2}. \tag{72}$$

Since the bound in (71) is a monotone increasing function in a,b, and c, we have that

$$\hat{\kappa} \geq \kappa \geq \|\bar{L}^{-1}\|.$$

The value of c results from a backsolve and can be computed exactly in $(k-1)^2$ flops. An $O(k)$ flop bound for c is

$$c \leq b\|\bar{L}(k, 1\colon k-1)\|\|L^{-1}\|.$$

Since

$$\bar{L}^{-T}\mathbf{e}_k = \bar{L}^{-1}(k,k) \begin{pmatrix} -\bar{L}^{-T}(1\colon k-1, 1\colon k-1)\bar{L}(k, 1\colon k-1)^T \\ 1 \end{pmatrix},$$

this computation comes out of inverse power or Lanczos methods with $\mathbf{e}_k$ as a starting vector. If $\|L^{-T}\mathbf{e}_k\|$ makes a substantial contribution to (68), then we would expect $\mathbf{e}_k$ to be a good starting vector for these iterations.

If $\hat{\kappa} \leq 1/\epsilon$, Algorithm 4 is done and $\hat{\kappa}$ is kept as our bound on $a = \|L^{-1}\|$ for the next update or downdate. Otherwise, we use a procedure discussed in

§2.2 to compute $\|\bar{L}^{-1}\|$ more precisely. If $\|\bar{L}^{-1}\|^{-1} \geq \epsilon$, again we are done and we keep $a = \|L^{-1}\|$ for the next update or downdate.

If $\|\bar{L}^{-1}\|^{-1} < \epsilon$, then step 5 of Algorithms 4 is completed by computing the ULVD of $\bar{L}$. That is, we find orthogonal matrices $\tilde{U}$ and $\tilde{V}$, and an $k \times k$ lower triangular matrix $\tilde{L}$ such that

$$\bar{L} = \tilde{U} \begin{pmatrix} \tilde{L} & 0 \\ \hat{\mathbf{f}}_1^T & \hat{g}_{11} \end{pmatrix} \tilde{V}^T \tag{73}$$

where $\|\tilde{L}^{-1}\|^{-1} \geq \epsilon > \|(\hat{\mathbf{f}}_1^T, \hat{g}_{11})^T\|$. Appropriate adjustments are made to $\bar{F}$ and $\bar{G}$ which include making $\hat{\mathbf{f}}_1^T$ the first row of $\bar{F}$ and making $\hat{g}_{11}e_1^T$ the first row of $\bar{G}$. We then compute $a = \|\tilde{L}^{-1}\|$ for use in the next update or downdate.

4.4 Adding a Row

The work that did much to revive interest in two-sided orthogonal decompositions [Stewart, 1992; Stewart, 1993] is based on the notion that adding a row to these decompositions is only a little bit more work than adding a row to a Q–R decomposition. The update algorithm in [Stewart, 1993] does not allow for the easy norm updates or the condition number bound given in §4.3.

Algorithm 5 is based upon the downdating algorithms in [Barlow et al., 1996; Barlow et al., 1999]. This update algorithm does some orthogonal transformations to C that produce a "spike" in the $(k+1)st$ column, performs the update on the $(k+1)st$ row, then restores triangular form. We suppose that X has the ULVD (5)–(7), and we want the ULVD of

$$X_{new} = \begin{pmatrix} \mathbf{x}_0^T \\ X \end{pmatrix}.$$

This routine produces a decompostion

$$X_{new} = U_{new} C_{new} V_{new}^T \tag{74}$$

where

$$C_{new} = \begin{matrix} & \begin{matrix} s & n-s \end{matrix} \\ \begin{matrix} s \\ n-s \end{matrix} & \begin{pmatrix} L_{new} & 0 \\ F_{new} & G_{new} \end{pmatrix} \end{matrix} \tag{75}$$

where $s = k$ or $s = k+1$ depending upon the result of a condition estimation technique used at the end of the computation.

Algorithm 5 (ULVD Update Procedure).
Let V_1 and V_2 comes from the partition in (7) and let L, F and G be as defined in (7).

Step 1. *Compute*

$$\mathbf{z}_1 = V_1^T \mathbf{x}_0, \quad \mathbf{z}_2 = V_2^T \mathbf{x}_0.$$

Using a chasing procedure closely related to that in §2.3, find $V^{(1)} \in \Re^{(n-k)\times(n-k)}$ *such that*

$$[V^{(1)}]^T \mathbf{z}_2 = \rho_2 \mathbf{e}_1,$$

and find $U^{(1)} \in \Re^{(n-k)\times(n-k)}$ *such that*

$$G^{(1)} = [U^{(1)}]^T G V^{(1)}$$

remains lower triangular. Then compute

$$F^{(1)} = [U^{(1)}]^T F, \quad U(:, k+1\colon n) \leftarrow U(:, k+1\colon n) U^{(1)},$$

$$V(:, k+1\colon n) \leftarrow V(:, k+1\colon n) V^{(1)}.$$

Step 2. *Let* $\mathbf{f}_1 = [F^{(1)}]^T \mathbf{e}_1$, $g_{11} = \mathbf{e}_1^T [G^{(1)}] \mathbf{e}_1$. *Define* S *as*

$$S = \begin{pmatrix} L & 0 \\ \mathbf{f}_1^T & g_{11} \end{pmatrix}.$$

Redefine F *and* G *as*

$$\begin{aligned} F^{(2)} &= \begin{pmatrix} F^{(1)}(2\colon n-k, :) & G^{(1)}(2\colon n-k, 1) \end{pmatrix}, \\ G^{(2)} &= G^{(1)}(2\colon n-k, 2\colon n-k). \end{aligned}$$

Using the chasing procedures in §2.3, construct $U^{(2)}, V^{(2)} \in \Re^{(k+1)\times(k+1)}$ *to be orthogonal matrices such that*

$$[V^{(2)}]^T \begin{pmatrix} \mathbf{z}_1 \\ \rho_2 \end{pmatrix} = \rho_1 \mathbf{e}_{k+1}$$

and such that

$$S^{(1)} = [U^{(2)}]^T S V^{(2)}$$

remains lower triangular.

$$F^{(3)} = F^{(2)} V^{(2)}.$$

Step 3. *Find orthogonal* $U^{(3)} \in \Re^{(k+1)\times(k+1)}$ *such that*

$$S^{(2)} = \begin{pmatrix} L^{(2)} & \mathbf{h} \\ 0 & s_{k+1,k+1}^{(2)} \end{pmatrix} = [U^{(3)}]^T S^{(1)}$$

where $L^{(2)}$ remains lower triangular. The vector $\mathbf{h}$ *is the "spike" referenced above.*

Compute

$$U(:,1:k+1) \leftarrow U(:,1:k+1)U^{(3)}$$

Let $U^{(4)} \in \Re^{(m+1)\times(m+1)}$ be a Givens rotation rotating rows $m+1$ and $k+1$ such that its nontrivial part computes

$$\begin{pmatrix} (cn) & -(sn) \\ (sn) & (cn) \end{pmatrix} \begin{pmatrix} \rho_2 \\ s^{(2)}_{k+1,k+1} \end{pmatrix} = \begin{pmatrix} 0 \\ s^{(3)}_{k+1,k+1} \end{pmatrix}.$$

Thereby replacing entry $s^{(2)}_{k+1,k+1}$ with $s^{(3)}_{k+1,k+1}$ to form $S^{(3)}$.

Let

$$U \leftarrow \frac{1}{m}\begin{pmatrix} 0 \\ U \end{pmatrix}.$$

Then the application of $U^{(4)}$ just changes column $k+1$ of U according to

$$U(:,k+1) \leftarrow \begin{pmatrix} (sn) \\ (cn)U(2:m+1,k+1) \end{pmatrix}.$$

Step 4. *Finally compute $V^{(3)} \in \Re^{(k+1)\times(k+1)}$ so that*

$$S^{(4)} = \begin{pmatrix} L^{(4)} & 0 \\ \bar{\mathbf{h}}^T & s^{(4)}_{k+1,k+1} \end{pmatrix} = \begin{pmatrix} L^{(3)} & \mathbf{h} \\ 0 & s^{(3)}_{k+1,k+1} \end{pmatrix} V^{(3)} = S^{(3)}V^{(3)}$$

$$F^{(4)} = F^{(3)}V^{(3)}.$$

Update V according to

$$V(:,1:k+1) = V(:,1:k+1)V^{(3)}$$

Step 5. *Compute or bound $\|[S^{(4)}]^{-1}\|$ using the procedure described in Remark 4 below. If $\|[S^{(4)}]^{-1}\|^{-1} \leq \epsilon$, accept $L_{new} = S^{(4)}$, $F_{new} = F^{(4)}$, $G_{new} = G^{(3)}$, and let $s = k+1$ in (75). Otherwise determine the ULVD of $S^{(4)}$ as*

$$S^{(4)} = U^{(5)} \begin{pmatrix} L_{new} & 0 \\ \hat{\mathbf{f}}_1^T & \hat{g}_{11} \end{pmatrix} V^{(4)}$$

using the techniques from §2.2 and §2.3. Then let

$$U(:,1:k+1) \leftarrow U(:,1:k+1)U^{(5)},$$
$$V(:,1:k+1) \leftarrow V(:,1:k+1)V^{(4)}.$$

Then let

$$F^{(5)} = F^{(4)}V^{(5)}$$

$$F_{new} = \begin{pmatrix} \hat{\mathbf{f}}_1^T \\ F^{(5)}(:,1:k) \end{pmatrix}, \quad G_{new} = \begin{pmatrix} \hat{g}_{11} & 0 \\ F^{(5)}(:,k+1) & G^{(3)} \end{pmatrix}.$$

In either case, accept $U_{new} = U$ and $V_{new} = V$ in (74).

Step one of this algorithm is $6(m+n)(n-k)+6n(n-k)+(m+n)$ flops. Step two is $6(m+n)k+6nk+O(m+n)$ flops. Step three is $6mk+3k^2+O(m)$ flops, and step four is $6nk + 3k^2 + O(n)$ flops. Thus steps one through four are $6(m+n)(n+2k)+6n^2+3k^2+O(m+n)$ flops. As in Algorithm 4, the last step may add as much as $6(m+n)k + (6 + c_{iter})k^2$ flops where c_{iter} is the parameter given in §2.2.

Remark 3. *The norms $\|F_{new}\|_F$, $\|G_{new}\|_F$ and $\|(F_{new} \quad G_{new})\|_F$ can be computed with $O(n)$ additional flops in the course of this computation in a very similar manner as in Algorithm 4. It is just an exploitation of orthogonal equivalence and the fact that the update is done through row $k+1$. Similarly, the norm $\|L_{new}^{-1}\|_F$ can be computed from $\|L^{-1}\|_F$ in $O(k^2)$ flops.*

Remark 4. *As shown by Yoon [Yoon and Barlow, 1998; Yoon, 1996], this algorithm can actually do a little better than the downdating algorithm in how it computes a bound on $\|[S^{(4)}]^{-1}\|$ for step 5. Suppose that*

$$\hat{a} \geq \|L^{-1}\|$$

is given. Then it is easily shown that

$$\|[S^{(4)}]^{-1}\| \leq \| \begin{pmatrix} a & 0 \\ b & c \end{pmatrix} \| \tag{76}$$

where

$$a = \|[L^{(2)}]^{-1}\|,$$

$$b = |s_{k+1,k+1}^{(3)}|^{-1}\|[L^{(2)}]^{-1}\mathbf{h}\|, \quad c = |s_{k+1,k+1}^{(3)}|^{-1}.$$

One can also show that

$$\hat{a} \geq \|L^{-1}\| \geq \|[L^{(2)}]^{-1}\|,$$

thus we just choose (72) to bound $\|[S^{(4)}]^{-1}\|$. Moreover, as shown in [Yoon and Barlow, 1998],

$$|s_{k+1,k+1}^{(4)}|^{-1} = \sqrt{b^2 + c^2}.$$

Thus, a little manipulation allows us to compute the bound in (76) using the formula in (72) in 11 flops. From there the procedure for obtaining the ULVD of $S^{(4)}$ is identical to that in §4.3.

5. Refining a ULVD

Refinement is a process of improving the accuracy of the ULVD. We discuss how to measure that improvement below.

The notion of refinement was popularized in [Mathias and Stewart, 1993] where the following was presented. It is equivalent to the QR algorithm without shifts.

Algorithm 6 (Mathias–Stewart Refinement Procedure).

Step 1. *Find orthogonal* $\tilde{U} \in \Re^{n\times n}$ *such that*

$$\tilde{U}^T C = \tilde{U}^T \begin{pmatrix} L & 0 \\ F & G \end{pmatrix} = \begin{pmatrix} L^{(1)} & F^{(1)} \\ 0 & G^{(1)} \end{pmatrix} = C^{(1)}.$$

$$U \leftarrow U\tilde{U}.$$

Step 2. *Find orthogonal* $\tilde{V} \in \Re^{n\times n}$ *such that*

$$C^{(1)}\tilde{V} = \begin{pmatrix} L^{(1)} & F^{(1)} \\ 0 & G^{(1)} \end{pmatrix} \tilde{V} = \begin{pmatrix} \bar{L} & 0 \\ \bar{F} & \bar{G} \end{pmatrix}$$

$$V \leftarrow V\tilde{V}.$$

We note three important properties of Algorithm 6. Other properties were shown in [Mathias and Stewart, 1993].

1 Since this is the QR algorithm without shifts, it follows that asymptotically

$$\|\bar{F}\| \approx \|F\| \frac{\sigma_{k+1}^2(C)}{\sigma_k^2(C)}.$$

2 One can show that

$$\|(\; \bar{F} \quad \bar{G} \;)\| \leq \|G\|, \quad \|(\; \bar{F} \quad \bar{G} \;)\|_F \leq \|G\|_F.$$

3 Repeated application of this algorithm will force $\|(\; \bar{F} \quad \bar{G} \;)\|$ to converge to $\sigma_{k+1}(C)$, but at about the rate of the shiftless QR algorithm.

This algorithm requires $O((m+n)(n-k)k)$ flops where the constant in $O(\cdot)$ depends upon the implementation. Thus, if $\min\{k, n-k\} = O(1)$, then this is an $O(mn)$ algorithm, but if $\min\{k, n-k\} = O(n)$, this is an $O(mn^2)$ algorithm.

An alternative to Algorithm 6, proposed in [Fierro and Bunch, 1995], is just to apply Algorithm 3 to C, and thereby produce a better ULVD. This costs about the same as Algorithm 6, it is $O(mn(n-k))$ flops, making it $O(mn)$ flops if $n-k=O(1)$ and $O(mn^2)$ flops if $n-k=O(n)$.

Below we give an algorithm that computes a refinement in $O(mn)$ flops no matter what the value of k is. It does not have a property similar to property (1) of Algorithm 6. Fortunately, it does have two properties similar to properties (2) and (3). Namely:

1 We establish

$$\|(\bar{F} \quad \bar{G})\|_F \leq \|(F \quad G)\|_F$$

where equality occurs if and only if $\|(F \quad G)\| = \sigma_{k+1}(C_0)$.

2 Repeated application of this algorithm will force $\|(\bar{F} \quad \bar{G})\|$ to converge to $\sigma_{k+1}(C)$. After one iteration, under reasonable assumptions, a very good approximation of $\sigma_{k+1}(C)$ will be produced.

Algorithm 7 (Alternative Refinement Algorithm).

Step 1. *Find unit vectors $\bar{\mathbf{u}}_1, \bar{\mathbf{v}}_1$ such that*

$$\begin{pmatrix} F & G \end{pmatrix} \bar{\mathbf{v}}_1 = \bar{\sigma}_1 \bar{\mathbf{u}}_1,$$

where $\bar{\sigma}_1$ is the largest singular value of $\begin{pmatrix} F & G \end{pmatrix}$.

Step 2. *Construct orthogonal matrices $U^{(1)}$ and $V^{(1)}$ such that*

$$[U^{(1)}]^T \bar{\mathbf{u}}_1 = \pm \mathbf{e}_1,$$

and

$$G^{(1)} = U^{(1)^T} G V^{(1)}$$

remains lower triangular. Also compute

$$F^{(1)} = [U^{(1)}]^T F,$$

$$U(:, k+1:n) \leftarrow U(:, k+1:n) U^{(1)},$$
$$V(:, k+1:n) \leftarrow V(:, k+1:n) V^{(1)}.$$

Note that

$$\begin{pmatrix} F^{(1)} & G^{(1)} \end{pmatrix} = \begin{matrix} 1 \\ n-k-1 \end{matrix} \overset{\begin{matrix} k & \quad n-k \end{matrix}}{\begin{pmatrix} [\mathbf{f}_1^{(1)}]^T & g_{11}^{(1)} \mathbf{e}_1^T \\ \tilde{F}^{(1)} & \tilde{G}^{(1)} \end{pmatrix}}$$

where

$$\|(\; \mathbf{f}_1^{(1)^T} \quad g_{11}^{(1)}\mathbf{e}_1^T \;)\| = \bar{\sigma}_1[(\; F \quad G \;)],$$
$$\|(\; \tilde{F}^{(1)} \quad \tilde{G}^{(1)} \;)\| = \sigma_2[(\; F \quad G \;)].$$

Step 3. *Let*

$$S = \begin{pmatrix} L & 0 \\ [\mathbf{f}_1^{(1)}]^T & g_{11}^{(1)} \end{pmatrix}$$

Find unit vectors $\bar{\mathbf{u}}_{k+1}, \bar{\mathbf{v}}_{k+1}$ *such that*

$$S\bar{\mathbf{v}}_{k+1} = \bar{\sigma}_{k+1}\bar{\mathbf{u}}_{k+1}$$

where $\bar{\sigma}_{k+1}$ *is the smallest singular value of* S.

Step 4. *Construct orthogonal matrices* $U^{(2)}, V^{(2)} \in \Re^{(k+1)\times(k+1)}$ *such that*

$$[U^{(2)}]^T\bar{\mathbf{u}}_{k+1} = \pm\mathbf{e}_{k+1}$$

$$S^{(1)} = \begin{pmatrix} L^{(1)} & 0 \\ 0 & \bar{\sigma}_{k+1} \end{pmatrix} = [U^{(2)}]^T S V^{(2)}$$

where $L^{(1)}$ *is lower triangular.*

$$(\; \tilde{F}^{(2)} \quad \tilde{G}^{(2)}(:,1) \;) = (\; \tilde{F}^{(1)} \quad \tilde{G}^{(1)}(:,1) \;)\, V^{(2)}$$

$$U(:,1\colon k+1) \leftarrow U(:,1\colon k+1)U^{(2)},$$
$$V(:,1\colon k+1) \leftarrow V(:,1\colon k+1)V^{(2)}.$$

If $\bar{\sigma}_{k+1} \geq \epsilon$, *let*

$$\bar{L} = S^{(1)}, \quad \bar{F} = (\; \tilde{F}^{(2)} \quad \tilde{G}(:,1) \;), \quad \bar{G} = \tilde{G}^{(2)}(:,2\colon n-k)$$

else

$$\bar{L} = L^{(1)}, \quad \bar{F} = \begin{pmatrix} 0 \\ \tilde{F}^{(2)} \end{pmatrix}, \quad \bar{G} = \begin{pmatrix} \bar{\sigma}_{k+1}\mathbf{e}_1^T \\ \tilde{G}^{(2)} \end{pmatrix}.$$

The refinement algorithm requires $6(m+n)n + 12nk + c_{iter}n^2$ flops. Thus, this is an $O(mn)$ algorithm for all k.

The following proposition exactly quantifies the decrease in $\|(\; F \quad G \;)\|_F$.

Proposition 1. *Assume the terminology of Algorithm 7. If* $\bar{\sigma}_{k+1} < \epsilon$ *then*

$$\|(\bar{F}\;\bar{G})\|_F^2 = \|(F\;G)\|_F^2 - \sigma_1^2[(F\;G)] + \sigma_{k+1}^2(S) \tag{77}$$

$$= \|(F\;G)\|_F^2 - \|([\mathbf{f}_1^{(1)}]^T, g_{11}^{(1)})\|^2 + \bar{\sigma}_{k+1}^2$$

If $\bar{\sigma}_{k+1} \geq \epsilon$ *then*

$$\|(\bar{F}\ \bar{G})\|_F^2 = \|(F\ G)\|_F^2 - \sigma_1^2[(F\ G)] \tag{78}$$

and

$$\|(\ \bar{F} \quad \bar{G}\ )\| = \sigma_2(\ F \quad G\ ) \leq \|(\ F \quad G\ )\|.$$

The computation of equation (77) requires $O(n)$ flops.

The following theorem shows that repeated application of Algorithm 7 forces $\|(\ F \quad G\ )\|$ to converge to $\sigma_{k+1}(C)$.

Theorem 2. *Suppose we apply Algorithm 7 to* $C_0 = \begin{pmatrix} L_0 & 0 \\ F_0 & G_0 \end{pmatrix}$ *t times to obtain the matrix* $C_t = \begin{pmatrix} L_t & 0 \\ F_t & G_t \end{pmatrix}$. *Then*

$$\lim_{t\to\infty} \|(\ F_t \quad G_t\ )\| = \sigma_{k+1}(C_0). \tag{79}$$

The next theorem shows that just one step of the refinement procedure can yield a value of $\bar{\sigma}_{k+1}$ that is a very good approximation of the $(k+1)st$ singular value of C.

Theorem 3. *For a given matrix C of the form*

$$C = \begin{matrix} \\ k \\ n-k \end{matrix} \begin{matrix} k & n-k \\ \left(\begin{matrix} L \\ F \end{matrix}\right. & \left.\begin{matrix} 0 \\ G \end{matrix}\right) \end{matrix}, \tag{80}$$

Algorithm 7 produces a matrix of the form

$$\bar{C} = \begin{matrix} \\ k \\ 1 \\ n-k-1 \end{matrix} \begin{matrix} k & n-k \\ \left(\begin{matrix} \bar{L} \\ 0 \\ \tilde{F} \end{matrix}\right. & \left.\begin{matrix} 0 \\ \bar{\sigma}_{k+1}\mathbf{e}_1^T \\ \tilde{G} \end{matrix}\right) \end{matrix} \tag{81}$$

such that

$$0 \leq \sigma_{k+1}(C) - \bar{\sigma}_{k+1} \leq \frac{\sigma_2^2[(\ F \quad G\ )]}{\sigma_{k+1}(C) + \bar{\sigma}_{k+1}}. \tag{82}$$

Theorems 2 and 3 with Proposition 1 imply the Algorithm 7 is an inexpensive method to decrease $\|(\ F \quad G\ )\|_F$ and to bring $\|(\ F \quad G\ )\|$ closer to $\sigma_{k+1}(C)$. This procedure has the advantage over Algorithm 6 that it is always $O(mn)$ flops rather that $O(mn(n-k))$ flops.

6. Conclusion

The ULVD is a useful decomposition for many if not most of the important applications of the SVD. Although its formation has about the same complexity as the SVD, it can be updated faster (in $O(mn)$ flops), provides good subspace approximations, and these approximations are easy to improve (also in $O(mn)$ flops). This makes it a convenient tool for constructing fast algorithms for recursive least squares problems and total least squares problems, principal component analysis, and subspace tracking applications.

References

Anderson, E., Bai, Z., Bischof, C., Demmel, J., Dongarra, J., DuCroz, J., Greenbaum, A., Hammarling, S., McKenney, A., Ostrouchov, S., and Sorensen, D. (1992). *LAPACK User's Guide*. SIAM Publications, Philadelphia.

Barlow, J., Erbay, H., and Slapničar, I. (2001a). Analysis of an alternative refinement algorithm. in preparation.

Barlow, J., Erbay, H., and Zhang, Z. (1999). A modified Gram-Schmidt based downdating technique for the ULV decompositions with applications to recursive TLS problems. In Luk, F., editor, *Advanced Signal Processing Algorithms, Architectures and Implementations IX*, pages 247–257, Bellingham, WA. to appear, in different form, in Computational Statistics and Data Analysis.

Barlow, J., Smoktunowicz, A., and Erbay, H. (2001b). An evaluation of Gram–Schmidt based downdating methods. to appear, Advanced Signal Processing Algorithms, Architectures and Implementations X, 46th SPIE Symposium.

Barlow, J. and Yoon, P. (1997). Solving recursive TLS problems using rank-revealing ULV decompositions. In Huffel, S. V., editor, *Recent Advances in Total Least Squares Techniques and Errors–In–Variables Modeling*, pages 117–126, Philadelphia, PA. SIAM Publications.

Barlow, J., Yoon, P., and Zha, H. (1996). An algorithm and a stability theory for downdating the ULV decomposition. *BIT*, 36:14–40.

Björck, A. (1994). Numerics of Gram–Schmidt orthogonalization. *Linear Alg. Appl.*, 197/198: 297–316.

Businger, P. and Golub, G. (1965). Linear least squares solutions by Householder transformations. *Numerische Mathematik*, 7:269–278.

Chan, T. (1987). Rank revealing QR factorization. *Lin. Alg. Appl.*, 88/89:67–82.

Cline, A., Conn, A., and Loan, C. V. (1982). Generalizing the LINPACK condition estimator. In Hennart, J., editor, *Numerical Analysis, Proceedings, Cocoyoc, Mexico*, volume 909 of *Lecture Notes in Mathematics*, pages 73–83, Berlin. Springer–Verlag.

Daniel, J. W., Gragg, W. B., Kaufman, L., and Stewart, G. W. (1976). Reorthogonalization and stable algorithms for updating the Gram-Schmidt QR factorization. *Math. Comp.*, 30(136): 772–795.

Davis, C. and Kahan, W. (1970). The rotation of eigenvectors by a perturbation III. *SIAM J. Num. Anal.*, 7:1–46.

Demmel, J. and Kahan, W. (1990). Accurate singular values of bidiagonal matrices. *SIAM J. Sci. Stat. Computing*, 11:873–912.

Erbay, H. and Barlow, J. (2000). Recursive ULV decomposition and an alternative refinement algorithm. In Luk, F., editor, *Advanced Signal Processing Algorithms, Architectures and*

Implementations X, volume 4116 of *45th SPIE Symposium*, pages 157–166, Bellingham, WA.

Faddeev, D., Kublanovskaya, V., and Faddeeva, V. (1968). *Sur les Systémes Linéires Algébriques de Matrices Rectangularies et Mal-Conditionnées*, volume VII, pages 161–170. Editions Centre Nat. Recherche Sci., Paris.

Fernando, K. (1999). Accurately counting singular values of bidiagonal matrices and eigenvalues of skew–symmetric tridiagonal matrices. *SIAM J. Matrix Anal. Appl.*, 20:373–399.

Fernando, K. and Parlett, B. (1994). Accurate singular values and differential qd algorithms. *Numer. Math.*, 67:191–229.

Fierro, R. and Bunch, J. (1995). Bounding the subspaces from rank revealing two-sided orthogonal decompositions. *SIAM J. Matrix Anal. Appl.*, 16:743–759.

Fierro, R., Vanhamme, L., and Van Huffel, S. (1997). Total least squares algorithms based on rank-revealing complete orthogonal decompositions. In Huffel, S. V., editor, *Recent Advances in Total Least Squares Techniques and Errors–In–Variabls Modeling*, pages 99–116, Philadelphia, PA. SIAM Publications.

Fierro, R. D., Hansen, P., and Kirk, P. S. (1999). UTV tools: Matlab templates for rank-revealing UTV decompositions. *Numer. Algorithms*, 20(2-3):165–194.

Foster, L. (1986). Rank and null space calculations using matrix decompositions without column pivoting. *Lin. Alg. Appl.*, 74:47–72.

Golub, G. and Kahan, W. (1965). Calculating the singular values and pseudoinverse of a matrix. *SIAM J. Num. Anal. Ser. B*, 2:205–224.

Golub, G. and Loan, C. V. (1996). *Matrix Computations, Third Edition*. The Johns Hopkins Press, Baltimore,MD.

Golub, G. and Reinsch, C. (1970). Singular value decomposition and least squares solutions. *Numer. Math*, 14:403–20.

Gu, M. and Eisenstat, S. (1995). Downdating the singular value decomposition. *SIAM J. Matrix Anal. Appl.*, 16:793–810.

Hanson, R. and Lawson, C. (1969). Extensions and applications of the Householder algorithm for solving linear least squares problems. *Math. Comp.*, 23:787–812.

Hoffman, W. (1989). Iterative algorithms for Gram-Schmidt orthogonalization. *Computing*, 41:353–367.

Mathias, R. and Stewart, G. (1993). A block QR algorithm and the singular value decomposition. *Linear Algebra and Its Applications*, 182:91–100.

Mathworks, Inc., T. (1996). *MATLAB: The language of technical computing – Using MATLAB*. The Mathworks, Inc., Natick, MA.

Parlett, B. and Marques, O. (2000). An implementation of the dqds algorithm (positive case). *Linear Alg. Appl.*, 309:217–259.

Ruhe, A. (1983). Numerical aspects of Gram–Schmidt orthogonalization of vectors. *Linear Alg. Appl.*, 52/53:591–601.

Stewart, G. (1992). An updating algorithm for subspace tracking. *IEEE Transactions on Signal Processing*, 40:1535–1541.

Stewart, G. (1993). Updating a rank-revealing ULV decomposition. *SIAM J. Matrix Anal. Appl.*, 14:494–499.

Van Huffel, S. and Vandewalle, J. (1991). *The Total Least Squares Problem: Computational Aspects and Analysis*. SIAM Publications, Philadelphia.

Wilkinson, J. (1965). *The Algebraic Eigenvalue Problem*. Oxford University Press, London.

Wilkinson, J. (1968). Global convergence of the tridiagonal QR algorithm with origin shifts. *Linear Alg. Appl.*, 1:409–20.

Yoo, K. and Park, H. (1996). Accurate downdating of a modified Gram–Schmidt QR decomposition. *BIT*, 36:166–181.

Yoon, P. (1996). *Modifying two-sided orthogonal decompositions: algorithms, implementation, and applications*. PhD thesis, The Pennsylvania State University, University Park, PA.

Yoon, P. and Barlow, J. (1998). An efficient rank detection procedure for modifying the ULV decomposition. *BIT*, 38:781–801.

ADVANCES IN JACOBI METHODS

Zlatko Drmač
drmac@math.hr
Department of Mathematics
University of Zagreb
Croatia

Vjeran Hari
hari@math.hr
Department of Mathematics
University of Zagreb
Croatia

Ivan Slapničar
ivan.slapnicar@fesb.hr
FESB
University of Split
Croatia

Abstract Two-sided and especially one-sided Jacobi methods for solving eigenproblems of Hermitian positive definite and symmetric indefinite matrices are reviewed. SVD one-sided methods are included in this survey. Special attention is paid to the latest results on accuracy and on asymptotic convergence of scaled iterates by these methods.

Keywords: Jacobi methods, one-sided methods, accuracy, scaled iterates, asymptotic convergence.

Introduction

It is known that diagonalization methods deliver output data with high relative accuracy, a property which is not shared with faster methods such as Divide and Conquer or QR methods which require tridiagonalization or bidiagonalization as a preprocessing step. After the pioneering research of Demmel and Veselić [6], accuracy of other methods for solving different matrix eigenproblems has been inspected (see [25, 41]). Closely related stopping criterions and measures of convergence have also been reinvestigated.

In this overview paper we present the latest research of Jacobi methods for solving Hermitian/symmetric eigenvalue problem and the singular value problem. The new results mostly address accuracy and the asymptotic convergence of scaled iterates.

The paper is divided into three sections. In Section 1 are briefly described two- and one-sided Jacobi methods for Hermitian/symmetric eigenvalue problem. A special attention is paid to J-symmetric Jacobi method of Veselić [51] and its one-sided version which is an excellent tool for accurate eigensolving of indefinite symmetric matrices. In Section 2, their asymptotic convergence is reviewed and several new results are shortly explained. Finally, in Section 3, the latest accuracy results concerning one-sided methods are presented.

Edited by Drmač *et al.*, Kluwer Academic/Plenum Publishers, 2002

1. Two-sided Methods

Here we first give a short introduction to Jacobi-type processes [36, 17]. Then we give a brief description of the the most important methods: Hermitian Jacobi method and J-symmetric Jacobi method. For each of the two methods, we introduce both the two- and one-sided versions.

1.1 Jacobi-type Processes

Jacobi-type methods are iterative processes of the form

$$A^{(k+1)} = P_k^* A^{(k)} Q_k, \quad k \geq 0, \tag{1.1}$$

where P_k, Q_k are nonsingular *elementary plane matrices*, P_k^* is the complex transpose of P_k and $A^{(0)} = A$ is the initial matrix of order n. An elementary plane matrix E is a nonsingular matrix which differs from identity matrix I_n in one principal submatrix of order two, denote it by $\hat{E}$, which is called *pivot submatrix* or the (i, j)-*restriction* of E. The pair of indices i, j, $i < j$ which determine position of $\hat{E}$ within E are *pivot indices* and (i, j) is *pivot pair*. In (1.1) pivot indices depend on k, so $i = i(k)$, $j = j(k)$. When emphasis is on pivot indices we shall write $P_{i(k)j(k)}$ instead of P_k (similar for Q_k) and when k is understood just P_{ij}. Transition from $A^{(k)}$ to $A^{(k+1)}$ is called the kth step or iteration of the method.

Jacobi methods are characterized by the requirement

$$a_{i(k)j(k)}^{(k+1)} = a_{j(k)i(k)}^{(k+1)} = 0, \quad k \geq 0,$$

which implies that for each k, $\hat{P}_k$ and $\hat{Q}_k$ are computed from $\hat{A}_k$.

Pivot strategy is a rule for selecting pivot pairs. We identify each pivot strategy with a function $\mathrm{I} : \mathbf{N}_0 \to \mathbf{P}_n$, where $\mathbf{N}_0 = \{0, 1, 2, \ldots\}$ and $\mathbf{P}_n = \{(l, m); 1 \leq l < m \leq n\}, n \geq 2$. Among different classes of (pivot) strategies we consider here only *periodic strategies* which are periodic functions. This means $\mathrm{I}(k + M) = \mathrm{I}(k)$, $k \geq 0$ for a positive integer M, the *period* of I. A periodic strategy I is *quasi-cyclic* if $\{\mathrm{I}(k); 0 \leq k \leq M - 1\} = \mathbf{P}_n$, and a quasi-cyclic strategy is *cyclic* if $M = N \stackrel{\text{def}}{=} n(n-1)/2$. The set of M successive iterations starting with k which is a multiple of M, is usually called a *quasi-cycle* (*cycle* for cyclic strategies). In the sequel the term strategy means periodic strategy. The most common are the row-cyclic and column-cyclic strategies (often referred to as serial) and the parallel ones.

The column-cyclic strategy is defined by $\mathrm{I}_C(k) = (i(k), j(k))$, $k \geq 0$, where as k increases, the pivot pair runs through the column-wise ordering of $\mathbf{P}_n$: $(1, 2)$, $(1, 3)$, $(2, 3)$,...,$(1, n)$, $(2, n)$, ...,$(n-1, n)$ of $\mathbf{P}_n$. The row-cyclic strategy I_R is defined in a similar way by the row-wise ordering of $\mathbf{P}_n$:

$(1,2)$, $(1,3)$, ..., $(1,n)$, $(2,3)$, ..., $(n-1,n)$. By parallel strategy we mean a cyclic strategy for which the defining ordering of $\mathbf{P}_n$ can be divided into p subsequences each containing mutually disjoint pairs (pairs (l,m) and (p,q) are disjoint if $\{l,m\} \cap \{p,q\} = \emptyset$). The plane matrices associated with each such subsequence mutually commute, they all can be computed in the same time and all can be applied simultaneously. Best efficiency is obtained when $p \approx n$. Then each subsequence contains around $n/2$ pairs (see [3, 28, 18, 29]). A cycle then consists of p parallel steps and each parallel step comprises $\approx n/2$ sequential steps.

Another interesting periodic strategy is the quasi-cyclic strategy of Mascarenhas [30, 31, 37] which enables cubic asymptotic convergence per quasi-cycle and for which $M \approx 1.25N$.

The notion of convergence depends on the method. For two-sided methods it usually means convergence of the iterated matrix to a diagonal matrix.

1.2 Hermitian Jacobi Method

Let $H = (h_{lm})$ be a (complex) Hermitian matrix of order n. Jacobi method for computing the eigendecomposition of Hermitian matrices generates sequence of Hermitian matrices by the rule (1.1) with $P_k = Q_k$, $k \geq 0$ being unitary matrices. Rewriting (1.1) with these assumptions, yields

$$H^{(k+1)} = V_k^* H^{(k)} V_k, \quad k \geq 0, \quad H^{(0)} = H,$$

where V_k, $k \geq 0$ are unitary plane matrices. If eigenvectors are wanted, then $V^{(k)} = V_0 V_1 \cdots V_{k-1}$ is computed by the rule $V^{(k)} = V^{(k-1)} V_{k-1}$, $k \geq 1$, $V^{(0)} = I_n$. For each $k \geq 0$ pivot submatrix of V_k,

$$\hat{V}_k = \begin{bmatrix} \cos\varphi^{(k)} & -\sin\varphi^{(k)} e^{\imath\omega_k} \\ \sin\varphi^{(k)} e^{-\imath\omega_k} & \cos\varphi^{(k)} \end{bmatrix}$$

is chosen to diagonalize the pivot submatrix $\hat{H}^{(k)}$ of $H^{(k)} = (h_{lm}^{(k)})$. Here $\imath$ denotes the imaginary unit and $\overline{z}$ denotes the complex conjugate of $z \in \mathbf{C}$. Hence, $\varphi^{(k)}$ and ω_k are chosen to produce $h_{ij}^{(k+1)} = 0$. For $h_{ij}^{(k)} = 0$, the k'th step is skipped i.e. $\omega_k = \varphi^{(k)} = 0$ is presumed. Otherwise, the usual choice

$$e^{\imath\omega_k} = \frac{h_{ij}^{(k)}}{|h_{ij}^{(k)}|}, \quad \tan 2\varphi^{(k)} = \frac{2|h_{ij}^{(k)}|}{h_{ii}^{(k)} - h_{jj}^{(k)}}, \quad \varphi^{(k)} \in [-\pi/4, \pi/4], \tag{1.2}$$

is assumed. If H is real symmetric, ω_k is set zero, so that all $V^{(k)}$ are orthogonal and all $H^{(k)}$ are real symmetric. In this case $|h_{ij}^{(k)}|$ in the relations (1.2) and (1.3) are replaced by $h_{ij}^{(k)}$. Using $c^{(k)} = \cos\varphi^{(k)}$, $s^{(k)} = \sin\varphi^{(k)}$, $t^{(k)} = \tan\varphi^{(k)}$, the transformation formulas read

$$\begin{aligned} h_{li}^{(k+1)} &= c^{(k)} h_{li}^{(k)} + e^{-\imath\omega_k} s^{(k)} h_{lj}^{(k)} = \overline{h_{il}^{(k+1)}}, \quad l \notin \{i,j\} \\ h_{lj}^{(k+1)} &= c^{(k)} h_{lj}^{(k)} - e^{\imath\omega_k} s^{(k)} h_{li}^{(k)} = \overline{h_{jl}^{(k+1)}}, \quad l \notin \{i,j\} \\ h_{ii}^{(k+1)} &= h_{ii}^{(k)} + |h_{ij}^{(k)}|\, t^{(k)}, \quad h_{jj}^{(k+1)} = h_{jj}^{(k)} - |h_{ij}^{(k)}|\, t^{(k)} \\ h_{ij}^{(k+1)} &= 0, \qquad h_{lm}^{(k+1)} = h_{lm}^{(k)} \quad \text{whenever} \quad l,m \notin \{i,j\}. \end{aligned} \tag{1.3}$$

As a consequence of the choice of transformation parameters, we have

$$\|\Omega(H^{(k+1)})\|_F^2 = \|\Omega(H^{(k)})\|_F^2 - 2\,|\, h_{ij}^{(k)}\,|^2, \quad k \geq 0, \tag{1.4}$$

where generally, $\Omega(X) = X - \text{diag}(X)$ stands for the off-diagonal part of X and $\|\cdot\|_F$ is the Frobenius (or Euclidean) matrix norm. By $\|\cdot\|_2$ is denoted the spectral or matrix 2-norm. The quantity $\|\Omega(X)\|_F$ is sometimes referred to as (see [20]) the *off-norm* of X. The quantity $\|\Omega(H^{(k)})\|_F$ can be used to measure the progress of the process.

The serial Jacobi methods are globally convergent, i.e. the sequence of matrices $H^{(k)}$ is convergent to diagonal matrix for any initial Hermitian H (see [16, 24]). The asymptotically faster quasi-cyclic method defined by Mascarenhas strategy is also globally convergent [37].

If $H \in \mathbf{C}^{n\times n}$ is positive definite, then all $H^{(k)}$, $k \geq 0$ are positive definite and Jacobi method delivers relatively accurate eigenvalues and eigenvectors (almost as accurate as H allows, see [6]). Scaled matrices

$$H_S^{(k)} = [\text{diag}(H^{(k)})]^{-1/2} H^{(k)} [\text{diag}(H^{(k)})]^{-1/2}, \quad k \geq 0. \tag{1.5}$$

play important role concerning accuracy. Their off-diagonal parts, and the corresponding norms

$$A^{(k)} = \Omega(H_S^{(k)}) = H_S^{(k)} - I, \quad \alpha_k = \|A^{(k)}\|_F \;\; k \geq 0,$$

are used in terminating of the process (see (2.8)). This will be discussed later. Note that the diagonal elements of $A^{(k)} = (a_{lm}^{(k)})$, $k \geq 0$ are zeros and the off-diagonal elements are given by

$$a_{lm}^{(k)} = h_{lm}^{(k)} / \sqrt{h_{ll}^{(k)} h_{mm}^{(k)}}, \quad l \neq m, \quad k \geq 0. \tag{1.6}$$

The property (1.4) is not shared with the sequence of scaled matrices since

$$H_S^{(k+1)} = (\Delta^{(k+1)})^{-1} (V^{(k)})^* \Delta^{(k)} H_S^{(k)} \Delta^{(k)} V^{(k)} (\Delta^{(k+1)})^{-1}$$

is not a unitary transformation. Actually, the off-norm of scaled matrices can temporarily increase during the process. In [34] it has been proven

$$
\begin{aligned}
|a_{il}^{(k+1)}|^2 + |a_{jl}^{(k+1)}|^2 &\leq \frac{|a_{il}^{(k)}|^2+|a_{jl}^{(k)}|^2}{1-|a_{ij}^{(k)}|}, \quad l \neq i,j\,, \\
\|A^{(k+1)}\|_F^2 - \|A^{(k)}\|_F^2 &\leq |a_{ij}^{(k)}|\frac{\|A^{(k)}\|_F^2-2|a_{ij}^{(k)}|}{1-|a_{ij}^{(k)}|}\,.
\end{aligned}
\tag{1.7}
$$

Hence, if $\|A^{(k+1)}\|_F > \|A^{(k)}\|_F$, then $0 < |a_{ij}^{(k)}| \leq \frac{1}{2}\|A^{(k)}\|_F^2$. We see that only quadratically small (in the scaled sense) pivot element can cause the growth of $\|A^{(k)}\|_F$. The above estimates make it possible to find an upper bound for the finite sequence $\alpha_0, \alpha_1, \ldots, \alpha_N$ provided that α_0 is small enough.

Scaled iterates can be defined in indefinite case, whenever diagonal elements are nonzero. Then the scaling matrix $[\mathrm{diag}(H^{(k)})]^{-1/2}$ should be replaced by $[|\ \mathrm{diag}(H^{(k)})\ |]^{-1/2} = [\mathrm{diag}(|H^{(k)}|)]^{-1/2}$, the diagonal elements of $H_S^{(k)}$ are ± 1, $A^{(k)} = \Omega(H_S^{(k)})$ and $a_{lm}^{(k)} = h_{lm}^{(k)}/\sqrt{|\ h_{ll}^{(k)}h_{mm}^{(k)}\ |}$, $l \neq m$, $k \geq 0$. Here $|X| = (|x_{lm}|)$ provided that $X = (x_{lm})$.

One-Sided Jacobi Method Let $H = LL^T$ be the Cholesky factorization of positive definite H, and let $L = U\Sigma V^T$ be the SVD of L. Then, since U is orthogonal and Σ diagonal, the decomposition $H = U\Sigma^2U^T$ is the spectral decomposition of H. Hence, we can diagonalize H in two steps: (1) compute the Cholesky factorization to get L and (2) compute the SVD of L. For the second task we can use the so called *one-sided* (or *right-handed*) Jacobi method. This method appears naturally, when two-sided Jacobi method is applied to L^TL. After k steps one obtains the matrix $(LV_0 \cdots V_{k-1})^T LV_0 \cdots V_{k-1}$, hence it is sufficient to iterate the process $G^{(k+1)} = G^{(k)}V_k$, $k \geq 0$, $G^{(0)} = L$. To compute the parameters of $\hat{V}_k$ one needs (see (1.2)) to compute scalar product of pivot columns $(G^{(k)}e_i \mid G^{(k)}e_j)$ and squares of pivot column norms $\|G^{(k)}e_i\|^2$ and $\|G^{(k)}e_j\|^2$ (e_l is the lth column of I_n). Fortunately, using formulas for updating the diagonal elements from (1.3) the squares of norms have to be computed once or twice during the whole process. The effect of kth transformation is to orthogonalize the pivot columns of $G^{(k)}$. Assuming a "convergent" (e.g. the serial) pivot strategy, the sequence $G^{(k)}$ approaches a set of matrices with orthogonal columns. Let $G^{(k)} = U^{(k)}\Sigma^{(k)}$, where $U^{(k)}$ has normalized columns and $\Sigma^{(k)}$ is diagonal positive definite. We can choose any convergent subsequence of $V^{(k)} = V_0 \cdots V_k$ and an appropriate convergent subsequence of $U^{(k)}$ to obtain in the limit $U\Sigma = LV$, the SVD of L. Left singular vectors of L are eigenvectors of H and squares of the singular values of L are the eigenvalues of H. In the context of computing the spectral decomposition of H, right singular vectors are not needed, hence accumulation of rotations can be skipped. This and several other attractive features of the one-sided method are first noted, analyzed and exploited by Veselić and Hari [52]. We shall address some of these features in Subsection 3.1.

1.3 J-Symmetric Jacobi Method

J-symmetric Jacobi method, introduced by Veselić in [51], is a diagonalization method for the generalized eigenvalue problem $Hx = \lambda Jx$ where $H \in \mathbf{R}^{n\times n}$ is symmetric matrix and $J = \mathrm{diag}(I_m, I_{n-m})$ is direct sum of I_m and $-I_{n-m}$. The method generates the sequence of matrices by the rule (1.1) with $P_k = Q_k$, $k \geq 0$. Since transformation matrices need not be orthogonal, we denote them by C_k, $k \geq 0$ so that (1.1) takes form

$$H^{(k+1)} = C_k^T H^{(k)} C_k, \quad k \geq 0, \quad H^{(0)} = H.$$

The method is so designed that $C_k^T J C_k = J$ holds for all $k \geq 0$. We see that transition from $H^{(k)}$ to $H^{(k+1)}$ is made by congruence transformation which preserves the symmetry and inertia of matrices $H^{(k)}$ as well as the eigenvalues of the initial pair (H, J). Since all C_k are $J-orthogonal$, the method is sometimes called J-orthogonal (see [43]). Although the method can be generalized to complex matrices, all known results refer to real matrices.

For $1 \leq i < j \leq m$ and $m+1 \leq i < j \leq n$, C_k is plane rotation and for $1 \leq i \leq m < j \leq n$ it is J-orthogonal plane matrix (*hyperbolic rotation*). Their pivot submatrices have form

$$\hat{C}^{(k)} = \begin{bmatrix} \cos\varphi^{(k)} & -\sin\varphi^{(k)} \\ \sin\varphi^{(k)} & \cos\varphi^{(k)} \end{bmatrix} \quad \text{and} \quad \hat{C}^{(k)} = \begin{bmatrix} \cosh\varphi^{(k)} & \sinh\varphi^{(k)} \\ \sinh\varphi^{(k)} & \cosh\varphi^{(k)} \end{bmatrix},$$

respectively. The angle $\varphi^{(k)}$ is in either case chosen to annihilate the pivot element $h^{(k)}_{i(k)j(k)}$, which leads to very simple angle formulas

$$\tan 2\varphi^{(k)} = \frac{2h_{ij}^{(k)}}{h_{ii}^{(k)} - h_{jj}^{(k)}}, \qquad \tanh 2\varphi^{(k)} = \frac{-2h_{ij}^{(k)}}{h_{ii}^{(k)} + h_{jj}^{(k)}},$$

with $\varphi^{(k)} \in [-\pi/4, \ \pi/4]$ for orthogonal rotation. Using $c^{(k)}$, $s^{(k)}$ and $t^{(k)}$ for both the trigonometric and hyperbolic functions and $\tau = 1$ ($\tau = -1$) for hyperbolic (orthogonal) rotation, we obtain the following transformation formulas (compare to the relation (1.3))

$$\begin{aligned}
h_{il}^{(k+1)} &= c^{(k)} h_{il}^{(k)} + s^{(k)} h_{jl}^{(k)}, \quad h_{li}^{(k+1)} = h_{il}^{(k+1)}, \quad l \notin \{i,j\} \\
h_{jl}^{(k+1)} &= c^{(k)} h_{jl}^{(k)} + \tau \cdot s^{(k)} h_{il}^{(k)}, \quad h_{lj}^{(k+1)} = h_{jl}^{(k+1)}, \quad l \notin \{i,j\} \\
h_{ii}^{(k+1)} &= h_{ii}^{(k)} + t^{(k)} h_{ij}^{(k)}, \quad h_{jj}^{(k+1)} = h_{jj}^{(k)} + \tau \cdot t^{(k)} h_{ij}^{(k)}, \\
h_{ij}^{(k+1)} &= 0, \qquad h_{pr}^{(k+1)} = h_{pr}^{(k)} \text{ if } \{p,r\} \cap \{i,j\} = \emptyset
\end{aligned} \tag{1.8}$$

The transformation defined by $\tau = -1$ ($\tau = 1$) is called kth *trigonometric (hyperbolic) step* or *iteration* of the method. If eigenvectors are wanted, the transformations are collected as product $C_0C_1\cdots$. After trigonometric step the relation (1.4) holds. To see what happens after hyperbolic step, let

$$H^{(k)} = \begin{bmatrix} H_{11}^{(k)} & H_{12}^{(k)} \\ H_{21}^{(k)} & H_{22}^{(k)} \end{bmatrix}, \quad J = \begin{bmatrix} I_m & \\ & -I_{n-m} \end{bmatrix}, \quad H_{11}^{(k)} \text{ is } m \text{ by } m,$$

$$S^2(H^{(k)}) = \|\Omega(H_{11}^{(k)})\|_F^2 + \|\Omega(H_{22}^{(k)})\|_F^2 - 2\|H_{12}^{(k)}\|_F^2 .$$

Then for the hyperbolic step holds (see [51])

$$S^2(H^{(k+1)}) = S^2(H^{(k)}) + 2(h_{ij}^{(k)})^2.$$

We can say that hyperbolic steps transfer the off-diagonal weight from the blocks $H_{12}^{(k)}$ and $H_{21}^{(k)}$ to diagonal blocks $H_{11}^{(k)}$ and $H_{22}^{(k)}$ and trigonometric steps diagonalize the diagonal blocks. Overall, $H^{(k)}$ converges to a diagonal matrix. In [51] it has been proved that the method converges globally under the serial strategies.

If H is n by n positive definite, then all $H^{(k)}$, $k \geq 0$ are such. Scaled iterate $H_S^{(k)}$ and its off-diagonal part $A^{(k)}$ are defined as above, so that relations (1.5) – (1.6) hold for this method too.

Resemblance with symmetric Jacobi method is also seen from the relations (1.7) which hold not only for trigonometric, but also for hyperbolic steps (see [35]). In particular, (1.7) has form

$$|a_{il}^{(k+1)}|^2 + |a_{jl}^{(k+1)}|^2 = \frac{|a_{il}^{(k)}|^2 + |a_{jl}^{(k)}|^2 - 2a_{ij}^{(k)}a_{il}^{(k)}a_{jl}^{(k)}}{1 - |a_{ij}^{(k)}|^2} \leq \frac{|a_{il}^{(k)}|^2 + |a_{jl}^{(k)}|^2}{1 - |a_{ij}|}.$$

One-Sided J-Symmetric Method J-symmetric Jacobi method also has its one-sided version which is an excellent method for accurate eigensolving of symmetric indefinite eigenproblem. Let us briefly derive the algorithm (see [40, 42, 46]).

Let $H = PG_1J_1G_1^TP^T$ be the Bunch-Parlett factorization [4] of a nonsingular, indefinite symmetric matrix H. Here P is permutation, G_1 lower block-triangular with diagonal blocks of order one or two and J_1 diagonal with ± 1 on its diagonal. Using the second permutation P_1 in

$$H = PG_1J_1G_1^TP^T = (PG_1P_1^T)(P_1J_1P_1^T)(P_1G_1^TP^T) = GJG^T$$

one can obtain $J = \text{diag}(I_m, I_{n-m})$. Now $Hx = \lambda x$ can be written as $GJG^T = \lambda x$ and since G is invertible, the last equation is equivalent to $G^TGJG^Tx = \lambda G^Tx$. This can be written as $G^TGz = \lambda Jz$ with $z = JG^Tx$.

Since $Gz = GJG^Tx = Hx = \lambda x$, one obtains $x = (Gz)/\lambda$. In conclusion, the initial indefinite symmetric eigenproblem $Hx = \lambda x$ is converted to *J-symmetric* eigenproblem with positive definite matrix G^TG. To solve the new problem one can apply J-symmetric Jacobi method. As will be seen later, this transition makes sense if accuracy of all eigenvalues and eigenvectors is important. A pleasant fact with the new eigenproblem is that G^TG need not be computed. One simply postmultiplies G by the J-orthogonal transformations,

$$G^{(k+1)} = G^{(k)}C_k\,, \quad k \geq 0, \qquad G^{(0)} = G\,.$$

Using additional vector to store the diagonal of $(G^{(k)})^TG^{(k)}$ and making benefit of simple updates of that vector described in (1.8), only one scalar product of the form $(G^{(k)}e_i \mid G^{(k)}e_j)$ is needed to compute the parameters of C_k. In addition, the matrices C_k need not be accumulated since $G_S^{(k)} = G^{(k)}\Delta_k^{-1}$, $\Delta_k = \operatorname{diag}(\|G^{(k)}e_1\|, \ldots, \|G^{(k)}e_n\|)$ approaches the set of eigenvector matrices of H and diagonal elements of Δ_k^2 approach the eigenvalues of H.

To see that, we have to assume a globally convergent pivot strategy, so that $(G^{(k)})^TG^{(k)} \mapsto \Lambda^2$ as k increases, where Λ is diagonal with positive diagonal elements. Next, note that the sequence $(G_S^{(k)})_k$ is bounded. So, let $(G_S^{(i_k)})_k$ be any convergent subsequence and let $\lim_k G_S^{(i_k)} = Q$. Since $Q^TQ = I_n + N$, where N has zero diagonal, we have

$$I_n + N \;=\; Q^TQ = \lim_k \Delta_{i_k}^{-1}(G^{(i_k)})^TG^{(i_k)}\Delta_{i_k}^{-1} = \Lambda^2 \lim_k \Delta_{i_k}^{-2}\,,$$

implying $\lim_k \Delta_{i_k} = \Lambda$ and orthogonality of Q. Since $G^{(i_k)} = G_S^{(i_k)}\Delta_{i_k}$, $(G^{(i_k)})_k$ is convergent. Because G is invertible, the product $C^{(i_k)} = C_0 \cdots C_{i_k}$ is also convergent. Let $\lim_{k\to\infty} C^{(i_k)} = C$. Since J-orthogonal matrices make a closed group in the group of regular matrices, C is J-orthogonal. In particular, this means that $CJC^T = J$. We have obtained $Q = GC\Lambda^{-1}$, so we can write

$$\begin{aligned} HQ &= (GJG^T)(GC\Lambda^{-1}) = G(CJC^T)G^TGC\Lambda^{-1} = GCJ\Lambda^2\Lambda^{-1} \\ &= GCJ\Lambda = GC\Lambda^{-1}\Lambda J\Lambda = QJ\Lambda^2\,. \end{aligned}$$

This proves that Q is an eigenvector matrix of H. Since we have the same conclusion for any convergent subsequence of $G_S^{(k)}$ we have proved that $G_S^{(k)}$ approaches the set of eigenvector matrices of H.

Finally, let us mention that the one-sided J-symmetric method is of interest on its own, not just as the part of the compound algorithm. For example, this method can be used to solve the downdating problem, which consists of finding the eigenvalue decomposition of the difference of outer products,

$$H = AA^T - BB^T.$$

The matrix H can be written in the product form

$$H = GJG^T, \quad G = [\ A \quad B\ ], \quad J = \text{diag}(I, -I),$$

and the latter problem can be solved by the the one-sided J-symmetric method. For more details on this problem see [58, 43].

2. Asymptotic Convergence

It is well-known that cyclic Jacobi-type methods converge, under standard conditions, asymptotically quadratically. This means that for large enough k,

$$\|\Omega(H^{(k+N)}\|_F \leq c\,\|\Omega(H^{(k)})\|_F^2\,, \tag{2.1}$$

where $N = n(n-1)/2$. The constant c does not depend on k. By standard conditions we mean either simple eigenvalues or multiple eigenvalues plus ordering of diagonal elements so that those ones converging to the same eigenvalue take successive positions on diagonal plus special (e.g. serial) pivot strategies. By an argument of Wilkinson [56], classical Jacobi method and all kinds of optimal Jacobi methods for symmetric/Hermitian matrices will converge quadratically or better. Since in praxis one uses serial strategies, the relation (2.1) is modified to

$$\|\Omega(H^{((k+1)N)}\|_F \leq c\,\|\Omega(H^{(kN)})\|_F^2\,, \quad k \geq k_0, \tag{2.2}$$

where k_0 is sufficiently large. The relation (2.2) shows that ultimately (in praxis after several cycles) the measure $\|\Omega(H)\|_F$ reduces quadratically per cycle.

The measure $\|\Omega(H^{(k)})\|_F$ appears naturally in asymptotic convergence results, since $\|\Omega(H^{(k)})\|_F = 0$ shows $H^{(k)}$ is diagonal. In addition, $\|\Omega(H^{(k)})\|_F$ is the upper bound for $\max_t |a_{tt}^{(k)} - \lambda_t|$ and therefore measures the absolute distance between diagonal elements and the affiliated eigenvalues.

Note however, that one of the most important properties of Jacobi methods is their relative accuracy. Hence, one would better look for a measure which bounds $\max_t |a_{tt}^{(k)} - \lambda_t|/|\lambda_t|$ or $\max_t |a_{tt}^{(k)} - \lambda_t|/|a_{tt}^{(k)}|$. That measure (see [22]) has form $\|\Omega(DHD)\|_F$, where $D = [\,|\,\text{diag}(D)|\,]^{-1/2}$ and it also appears in the accuracy estimates from [6]. This has given rise to recent research of the quadratic convergence of scaled iterates by Jacobi methods. The new results will be briefly presented here.

Since one-sided methods can be viewed as intelligent implementations of two-sided methods, their asymptotic convergence amounts to that of two-sided methods.

2.1 Hermitian Jacobi Method

Let us first show using ε notation, the cubic convergence of the row-cyclic Hermitian Jacobi method, when $n = 4$. Suppose the considered cycle is not the first one (hence $h_{34} = 0$). Suppose the eigenvalues are distinct and all off-diagonal elements are $O(\epsilon)$. We assume that ϵ is so small that all angles and their sines have the same order of magnitude as the corresponding pivot elements. Then, we have

$$
\begin{array}{ccccc}
 & \downarrow & \downarrow & & \\
\rightarrow & \bullet & \varepsilon & \varepsilon & \varepsilon \\
\rightarrow & \varepsilon & \bullet & \varepsilon & \varepsilon \\
 & \varepsilon & \varepsilon & \bullet & 0 \\
 & \varepsilon & \varepsilon & 0 & \bullet
\end{array}
\mapsto
\begin{array}{ccccc}
 & \downarrow & & \downarrow & \\
\rightarrow & \bullet & \circ & \varepsilon & \varepsilon \\
 & \circ & \bullet & \varepsilon & \varepsilon \\
\rightarrow & \varepsilon & \varepsilon & \bullet & 0 \\
 & \varepsilon & \varepsilon & 0 & \bullet
\end{array}
\mapsto
\begin{array}{ccccc}
 & \downarrow & & & \downarrow \\
\rightarrow & \bullet & \varepsilon^2 & \circ & \varepsilon \\
 & \varepsilon^2 & \bullet & \varepsilon & \varepsilon \\
 & \circ & \varepsilon & \bullet & \varepsilon^2 \\
\rightarrow & \varepsilon & \varepsilon & \varepsilon^2 & \bullet
\end{array}
\mapsto
$$

$$
\begin{array}{ccccc}
 & & \downarrow & \downarrow & \\
 & \bullet & \varepsilon^2 & \varepsilon^3 & \circ \\
\rightarrow & \varepsilon^2 & \bullet & \varepsilon & \varepsilon \\
\rightarrow & \varepsilon^3 & \varepsilon & \bullet & \varepsilon^2 \\
 & \circ & \varepsilon & \varepsilon^2 & \bullet
\end{array}
\begin{array}{ccccc}
 & & \downarrow & & \downarrow \\
 & \bullet & \varepsilon^2 & \varepsilon^3 & \circ \\
\rightarrow & \varepsilon^2 & \bullet & \circ & \varepsilon \\
 & \varepsilon^3 & \circ & \bullet & \varepsilon^2 \\
\rightarrow & \circ & \varepsilon & \varepsilon^2 & \bullet
\end{array}
\begin{array}{ccccc}
 & & & \downarrow & \downarrow \\
 & \bullet & \varepsilon^2 & \varepsilon^3 & \varepsilon^3 \\
 & \varepsilon^2 & \bullet & \varepsilon^3 & \circ \\
\rightarrow & \varepsilon^3 & \varepsilon^3 & \bullet & \varepsilon^2 \\
\rightarrow & \varepsilon^3 & \circ & \varepsilon^2 & \bullet
\end{array}
\mapsto
$$

$$
\begin{array}{ccccc}
 & \downarrow & \downarrow & & \\
\rightarrow & \bullet & \varepsilon^2 & \varepsilon^3 & \varepsilon^3 \\
\rightarrow & \varepsilon^2 & \bullet & \varepsilon^3 & \varepsilon^5 \\
 & \varepsilon^3 & \varepsilon^3 & \bullet & \circ \\
 & \varepsilon^3 & \varepsilon^5 & \circ & \bullet
\end{array}
\mapsto
\begin{array}{c}\text{after}\\ \text{the next}\\ \text{cycle}\end{array}
\begin{array}{ccccc}
 & \downarrow & \downarrow & & \\
\rightarrow & \bullet & \varepsilon^6 & \varepsilon^9 & \varepsilon^{11} \\
\rightarrow & \varepsilon^6 & \bullet & \varepsilon^{11} & \varepsilon^{17} \\
 & \varepsilon^9 & \varepsilon^{11} & \bullet & \circ \\
 & \varepsilon^{11} & \varepsilon^{17} & \circ & \bullet
\end{array}
\mapsto
$$

$$
\begin{array}{c}\text{after}\\ \text{the next}\\ \text{cycle}\end{array}
\begin{array}{ccccc}
 & \downarrow & \downarrow & & \\
\rightarrow & \bullet & \varepsilon^{20} & \varepsilon^{31} & \varepsilon^{37} \\
\rightarrow & \varepsilon^{20} & \bullet & \varepsilon^{37} & \varepsilon^{57} \\
 & \varepsilon^{31} & \varepsilon^{37} & \bullet & \circ \\
 & \varepsilon^{37} & \varepsilon^{57} & \circ & \bullet
\end{array}
\qquad
\begin{array}{ll}
\downarrow & \text{Pivot column} \\
\rightarrow & \text{Pivot row} \\
0 & \text{Initial zeros} \\
\circ & \text{Produced zeros} \\
\varepsilon^r & O(\epsilon^r)\ \text{element.}
\end{array}
$$

If the initial matrix has multiple eigenvalues, the above analysis is not correct since almost diagonal matrices with multiple eigenvalues have special structure which influences the asymptotic rate of convergence. Let

$$\lambda_1 = \cdots = \lambda_{s_1}\,,\ \lambda_{s_1+1} = \cdots = \lambda_{s_2}\,, \ldots,\ \lambda_{s_{p-1}+1} = \cdots = \lambda_{s_p} \tag{2.3}$$

be the eigenvalues of Hermitian H and let

$$\delta_i = \min_{\substack{j \\ j\neq i}} |\lambda_{s_i} - \lambda_{s_j}|\,, \quad 1 \le i \le p, \quad \delta = \min_{1\le i\le p} \delta_i$$

be the absolute gaps in the spectrum of H. Suppose that H satisfies

$$\|\Omega(H)\|_F \le \frac{\delta}{3}\,,$$

and the diagonal elements of H affiliated with the same multiple eigenvalue occupy successive positions on the diagonal. Then H can be partitioned as

(H_{ij}) where each block H_{ii} has order $n_i = s_i - s_{i-1}$, $1 \le i \le p$, $s_0 = 0$, $s_p = n$. Here n_i is the algebraic multiplicity of λ_{s_i}. We can assume that $p > 1$, since otherwise $H = \lambda_1 I_n$. Let

$$\begin{aligned} \pi_i(H) &= H_{ii}, \qquad \tau_i(H) = [H_{i1}, \ldots, H_{i,i-1}, H_{i,i+1}, \ldots, H_{ip}] \\ \pi(H) &= \operatorname{diag}(H_{11}, \ldots, H_{pp}), \;\; \tau(H) = H - \pi(H). \end{aligned}$$

The special structure of an almost diagonal Hermitian matrix (see [56], [26, 27] and [19, 20, 21, 23]) is revealed from the following result from [21]. For $1 \le i \le p$ holds

$$\|\Omega(H_{ii})\|_F \le \|H_{ii}\text{-}\lambda_{s_i} I_{n_i}\|_F \le \frac{1.32}{\delta_i} \sum_{\substack{j=1 \\ j \ne i}}^{p} \|H_{ij}\|_F^2 \le 0.66 \frac{\|\Omega(H)\|_F^2}{\delta_i}. \tag{2.4}$$

This relation implies that within diagonal blocks the off-diagonal elements are quadratically small compared to those outside the diagonal blocks. When a Jacobi method is applied to such H the relation (2.4) will hold for all iteration matrices i.e. the partition will be preserved (see [21, Lemma 2.2]). This implies that methods defined by the classical and other optimal strategies will never choose at that stage, pivot elements from diagonal blocks and therefore all angles will ultimately tend to zero.

Suppose, $n = 4$, $p = 2$, $n_1 = 3$, $n_2 = 1$. Using the analysis as before and taking into account (2.4), we obtain

$$\begin{array}{ccccc} & \downarrow & \downarrow & & \\ \rightarrow & \bullet & \varepsilon^2 & \varepsilon^2 & \varepsilon \\ \rightarrow & \varepsilon^2 & \bullet & \varepsilon^2 & \varepsilon \\ & \varepsilon^2 & \varepsilon^2 & \bullet & 0 \\ & \varepsilon & \varepsilon & 0 & \bullet \end{array} \quad \begin{array}{ccccc} & \downarrow & & \downarrow & \\ \rightarrow & \bullet & \circ & \varepsilon^2 & \varepsilon \\ & \circ & \bullet & \varepsilon^2 & \varepsilon \\ \rightarrow & \varepsilon^2 & \varepsilon^2 & \bullet & 0 \\ & \varepsilon & \varepsilon & 0 & \bullet \end{array} \quad \begin{array}{ccccc} & \downarrow & & & \downarrow \\ \rightarrow & \bullet & \varepsilon^2 & \circ & \varepsilon \\ & \varepsilon^2 & \bullet & \varepsilon^2 & \varepsilon \\ & \circ & \varepsilon^2 & \bullet & \varepsilon \\ \rightarrow & \varepsilon & \varepsilon & \varepsilon & \bullet \end{array} \mapsto$$

$$\begin{array}{ccccc} & & \downarrow & \downarrow & \\ & \bullet & \varepsilon^2 & \varepsilon^2 & \circ \\ \rightarrow & \varepsilon^2 & \bullet & \varepsilon^2 & \varepsilon \\ \rightarrow & \varepsilon^2 & \varepsilon^2 & \bullet & \varepsilon \\ & \circ & \varepsilon & \varepsilon & \bullet \end{array} \quad \begin{array}{ccccc} & & \downarrow & & \downarrow \\ & \bullet & \varepsilon^2 & \varepsilon^2 & \circ \\ \rightarrow & \varepsilon^2 & \bullet & \circ & \varepsilon \\ & \varepsilon^2 & \circ & \bullet & \varepsilon \\ \rightarrow & \circ & \varepsilon & \varepsilon & \bullet \end{array} \quad \begin{array}{ccccc} & & & \downarrow & \downarrow \\ & \bullet & \varepsilon^6 & \varepsilon^6 & \varepsilon^3 \\ & \varepsilon^6 & \bullet & \varepsilon^6 & \circ \\ \rightarrow & \varepsilon^6 & \varepsilon^6 & \bullet & \varepsilon \\ \rightarrow & \varepsilon^3 & \circ & \varepsilon & \bullet \end{array} \mapsto$$

$$\begin{array}{ccccc} & & & \downarrow & \downarrow \\ & \bullet & \varepsilon^6 & \varepsilon^6 & \varepsilon^3 \\ & \varepsilon^6 & \bullet & \varepsilon^6 & \varepsilon^7 \\ \rightarrow & \varepsilon^6 & \varepsilon^6 & \bullet & \circ \\ \rightarrow & \varepsilon^3 & \varepsilon^7 & \circ & \bullet \end{array}$$

showing again the cubic convergence. Note that cubic reduction inside H_{11} has occurred earlier than indicated above, probably after annihilating $(1, 4)$ – element.

In general, when $n > 4$, we have only quadratic convergence for the serial Jacobi methods.

Quadratic convergence considerations of symmetric Jacobi methods in the case of simple and double eigenvalues originated from early works of Wilkinson [55] and Schönhage [39] (see also [54] and [24]).

The harder case of multiple eigenvalues was first considered by Van Kempen [26, 27] and Wilkinson [56]. Later, Hari [21] gave the first complete proof of the quadratic convergence of the serial Jacobi methods. The proof in [21] broadens and completes the considerations of van Kempen from [27], who has not taken into account some quantities which influence the bound. By sophisticated estimates it has been verified in [21] that the result stated in [27] indeed holds. The main result from [21] has the form

$$\|\Omega(H)\|_F \leq \frac{\delta}{3} \quad \Rightarrow \quad \|\Omega(H^{(N)})\|_F \leq \frac{9}{5}\frac{\|\Omega(H)\|_F^2}{\delta}. \tag{2.5}$$

It has been shown in [20] that in the case of multiple eigenvalues, large angles can be expected during the process, irrespectively of how tiny the off-diagonal elements are. Hence, the condition on diagonal elements which ensures that the relation (2.4) holds cannot be omitted. If eigenvalues form clusters of small width, one can incorporate perturbation theory (see [21]) to prove the quadratic reduction of $\|\Omega(H^{(kN)}))\|_F$ per cycle. As $\|\Omega(H^{(kN)}))\|_F$ approaches the width of clusters, the asymptotic speed slows down, but further shrinking of $\|\Omega(H^{(kN)}))\|_F$ again increases the speed to reach the quadratic rate.

If δ is tiny however, the estimate (2.5) becomes useless. In praxis this sometimes happens when eigenvalues cluster around zero in such a way that it is difficult or impossible to bound the cluster.

Later, in 1990 Mascarenhas [30, 31] showed that using special quasi-cyclic strategies Jacobi method can perform cubic asymptotic convergence per quasi-cycle. Since his quasi-cycles consist of around 1.25 cycles, this corresponds to the asymptotic rate of order $3^{0.8} \approx 2.41$, thus, between quadratic and cubic speed per cycle. Rhee and Hari [37] proved the global and the cubic (per quasi-cycle) convergence of his method.

Recent Results Although recent research of Jacobi methods has been mostly concentrated on their accuracy properties, closely related are new measures of advancing of the processes. These measures bound the maximum relative distance between diagonal elements and the corresponding eigenvalues. Therefore, they are included in the stopping criterions of the methods. They involve *scaled diagonally dominant* (see [2], [22]) matrices and *relative gaps*.

Suppose $A = D + N$, where D is diagonal and N has zero diagonal. Then $A = (a_{ij})$ is referred to as *α–diagonally dominant* with respect to a norm $\|\cdot\|$ if $\|N\| \leq \alpha \min_{1\leq i\leq n} |a_{ii}|$, with $0 \leq \alpha < 1$. If $A = D + N$ with $|a_{ii}| = 1$, $1 \leq i \leq n$ and Δ_1, Δ_2 are arbitrary nonsingular diagonal matrices, then $B = \Delta_1 A \Delta_2$ is *α–scaled diagonally dominant* (α–s.d.d.) with respect to a given norm, provided that A is α–diagonally dominant with respect to that norm. Note that an α–s.d.d. matrix has nonzero diagonal elements. If A is Hermitian, it is presumed that $\Delta_1 = \Delta_2$ and Δ_1 is real. Such scaling, which is a congruence transformation, will be called symmetric.

Relative gaps are in applications often connected with s.d.d. matrices. Several definitions of relative gaps have been used (cf. [40, 46, 25]). Here we use the one from [22].

Let the eigenvalues of the Hermitian matrix H be ordered as in the relation (2.3). The relative gaps and the minimum relative gap in the spectrum of H are defined as follows

$$\gamma_i = \min_{\substack{1\leq j\leq p \\ j\neq i}} \frac{|\lambda_{s_i} - \lambda_{s_j}|}{|\lambda_{s_i}| + |\lambda_{s_j}|} \quad 1 \leq i \leq p; \qquad \gamma = \min_{1\leq i\leq p} \gamma_i\,. \tag{2.6}$$

The following result can be used in selecting a measure for convergence of Hermitian Jacobi methods.

Proposition 2.1. [22] *Let $H = H^*$ be α-s.d.d. and $H = \Delta_H H_S \Delta_H$, where $\Delta_H = [\,|diag(H)|\,]^{\frac{1}{2}}$. Let γ_i and γ be as in the relation (2.6). If $\alpha < \gamma/(\gamma+3)$, then*

$$\sum_{j=s_{i-1}+1}^{s_i} |1 - \frac{\lambda_{s_i}}{h_{jj}}|^2 + \|\Omega\,(\pi_i(H_S))\,\|_F^2 \leq \frac{16}{\gamma_i^2}\|\tau_i(H_S)\|_F^4,\ 1 \leq i \leq p. \tag{2.7}$$

If $H = H^*$ is α-s.d.d. and positive definite, then the relation (2.7) holds with constant 4 instead of 16 and under less stringent assumption $\alpha < \gamma/3$. ■

Summing up the equations (2.7) over i, one obtains, say for positive definite matrix H (see [22, Collorary 3.2(ii)]),

$$\sum_{j=1}^{n} |\,1 - \frac{\lambda_j}{h_{jj}}\,|^2 + \|\Omega\,(\pi(H_S))\,\|_F^2 \leq \frac{2}{\gamma^2}\|\tau(H_S)\|_F^4. \tag{2.8}$$

Note that $|1 - \frac{\lambda_{s_i}}{h_{jj}}| \leq \beta$ implies $|1 - \frac{h_{jj}}{\lambda_{s_i}}| \leq \frac{\beta}{1-\beta}$. Hence, if $\|\tau(H_S)\|_F^2/\gamma$ is small, we have accurate estimates for both $|(\lambda_{s_i} - h_{jj})/h_{jj}|$ and $|(h_{jj} - \lambda_{s_i})/\lambda_{s_i}|$. The relation (2.7) for positive definite matrices implies that $\log_{10}(2\|\tau_i(H_S)\|_F^2/\gamma_i)$ indicates how many significant decimal digits are correct in the appropriate diagonal elements as approximations of λ_{s_i}. As a global measure for all diagonals, one can also use $\sqrt{2}\|\tau(H_S)\|_F^2/\gamma$ or larger measures, $\|\tau(H_S)\|_F$ and $\|\Omega(H_S)\|_F$.

Returning to Jacobi methods, note that γ does not change with iteration matrix. Usually, in the first few cycles the partition $(H_{ij}^{(k)})$ is not yet recognized and one prefers to use $\|\Omega(H_S)\|_F$ a simpler although (in rare situations) larger measure than $\|\tau(H_S)\|_F$.

Suppose a serial Jacobi method is applied to a positive definite H producing the sequence $(H^{(k)})_k$. By the result [6] we know that the method computes the eigenvalues and eigenvectors almost with the accuracy that is warranted by the eigenproblem. After several cycles the diagonal elements become approximations of the eigenvalues and lower bounds of absolute or relative gaps can easily be computed from diagonal elements (see Example 3.3 from [22]). If the eigenvalues cluster around zero, some absolute gaps δ_i and therefore δ will be tiny and the result (2.5) becomes useless for practical purposes. The measure $\|\Omega(H^{(k)}\|_F$ alone bounds $\max_t |h_{tt}^{(k)} - \lambda_t|$. Hence, for tiny λ_{s_i}, $\max_{s_{i-1}+1\leq t\leq s_i} |h_{tt}^{(k)} - \lambda_{s_i}|/\lambda_{s_i}$ is bounded by $\|\Omega(H^{(k)}\|_F/\lambda_{s_i}$ which can be large. Thus, the measures involving $\|\Omega(H^{(k)})\|_F$ and absolute gaps will not be appropriate.

On the contrary, for such matrices, all relative gaps γ_i and γ can be large (close to one). As the process advances one can use measures $\|\Omega(H_S)\|_F$ and $\|\Omega(H_S)\|_F^2/\gamma$. This implies that after each cycle one has to compute $\|\Omega(H_S)\|_F$ and perhaps a lower bound of γ. For two-sided Jacobi methods this is appropriate since it requires only $O(n^2)$ flops. However, in the context of one-sided methods, computing $\|\Omega(H_S)\|_F$ requires $O(n^3)$ flops. Hence, during the whole process, one would not like to compute $\|\Omega(H_S)\|_F$ more than once. In such a situation it is necessary to exploit the knowledge that serial Jacobi methods converge asymptotically quadratically. As we have just explained, the quadratic convergence result (2.5) is useless. In addition, in described situation, the relation between $\|\Omega(H_S)\|_F$ and $\|\Omega(H)\|_F$ can be only very roughly estimated, so (2.5) would give even poorer result when translated in terms of $\|\Omega(H_S)\|_F$.

We need a quadratic convergence result for $\|\Omega(H_S)\|_F$ which involves relative instead of absolute gaps. Fortunately, such results can be formulated and derived in the context of two-sided methods and then directly applied to one-sided methods.

In his Ph. D. thesis [32] Matejaš solved this problem for the serial Jacobi methods for positive definite matrices. The easier case of distinct eigenvalues (see [33]) and real matrices assumes form

$$\alpha_N \leq 0.715\frac{\alpha_0^2}{\gamma} \qquad \text{whenever} \qquad \alpha_0 \leq \frac{1}{4}\min\{\frac{1}{n}, \gamma\}, \tag{2.9}$$

where $\alpha_k = \|A^{(k)}\|_F$ as defined by the relations (1.5) – (1.6). In (2.9) it is assumed that the diagonal elements are decreasingly ordered. The harder case of multiple eigenvalues and complex matrices takes similar form (see [34])

$$\alpha_N \leq \sqrt{\frac{5}{2}\frac{\alpha_0^2}{\gamma}} \qquad \text{whenever} \qquad \alpha_0 \leq \frac{1}{6}\min\{\frac{1}{n}, \gamma\}.$$

A similar result holds for the case of indefinite Hermitian matrix H. In the context of one-sided methods, these results can be used for predicting the number of cycles till convergence. If eigenvectors are wanted one can use appropriate eigenvector perturbation estimates (e.g. [6, Theorem 2.7]) which use condition of scaled matrix (which is in our case close to one) and appropriate relative gaps.

We end this subsection by a brief discussion on stopping criterions. Jacobi methods are predominantly used when accuracy of the output data is important. Therefore, one will probably choose the criterion $\|\Omega(H_S^{(k)}\|_F \leq \text{tol}$, where tolerance tol is chosen by the user. For one-sided methods, where N scalar products are needed to obtain $\|\Omega(H_S^{(k)})\|_F$, tol can be set $f(n)\epsilon$ where f is slowly increasing function of n and ϵ is machine epsilon. For two-sided methods however, one can use a nice stopping criterion of Rutishauser (see [57]) which is almost equivalent to $\|\Omega(H_S^{(k)})\|_F \leq \epsilon$.

2.2 J-Symmetric Jacobi Method

J-Symmetric Jacobi method has similar asymptotic properties. Quadratic convergence is again defined by the relation (2.1) or (2.2). The proof assumes that the diagonal elements of JH approximating the same eigenvalue of (H, J) take successive positions on the diagonal of JH. Absolute gaps δ_i and δ are defined as above. Drmač and Hari [10] have shown that

$$\|\Omega(H^{(N)})\|_F \leq 3\frac{\|\Omega(H)\|_F^2}{\delta} \quad \text{provided that} \quad \|\Omega(H)\|_F < \frac{\delta}{3m(n-m)},$$

where m is number of positive ones in J. Here the asymptotic assumption is stronger than for the symmetric Jacobi method. In estimating nonorthogonal (hyperbolic) transformations one uses mathematical induction. The number $m(n-m)$ in the denominator is used to compensate a possible gradual increase of $\|\Omega(H^{(k)})\|_F$ during the considered cycle.

J-symmetric method is very well suited to work with the accelerated strategy of Mascarenhas, which yields cubic rate of convergence per quasi-cycle. For the proof, one could combine ideas from [10] and [37].

Since the J-symmetric quadratic convergence result suffers the same shortcomings as its symmetric counterpart (2.5), one is challenged to find a similar remedy. First, one needs a bound for the relative distance between diagonal elements of JH and the eigenvalues of (H, J). This is done for the case of positive definite H. In the following result, nonincreasing ordering of the eigenvalues of the pair (H, J) is assumed. Relative gaps γ_i, $1 \leq i \leq p$ and γ are defined as in (2.6).

Proposition 2.2. [35, Theorem 1(ii)] *Let $H = H^*$ be α-s.d.d. positive definite. Let $H = \Delta H_S \Delta$, where $\Delta = [\,diag(H)\,]^{\frac{1}{2}}$. Let $\alpha < \gamma/3$ and let the diagonal* *elements of JH affiliated with the same eigenvalue of (H, J) occupy successive positions on the diagonal. Then for the same partition of $A = H_S - I_n$ and Δ holds*

$$\|A_{ii} - |\lambda_{s_i}|\Delta_{ii}^{-2}\|_F \leq \frac{2}{\gamma_i}\|\tau_i(A)\|_F^2 = \frac{2}{\gamma_i}\sum_{\substack{j=1\\ j\neq i}}^{p}\|A_{ij}\|_F^2, \qquad 1 \leq i \leq p.$$

■

As earlier, one can deduce

$$\sum_{r=1}^{n}\left|1 - \frac{|\lambda_r|}{h_{rr}}\right|^2 + \|\pi(A)\|_F^2 \leq \frac{2}{\gamma^2}\|\tau(A)\|_F^4 .$$

Thus, scaled diagonally dominant pair (H, J) has the same structure as the pair (H, I_n) which has been discussed earlier. Hence the rotation angles can be large if pivot pair happens to be inside a diagonal block. On the contrary, hyperbolic angles ultimately tend to zero as k increases. This follows from the fact that $m = n_1 + \cdots + n_{r_o}$ for some $2 \leq r_o \leq p-1$ and that is a consequence of the assumption that (H, J) is positive definite. We see that the measure $\|\tau(A)\|_F$ or $\|\Omega(A)\|_F$ should be used for stopping of the process.

The new quadratic convergence result from [35] has form

$$\alpha_N \leq 3.5\frac{\alpha_0^2}{\gamma} \qquad \text{whenever} \qquad \alpha_0 \leq \frac{1}{6}\min\{\frac{1}{n}, \gamma\}.$$

Here $\alpha_k = \|\Omega(A^{(k)})\|_F$, $k \geq 0$ and the result assumes that $\lambda_1 \geq \cdots \geq \lambda_m$, $|\lambda_{m+1}| \geq \cdots \geq |\lambda_n|$, which requires block-permutational similarity of the partition (2.3) and renumbering of the relative gaps.

3. Accuracy

As noticed by Rosanoff et al [38], and theoretically explained by Demmel and Veselić [6], the Jacobi algorithm is more accurate than any algorithm that starts with tridiagonalization of the symmetric matrix (bidiagonalization in the case of SVD computation). In this section we explain this important fact, using the results of Demmel, Veselić, Hari and Drmač. As we will see, if the objective is to compute all eigenvalues with small relative error, the definite and the indefinite case must be treated differently. We first analyze the

3.1 Symmetric Definite Case

Numerical analysis of the two–stage diagonalization procedure is simple but with far reaching consequences. We start with the analysis of the Cholesky factorization.

In floating point computation, the computed approximation of L is $\tilde{L}$ and we need an estimate for the backward error $\tilde{L}\tilde{L}^T - H$. The following proposition is due to Demmel [7].

Proposition 3.1. *Suppose the Cholesky factorization algorithm has successfully completed all steps in floating point arithmetic with unit roundoff ϵ. If $\tilde{L}$ is the computed lover triangular matrix, then $\tilde{L}\tilde{L}^T = H+\delta H$, where δH is symmetric matrix such that for all $1 \le i, j \le n$*

$$|\delta H_{ij}| \le \eta_C \sqrt{H_{ii}H_{jj}}, \quad \eta_C = \frac{c(n)\epsilon}{1-2c(n)\epsilon}, \quad c(n) = \max\{3, n\}. \qquad (3.1)$$

■

Note that the backward error δH is bounded entry–wise, rather than norm–wise.

Now, we apply the one-sided Jacobi algorithm to the matrix $\tilde{L}$, that is, we implicitly run the symmetric Jacobi on the matrix $\tilde{L}^T\tilde{L}$. As has been shown in subsection 1.2, this iteration process has form $G^{(k+1)} = G^{(k)}V^{(k)}$, $k \ge 0$, where $V^{(k)}$ is plane Jacobi rotation and $G^{(0)} = \tilde{L}$. The process terminates at index ℓ where the normalized columns of $G^{(\ell)}$ are orthogonal up to a tolerance $O(n\epsilon)$. The following proposition from [9] describes the numerical behaviour of the right–handed Jacobi SVD algorithm.

Proposition 3.2. *Let the cyclic one-sided Jacobi algorithm be applied to $\tilde{L}$ in floating point arithmetic with roundoff ϵ. Let each cycle comprise p parallel steps and let the stopping criterion be satisfied after s cycles. Let $\tilde{G}^{(k)}$, $k = 0,\ldots,\ell = p\cdot s$, $\tilde{G}^{(0)} = \tilde{L}$ be the generated matrices. Then there exists an orthogonal matrix $\hat{V}$ and a backward error $\delta\tilde{L}$ such that $\tilde{G}^{(\ell)} = (\tilde{L}+\delta\tilde{L})\hat{V}$ and*

$$\|e_i^T\delta\tilde{L}\|_2 \le \eta_J\|e_i^T\tilde{L}\|_2, \quad 1 \le i \le n, \quad \eta_J \le (1+6\epsilon)^\ell - 1.$$

Further, due to the stopping criterion, the columns of $\tilde{G}^{(\ell)}$ are numerically orthogonal, that is

$$\max_{i,j} \cos\angle(\tilde{G}^{(\ell)}e_i\,,\, \tilde{G}^{(\ell)}e_j) \le O(n\epsilon).$$

■

It is important to note that the error analysis is done row–wise, while the convergence is defined column–wise.

From Proposition 3.2 it follows that $\tilde{G}^{(\ell)}$ can be written as $\tilde{G}^{(\ell)} = \tilde{U}\tilde{\Sigma}$, where $\tilde{\Sigma}$ is diagonal with column norms of $\tilde{G}^{(\ell)}$ along its diagonal, and $\tilde{U}$ is numerically orthogonal, $|\tilde{U}^T\tilde{U} - I_n|_{ij} \le O(n\epsilon)$.

Combining Propositions 3.1 and 3.2, we get

$$\begin{aligned}\tilde{U}\tilde{\Sigma}^2\tilde{U}^T &= (\tilde{L}+\delta\tilde{L})(\tilde{L}+\delta\tilde{L})^T = \tilde{L}\tilde{L}^T + \tilde{L}\delta\tilde{L}^T + \delta\tilde{L}\tilde{L}^T + \delta\tilde{L}\delta\tilde{L}^T \\ &= H + \delta H + E, \quad E = \tilde{L}\delta\tilde{L}^T + \delta\tilde{L}\tilde{L}^T + \delta\tilde{L}\delta\tilde{L}^T.\end{aligned}$$

The perturbation matrix $\Delta H = \delta H + E$ is symmetric and it holds

$$\max_{i,j} \frac{|\Delta H_{ij}|}{\sqrt{H_{ii}H_{jj}}} \leq \eta \equiv \eta_C + 2\eta_J + O(\eta_J^2).$$

If we set $\tilde{\Lambda} = \tilde{\Sigma}^2$, then we have $H + \Delta H = \tilde{U}\tilde{\Lambda}\tilde{U}^T$. This means that *this variant of the Jacobi diagonalization method computes the eigenvalues and eigenvectors with entry–wise small backward error. The methods that first tridiagonalize the matrix do not share this important property.*

Let us estimate the forward error in the computed approximations $\tilde{\Lambda}_{ii}$ of the eigenvalues λ_i of H. Let $\tilde{\lambda}_1 \geq \cdots \geq \tilde{\lambda}_n$ be the eigenvalues of $H + \Delta H$ and let the eigenvalues λ_i of H as well as the diagonals of $\tilde{\Lambda}$ be nonincreasingly ordered. First we estimate the maximum relative distance between λ_i and $\tilde{\lambda}_i$.

We can presume that η is small enough so that $\|L^{-1}\Delta H L^{-T}\|_2 < 1$ holds, implying that the positive square root of $I_n + L^{-1}\Delta H L^{-T}$ is well defined. Since the matrix

$$H + \Delta H = L\sqrt{I + L^{-1}\Delta H L^{-T}}\sqrt{I + L^{-1}\Delta H L^{-T}}L^T$$

is similar to

$$\sqrt{I + L^{-1}\Delta H L^{-T}}L^T L\sqrt{I + L^{-1}\Delta H L^{-T}},$$

and $L^T L$ is similar to $H = LL^T$, an application of Ostrowsky's theorem yields $\tilde{\lambda}_i = \lambda_i(1 + \theta_i)$, $|\theta_i| \leq \|L^{-1}\Delta H L^{-T}\|_2$, $1 \leq i \leq n$. The key observation in this estimate is as follows. If we set $D = [\mathrm{diag}(H)]^{1/2}$, then $L^{-1}\Delta H L^{-T} = L^{-1}D(D^{-1}\Delta H D^{-1})DL^{-T}$ and

$$\begin{aligned} \|L^{-1}\Delta H L^{-T}\|_2 &\leq \|L^{-1}D\|_2^2 \|D^{-1}\Delta H D^{-1}\|_2 \\ &= \|(D^{-1}HD^{-1})^{-1}\|_2 \|D^{-1}\Delta H D^{-1}\|_2 \\ &\leq n\eta\|(D^{-1}HD^{-1})^{-1}\|_2. \end{aligned}$$

The matrix $H_S = D^{-1}HD^{-1}$ has unit diagonal, off-diagonals less than one in modulus, and by the well–known result of van der Sluis [50] it holds

$$\|H_S^{-1}\|_2 \leq \kappa_2(H_S) \leq n \min_{S=\mathrm{diag},\det(S)\neq 0} \kappa_2(SHS) \leq n\kappa_2(H).$$

Here $\kappa_2(X) = \|X\|_2\|X^{-1}\|_2$ is the spectral condition number. Therefore, we can write

$$\tilde{\lambda}_i = \lambda_i(1 + \theta_i), \quad |\theta_i| \leq n\eta\|H_S^{-1}\|_2, \quad 1 \leq i \leq n. \tag{3.2}$$

Next, we estimate the relative distance between $\tilde{\lambda}_i$ and $\tilde{\Lambda}_{ii}$. The stopping criterion ensures that $\tilde{U}^T\tilde{U} = I + X$ with $\max_{i,j}|x_{ij}| \leq O(n\epsilon)$, where $X =$

(x_{ij}). An easy calculus shows that there exists an orthogonal matrix $\hat{U}$ such that $\tilde{U} = (I+Y)^{1/2}\hat{U}$, where Y is symmetric and $\|Y\|_2 = \|X\|_2 \leq O(n^2\epsilon)$. Hence,

$$H + \Delta H = \tilde{U}\tilde{\Lambda}\tilde{U}^T = (I+Y)^{1/2}\hat{U}\tilde{\Lambda}\hat{U}^T(I+Y)^{1/2},$$

and we can again apply the Ostrowsky's theorem to obtain $\tilde{\lambda}_i = \tilde{\Lambda}_{ii}(1+\theta_i')$, $|\theta_i'| \leq O(n^2\epsilon))$. This together with (3.2) implies

$$\tilde{\Lambda}_{ii} = \lambda_i \frac{1+\theta_i}{1+\theta_i'} = \lambda_i(1+\eta_i), \quad |\eta_i| \leq O(n^2\epsilon)\|H_S^{-1}\|_2, \quad 1 \leq i \leq n.$$

This bound is nearly the best one can hope for in computing with floating point positive definite matrices. For, Demmel [7] has shown that relative entry–wise perturbations of size $1/\|H_S^{-1}\|_2$ can make H exactly singular, and that H can be considered numerically positive definite if $\|H_S^{-1}\|_2 < 1/(n\eta_C)$. Moreover, Veselić and Slapničar [53] have shown that the spectrum of positive definite H is stable under entry–wise floating point perturbations if and only if $\|H_S^{-1}\|_2$ is moderate.

We can conclude that *the forward error in the computed eigenvalues depends on $\kappa_2(H_S)$, and not on $\kappa_2(H)$, as is in the case of methods based on tridiagonalization. Since $\kappa_2(H_S) \leq n\kappa_2(H)$, and since it is possible $\kappa_2(H_S) \ll \kappa_2(H)$, the Jacobi method is numerically clearly superior to any method that first tridiagonalizes the matrix. This method is method of choice for computing the eigenvalues of positive definite matrices with high relative accuracy.*

The diagonalization procedure just described (Cholesky factorization followed by the right–handed Jacobi on L) implicitly diagonalizes $L^T L$. This is more than just a nice observation. It actually means that this Jacobi method is preconditioned using one implicit (cost free) step of the Rutishauser LR method. Because of diagonalizing effect of preconditioning, the Jacobi method converges faster, especially if the Cholesky factorization is computed with pivoting. For more discussion see [52] and [14].

3.2 Indefinite Case

Here we provide elements of perturbation analysis and error bounds for the J-symmetric Jacobi method and one-sided J-symmetric compound Jacobi method described in Subsection 1.3.

Let us first consider the J-symmetric eigenvalue problem $Hx = \lambda Jx$ with positive definite H,

$$Q^T HQ = \Lambda, \quad Q^T JQ = J, \quad \Lambda = \mathrm{diag}(\lambda_i).$$

Let $H = DAD$, where $D = \mathrm{diag}(d_{11}, \ldots, d_{nn})$ is diagonal and A is positive definite with unit diagonal. Further, let $\delta H = D\delta AD$. By the result of Veselić and Slapničar [53, 46], if $|\delta H_{ij}| \leq \eta d_{ii} d_{jj}$, and if $\tilde{\eta} \equiv \|\delta A\|_2 \|A^{-1}\|_2 \leq n\eta \|A^{-1}\|_2 < 1$, then

$$1 - \tilde{\eta} \leq \frac{\tilde{\lambda}_i}{\lambda_i} \leq 1 + \tilde{\eta}. \tag{3.3}$$

The perturbation result for the invariant subspaces is given by Slapničar and Truhar in [45]. Let us partition the eigenvalue problem as

$$\begin{bmatrix} Q_1^T \\ Q_2^T \end{bmatrix} H \, [Q_1 \, Q_2] = \begin{bmatrix} \Lambda_1 & \\ & \Lambda_2 \end{bmatrix}, \quad \begin{bmatrix} Q_1^T \\ Q_2^T \end{bmatrix} J \, [Q_1 \, Q_2] = \begin{bmatrix} J_1 & \\ & J_2 \end{bmatrix},$$

and let the perturbed problem be partitioned accordingly. Let $\tilde{X}_1$ and X_2 be orthogonal bases for subspaces spanned by the columns of $\tilde{Q}_1$ and JQ_2J_2, respectively. Let $U\Sigma V^*$ be a singular value decomposition of the matrix $X_2^T \tilde{X}_1$. The diagonal entries of the matrix $\sin \Theta(Q_1, \tilde{Q}_1) \equiv \Sigma$ are the sines of canonical angles between subspaces spanned by the columns of Q_1 and $\tilde{Q}_1$ (see [48]). The relative gap is in this case defined by

$$\gamma(\tilde{\Lambda}_1, \Lambda_2) = \min_{i,j} \frac{|\, [\tilde{\Lambda}_1]_{ii}[J_1]_{ii} - [\Lambda_2]_{jj}[J_2]_{jj}|}{\sqrt{[\tilde{\Lambda}_1]_{ii} \cdot [\Lambda_2]_{jj}}}. \tag{3.4}$$

Then, if $\|\delta A\|_2 \|A^{-1}\|_2 < 1$, we have

$$\| \sin \Theta(Q_1, \tilde{Q}_1)\|_F \leq \|Q\|_2^2 \left(\frac{1}{2}\psi + \sqrt{1 + \frac{1}{4}\psi^2} \right) \frac{\psi}{\gamma(\tilde{\Lambda}_1, \Lambda_2)}, \tag{3.5}$$

where

$$\psi = \frac{\|\delta A\|_2 \|A^{-1}\|_F}{\sqrt{1 - \|\delta A\|_2 \|A^{-1}\|_2}}.$$

Clearly, when $J = I$, then Q is orthogonal, and (3.5) is a subspace version of the corresponding eigenvector bounds from [6]. However, when $J \neq I$, then $\|Q\|_2^2 \equiv \kappa_2(Q)$ is the spectral condition number. It is a remarkable fact that $\kappa_2(Q)$ is bounded by the condition number of A, the same quantity that governs the accuracy of the computation. In [47], Slapničar and Veselić proved that

$$\kappa(Q) \leq \sqrt{\min_{\Delta J = J\Delta} \kappa_2(\Delta^T A \Delta)} \leq \sqrt{\kappa_2(A)}.$$

We now describe results of numerical analysis. Error analysis of a single hyperbolic rotation is technically more complicated because such transformations are nonorthogonal and possibly of large norm. One step of the method, $H^{(k+1)} = C_k^T H^{(k)} C_k$, in floating point computation is of the form

$$\tilde{H}^{(k+1)} = \tilde{C}_k^T (\tilde{H}^{(k)} + \delta \tilde{H}^{(k)}) \tilde{C}_k$$

where $\tilde{C}_k$ is hyperbolic rotation, and the backward perturbation $\delta\tilde{H}^{(k)}$ is bounded as follows: If $\tilde{H}^{(k)} = D_k A_k D_k$, where D_k is diagonal and A_k has unit diagonal, then we can write $\delta\tilde{H}^{(k)}$ as $\delta\tilde{H}^{(k)} = D_k \delta A_k D_k$, where $\|\delta A_k\|_2 \leq \alpha_k \varepsilon$. Here $\alpha_k = O(\sqrt{\kappa_2(A_k)}\,\sqrt{n})$. Thus, according to (3.3), the perturbation of the eigenvalues due to single floating point hyperbolic rotation is determined by the value

$$\|\delta A_k\|_2 \, \|A_k^{-1}\|_2 \leq \alpha_k \varepsilon \|A_k^{-1}\|_2.$$

Accumulating the effect of a total of r rotations, gives the relative error bound for the computed eigenvalues of the order of

$$\varepsilon \sum_{k=0}^{r} \alpha_k \|A_k^{-1}\|_2 + O(n^2)\varepsilon.$$

The error bound for the computed eigenspaces follows by plugging the above bound for $\|\delta A_k\|_2$ into (3.5) and accumulating the effect of a total of r rotations in a similar manner. For details see [43].

During the process, $\|A_k^{-1}\|_2$ and $\kappa_2(A_k)$ tend to one, starting with $\|A^{-1}\|_2$ and $\kappa_2(A)$, respectively. While the theoretical bound for the values of $\|A_k^{-1}\|_2$ and $\kappa_2(A_k)$, $k = 1, 2, \ldots$ is pessimistic, numerical evidence indicates that these values never grow too much above the initial values. For more discussion see [40, 43, 41, 44].

Next, we consider the symmetric indefinite eigenvalue problem $Hx = \lambda x$,

$$Q^T H Q = \Lambda, \quad Q^T Q = I, \quad \Lambda = \text{diag}(\lambda_i) \text{ nonsingular},$$

and the appropriate compound method.

The perturbation bound for the eigenvalues is given by Veselić and Slapničar [53, 46]. Let $|H|_s = \sqrt{H^2}$ be the spectral absolute value of H, and let $|H|_s = DAD$, where $D = \text{diag}(d_{11}, \ldots, d_{nn})$ is diagonal and A is positive definite with unit diagonal. Further, let $\delta H = D\delta A D$. If $|\delta H_{ij}| \leq \eta d_{ii} d_{jj}$, and if $\tilde{\eta} \equiv \|\delta A\|_2 \, \|A^{-1}\|_2 \leq n\eta \|A^{-1}\|_2 < 1$, then the perturbation of the eigenvalues is again bounded by (3.3).

The perturbation theory for the invariant subspaces, given by Truhar and Slapničar [49], assumes the following partition of the eigenvalue decomposition

$$\begin{bmatrix} Q_1^T \\ Q_2^T \end{bmatrix} H \begin{bmatrix} Q_1 & Q_2 \end{bmatrix} = \begin{bmatrix} \Lambda_1 & \\ & \Lambda_2 \end{bmatrix}.$$

Let the perturbed problem be partitioned accordingly. Similarly to (3.5), if $\|\delta A\|_2 \, \|A^{-1}\|_2 < 1$, then

$$\|\sin\Theta(Q_1, \tilde{Q}_1)\|_F \leq \|V\|_2^2 \frac{\|\delta A\|_2 \, \|A^{-1}\|_F}{\sqrt{1 - \|\delta A\|_2 \, \|A^{-1}\|_2}} \frac{1}{\gamma(\tilde{\Lambda}_1, \Lambda_2)}. \tag{3.6}$$

Here $\gamma(\tilde{\Lambda}_1, \Lambda_2)$ is again defined as in (3.4), but without J_1 and J_2, and V is the hyperbolic eigenvector matrix of the pair $(G^T G, J)$.

We now describe numerical analysis of the factorization. For the sake of simplicity, we assume that the matrix is already pivoted so that the Bunch–Parlett symmetric indefinite factorization $PHP^T = GJG^T$ runs with $P = I$. Slapničar [42] has shown that in floating point, the computed matrix G satisfies

$$GJG^T = H + \delta H$$

with symmetric backward perturbation δH bounded entry–wise by

$$|\delta H| \leq O(n)\varepsilon(|H| + |G||G|^T).$$

The effect of this backward error to the eigenvalues of H is given by (3.3) with

$$\tilde{\eta} \leq O(n)\varepsilon\|(\hat{D}^{-1}GVV^TG^T\hat{D}^{-1})^{-1}\|_2, \tag{3.7}$$

where $\hat{D}$ is a diagonal scaling matrix and the rows of the matrix $\hat{D}^{-1}G$ are of unit Euclidean norm. The details of this estimation can be found in [40, 43]. Also, by using (3.6) it can be shown that the error in the computed invariant subspaces, which is due to factorization is bounded by (see [40, 43])

$$\|\sin\Theta(Q_1, \tilde{Q}_1)\|_F \leq O(\|V\|_2^2)\frac{\tilde{\eta}}{\gamma(\tilde{\Lambda}_1, \Lambda_2)}. \tag{3.8}$$

The factorization $PHP^T = GJG^T$ has two more remarkable properties worth mentioning (see [42] for details). First, if the computed factor $\tilde{G}$ is lower triangular (meaning that only 1×1 pivots took place), then the factorization is also forward stable,

$$|\tilde{G} - G| \leq 3n|G|\text{tril}(|G^{-1}|(|H| + |G|\,|G|^T)|G^{-1}|^T)\varepsilon + O(\varepsilon^2).$$

Second, let $G = BD$, where D is diagonal scaling and the columns of B are of unit Euclidean norm. Then the condition number of B is bounded by a function of n, irrespective of the condition number of G,

$$\kappa_2(B) \leq 3.781^n\sqrt{15\,n^2 + n}\,.$$

Let us now consider the iterative part of the method. Let $\tilde{H} = \tilde{G}J\tilde{G}^T$ and $\tilde{G} = (B + \delta B)D$. Then the eigenvalue perturbation can be bounded by (see [53])

$$(1-\eta)^2 \leq \frac{\tilde{\lambda}_i}{\lambda_i} \leq (1+\eta)^2, \quad \eta = \|\delta B B^\dagger\|_2. \tag{3.9}$$

One step of the one–sided method, $G^{(k+1)} = G^{(k)}C_k$, in floating point computation is of the form

$$\tilde{G}^{(k+1)} = (\tilde{G}^{(k)} + \delta\tilde{G}^{(k)})\tilde{C}_k$$

where $\tilde{C}_k$ is hyperbolic rotation, and the backward perturbation $\delta\tilde{G}^{(k)}$ is bounded as follows: If $\tilde{G}^{(k)} = B_k D_k$, where D_k is diagonal and B_k has unit columns, then we can write $\delta\tilde{G}^{(k)}$ as $\delta\tilde{G}^{(k)} = \delta B_k D_k$, where $\|\delta B_k\|_2 \leq \beta_k \varepsilon$. Here β_k is moderate constant. Thus, the perturbation of the eigenvalues due to a single floating point rotation is determined by the value of $\|\delta B_k B_k^\dagger\|_2 \leq \beta_k \varepsilon \|B_k^\dagger\|_2$. Accumulating the effect of a total of r rotations, where $\tilde{G}^{(r)}$ has columns orthogonal up to $O(n\varepsilon)$, gives the relative error bound for the computed eigenvalues of the order of

$$\tilde{\beta} = \varepsilon \sum_{k=0}^{r} \beta_k \|B_k^\dagger\|_2 + O(n^2)\varepsilon. \tag{3.10}$$

By combining (3.3), (3.7), (3.9) and (3.10), it follows that the relative error in the eigenvalues $\tilde{\lambda}_i$ computed by the compound method (indefinite factorization followed by the one-sided J-symmetric Jacobi method) is bounded by

$$\frac{|\tilde{\lambda}_i - \lambda_i|}{|\lambda_i|} \leq \tilde{\eta} + \tilde{\beta} + O(\varepsilon^2). \tag{3.11}$$

Error bound for the computed invariant subspaces is obtained by combining (3.8) with the bound for the errors due to iterative part of the algorithm from [43], essentially giving

$$\|\sin\Theta(Q_1, \tilde{Q}_1)\|_F \leq O(\|V\|_2^2) \frac{\tilde{\eta} + \tilde{\beta}}{\gamma(\tilde{\Lambda}_1, \Lambda_2)}. \tag{3.12}$$

Similarly to the two-sided method, during the process, $\|B_k^\dagger\|_2$ tends to one, starting with $\|B^\dagger\|_2$. While the theoretical bound for the values of $\|B_k^\dagger\|_2$, $k = 1, 2, \ldots$ is again pessimistic, numerical evidence indicates that these values never grow too much above the initial value $\|B^\dagger\|_2$. Moreover, it has been observed in [40, 43], that $\|B^\dagger\|_2$ is in general very low, primarily due to the rank revealing property of the factorization $PHP^T = GJG^T$. Consequently, the final errors in (3.11) and (3.12) are mainly due to the factorization part of the algorithm, that is, to $\tilde{\eta}$. For more discussion see [40, 43, 41, 44].

3.3 Computing SVD

Here we discuss relative accuracy issues in one-sided Jacobi method for computing singular value decomposition of a general $m \times n$ matrix A, $m \geq n$. The simplest variant of the method is simply implicit form of the symmetric Jacobi on $H = A^T A$. More sophisticated versions use preconditioning to enhance numerical accuracy and efficiency (speed). For instance, the QR factorization with pivoting is an excellent preconditioner in the following sense. If $A = QR$ is the QR factorization with standard Businger–Golub column pivoting ($A \leftarrow AP$, P permutation matrix), then the SVD Jacobi applied to R^T

converges much faster than applied to A or R. This is because of implicitly performed step of Rutishauser's LR method (transition from $R^T R$ to RR^T). Moreover, the computed $\tilde{R}$ satisfies $A + \delta A = \hat{Q}\tilde{R}$, where $\hat{Q}$ is orthonormal and $\|\delta A e_i\|_2 \le \eta_{QR}\|Ae_i\|_2$, $1 \le i \le n$. Here η_{QR} is bounded by a modest polynomial times the roundoff unit ϵ. Note that the relative backward error is small in each column of A. For more details see [9].

If we apply the SVD Jacobi on $\tilde{L} = \tilde{R}^T$, then by Proposition 3.2, for some orthogonal $\hat{V}$ it holds $\tilde{R} + \delta\tilde{R} = \hat{V}\tilde{\Sigma}\tilde{U}^T$. Here $\tilde{U}$ is numerically orthogonal, $\tilde{\Sigma}$ is diagonal, and $\|\delta\tilde{R}e_i\|_2 \le \eta_J\|\tilde{R}e_i\|$, $1 \le i \le n$. Combining the results we obtain the relation

$$A + \Delta A = \hat{Q}\hat{V}\tilde{\Sigma}\tilde{U}^T, \quad \Delta A = \delta A + \hat{Q}\delta\tilde{R},$$

where

$$\|\Delta A e_i\|_2 \le (\eta_{QR} + \eta_J + \eta_{QR}\eta_J)\|Ae_i\|_2, \quad 1 \le i \le n.$$

Here we note that the angle of Jacobi rotation underflows if the condition number $\kappa_2(A)$ overflows, and that standard construction of Jacobi rotation can lead to misconvergence of the algorithm. To avoid this, Jacobi rotation must be modified as shown in [11]. Also, instead of $\hat{Q}$, $\hat{V}$ we will have computed numerically orthogonal matrices $\tilde{Q}$, $\tilde{V}$ such that $\|\tilde{Q} - \hat{Q}\|_2$ and $\|\tilde{V} - \hat{V}\|_2$ are bounded by moderate polynomials of the dimensions times the roundoff ϵ.

Let $\sigma_1 \ge \cdots \ge \sigma_n > 0$ and $\tilde{\sigma}_1 \ge \cdots \ge \tilde{\sigma}_n$ be the eigenvalues of A and $A + \Delta A$, respectively. Write $D = \mathrm{diag}(\|Ae_1\|_2, \ldots, \|Ae_1\|_2)$ and

$$A + \Delta A = (I + \Delta A A^\dagger)A = (I + (\Delta A D^{-1})(AD^{-1})^\dagger)A\,.$$

From the variational characterization of the singular values, we immediately conclude

$$\max_i \frac{|\tilde{\sigma}_i - \sigma_i|}{\sigma_i} \le \eta \equiv \|\Delta A A^\dagger\|_2 \le \sqrt{n}(\eta_{QR} + \eta_J + \eta_{QR}\eta_J)\|A_s^\dagger\|_2,$$

where $A_s = AD^{-1}$. On the other hand, one can show that, for all i, $\tilde{\sigma}_i = \tilde{\Sigma}_{ii}(1 + O(n^2\epsilon))$.

By a theorem due to van der Sluis [50], we know that

$$\kappa_2(A_s) \equiv \|A_s\|_2\|A_s^\dagger\|_2 \le \sqrt{n} \min_{S=diag,\det(S)\neq 0} \kappa_2(AS) \le \sqrt{n}\kappa_2(A).$$

Thus, the SVD Jacobi algorithm computes the singular values with small relative backward error in each column of A. This means that small columns are preserved. The relative error in the computed singular values depends on the condition number of the column equilibrated matrix A_s, and not on the condition number of the initial A. These properties are not shared by bidiagonalization based methods. (Recent modification of the bidiagonalization process, due to

Barlow [1], improves the accuracy of the bidiagonalization, but not to the level of the Jacobi SVD algorithm.)

The backward error in the Jacobi algorithm can be put into multiplicative form $(I + \Delta AA^{\dagger})A$ with small $\eta = \|\Delta AA^{\dagger}\|$ if $\kappa_2(A_s)$ is moderate. This fact also has important implications to the accuracy of the computed singular vectors. Let σ_i be simple with singular vectors u_i and v_i. If $u_i + \delta u_i$, $v_i + \delta v_i$ are the singular vectors of $A + \Delta A$, corresponding to $\tilde{\sigma}_i$, then

$$\max\{\sin\angle(u_i, u_i + \delta u_i), \sin\angle(v_i, v_i + \delta v_i)\} \leq \frac{O(\eta)}{\text{gap}_i} \tag{3.13}$$

where

$$\text{gap}_i = \min\left\{\min_{j\neq i} \frac{|\sigma_i - \tilde{\sigma}_j|}{\sigma_i}, 2\right\}.$$

Thus, the approximation error for the singular vectors of $A + \Delta A$ depends on the condition number $\kappa_2(A_s)$ and the relative separation of the singular values. Since our computed vectors are close to those of $A + \Delta A$, we can conclude that the SVD Jacobi computes the singular vectors with a bound like (3.13). The same conclusion then holds for the eigenvectors computed by the previously explained diagonalization procedure of symmetric positive definite matrix.

From the above analysis we can see that the SVD Jacobi can compute with high relative accuracy the SVD of any matrix A of the form $A = BS$, where S is any diagonal matrix, and B is well conditioned ($\kappa_2(B)$ moderate). A simple device can preserve this property if A is more general, for instance if $A = S_1 B S_2$, where S_1, S_2 are arbitrary diagonal scalings, and B is well conditioned. In that case, the QR factorization in the first step should be computed with column and row pivoting. For the details see [14, 5]. We only note that our theoretical understanding is one step behind numerical experience.

The SVD Jacobi method can be generalized e.g. to the SVD of the product of two matrices. For instance, if $A = BS$, $C = GD$ with well conditioned full column rank B, G with equilibrated columns, and arbitrary diagonal matrices S, D, then the SVD of $AC^T = BSDG^T$ can be computed as follows. First, we compute QR factorization with column pivoting of GDS, $(GDS)P = QR$. Due to pivoting, the matrix R is structured as $R = D_1 R_r$, where D_1 is diagonal and R_r is well conditioned. In fact, $\kappa_2(R_r)$ is bounded by a function of the dimension, for any initial matrix. Then, $AC^T = BPR^TQ^T$, and the product BPR^T is $BPR_r^T D_1$, where BPR_r^T is again well conditioned. Because of that, explicit computation of the product BPR^T will cause no loss of information and the Jacobi SVD will compute accurate SVD of the explicitly computed matrix product. For more details see [13, 12, 15].

In some cases, the matrix A is rather ill conditioned, but with special structure that allows accurate LU decomposition with complete pivoting, $P_1 A P_2 =$

LDU, where P_1, P_2 are permutations, D is diagonal, and L, U are well conditioned. This means that computed matrices $\tilde{L}$, $\tilde{D}$, $\tilde{U}$ are such that the SVD of the product $P_1^T \tilde{L}\tilde{D}\tilde{U} P_2^T$ is highly accurate approximation of the SVD of A. On the other hand, $(P_1^T \tilde{L})(\tilde{D}\tilde{U} P_2^T)$ has the structure of the product of two matrices that allow accurate SVD by the Jacobi method. For more details see [8].

References

[1] Barlow J.: More Accurate Bidiagonal Reduction for Computing the Singular Value Decomposition. SIAM J. Matrix. Anal. Appl. To appear.

[2] Barlow J. and Demmel J.: Computing accurate eigensystems of scaled diagonally dom inant matrices. SIAM J. Num. Anal., 27 (1990) 762–791.

[3] Brent R. and Luk F.: The Solution of Singular-value and Symmetric Eigenvalue Problems on Multiprocessor Arrays, SIAM Journal of Scientific and Statistical Computing 6 (1985) 69–84.

[4] Bunch J. and Parlett B.: Direct Methods for solving Symmetric indefinite systems of linear equations, SIAM J. Num. Anal. Vol. 8 No. 4 (1971) 639-655.

[5] Cox A. and Higham N. : Stability of Householder QR Factorization for Weighted Least Squares Problems. Proceedings of the 17th Dundee Biennial Conference Numerical Analysis 1997, Pitman Research Notes in Mathematics 380 (1998) 57-73. Ed. Griffiths D. F., Higham D. J. and Watson G. A. Pub. Addison Wesley Longman, Harlow, Essex, UK.

[6] Demmel J. and Veselić K.: Jacobi's method is more accurate than QR. SIAM J. Matrix Anal. Appl. 13 (1992) 1204–1245.

[7] Demmel J.: On floating point errors in Cholesky. Computer Science Department, University of Tennessee 1989. LAPACK Working Note 14.

[8] Demmel J., Gu M., Eisenstat S., Slapničar I., Veselić K. and Drmač Z.: Computing the singular value decomposition with high relative accuracy. Linear Algebra Appl. 299 (1999) 21-80.

[9] Drmač Z.: Computing the Singular and the Generalized Singular Values. Ph.D. thesis, Lehrgebiet Mathematische Physik, Fernuniversität Hagen, 1994.

[10] Drmač Z. and Hari V.: On the Quadratic Convergence of the J–symmetric Jacobi Method. Num. Math. 64 (1993) 147–180.

[11] Drmač Z.: Implementation of Jacobi rotations for accurate singular value computation in floating-point arithmetic. SIAM J. Sci. Comp. 18 (1997) 1200–1222.

[12] Drmač Z.: Accurate computation of the product induced singular value decomposition with applications. SIAM J. Numer. Anal. Vol 35, No. 5 (1998) 1969–1994.

[13] Drmač Z.: A tangent algorithm for computing the generalized singular value decomposition. SIAM J. Numer. Anal. Vol. 35, No. 5 (1998) 1804–1832.

[14] Drmač Z.: A posteriori computation of the singular vectors in a preconditioned Jacobi SVD algorithm. IMA J. Numer. Anal. 19 (1999) 191-213.

[15] Drmač Z.: New accurate algorithms for singular value decomposition of matrix triplets. SIAM J. Matrix Anal. Appl. 21 (2000) 1026–1050.

[16] Forsythe G. and Henrici P.: The Cyclic Jacobi Method for Computing the Principal Values of a Complex Matrix. Trans. Amer. Math. Soc. 94 (1960), 1–23.

[17] Golub G. and van Loan C. : Matrix Computations. The John Hopkins University Press, Baltimore and London 1989.

[18] Hari V. and Veselić K.: On Jacobi methods for singular value decompositions. SIAM J. Sci. Stat. Comput. Vol.8, No.5 (1987).

[19] Hari V.: On Almost Diagonal Square Matrices With Multiple Singular Values. Radovi Matematički 4 (1988) 209–225.

[20] Hari V.: On Pairs of Almost Diagonal Matrices. Linear Algebra Appl. 148 (1991) 193–223.

[21] Hari V.: On Sharp Quadratic Convergence Bounds for the Serial Jacobi Methods. Numer. Math. 60: 375–406, 1991.

[22] Hari V. and Drmač Z.: On Scaled Almost Diagonal Hermitian Matrix Pairs. SIAM J. Matrix Anal. Appl. Vol. 18. No. 4 (1997) 12pp.

[23] Hari V.: Structure of Almost Diagonal Matrices. Mathematical Communications 4 (1999), 135–158.

[24] Henrici P. and Zimmermann K.: An Estimate for the Norms of Certain Cyclic Jacobi Operators. Linear Algebra Appl. 1 (1968) 289–501.

[25] Ipsen I.: Relative Perturbation Results for Matrix Eigenvalues and Singular Values. Acta Numerica 8 (1998) 151–201.

[26] van Kempen H.: On Quadratic Convergence of the Classical Jacobi Method for Real Symmetric Matrices with Nondinstinct Eigenvalues. Numer. Math. 9 (1966) 11–18.

[27] van Kempen H.: On Quadratic Convergence of the Special Cyclic Jacobi Method. Numer. Math. 9 (1966) 19–22.

[28] Luk F. and Park H.: On the equivalence and convergence of parallel Jacobi SVD algorithms. IEEE Computer (1987).

[29] Luk F. and Park H.: On parallel Jacobi orderings. SIAM J. Sci. Statist. Comput. 10 (1989) 18–26.

[30] Mascarenhas W.: On the Convergence of the Jacobi Method, Poster Presentation, Fourth SIAM Conference on Parallel Processing for Scientific Computing, Chicago, Illinois, December, 1989.

[31] Mascarenhas W.: On the Convergence of the Jacobi Method for Arbitrary Orderings I, SIAM Journal of Scientific and Statistical Computing, submitted, 1990.

[32] Matejaš J.: Quadratic Convergence of Scaled Iterates by Diagonalization Methods. Ph. D. thesis, University of Zagreb, 1999. Croatian language.

[33] Matejaš J.: Quadratic Convergence of Scaled Matrices in Jacobi Method. Numer. Math. 87 (1999) 171–199.

[34] Matejaš J.: Convergence of scaled iterates by Jacobi method. To appear in Linear Algebra Appl. p.p. 1-37.

[35] Matejaš J. and Hari V.: Note on the quadratic convergence of scaled matrices by J-symmetric Jacobi method. Preprint, University of Zagreb, 2002.

[36] Parlett B.: The Symmetric Eigenvalue Problem, Prentice–Hall Inc., Englewood Cliffs, N. J. 1980.

[37] Rhee N. and Hari V.: On the Global and Cubic Convergence of a Quasy-cyclic Jacobi Method. Numer. Math. 66, (1993) 97–122.

[38] Rosanoff R., Gloudeman J. and Levy S.: Numerical Conditions of Stiffness Matrix Formulations for Frame Structures. Proc.of the Second Conference on Matrix Methods in Structural Mechanics, WPAFB Dayton, Ohio, 1968.

[39] Schönhage A.: On Quadratic Convergence of the Jacobi Process. Numer. Math. 6 (1964) 410–412.

[40] Slapničar I.: Accurate Symmetric Eigenreduction by a Jacobi Method. Ph. D. Fernuniversität Hagen, Germany, 1992.

[41] Slapničar I.: Accurate computation of singular values and eigenvalues of symmetric matrices, Mathematical Communications 1 (1996) 153–168.

[42] Slapničar I.: Componentwise Analysis of Direct Factorisation of Real Symmetric and Hermitian Matrices. Linear Algebra Appl. 272 (1997) 227–275.

[43] Slapničar I.: Error Analysis of J-orthogonal Jacobi Method. submitted to Numer. Math.

[44] Slapničar I.: Highly Accurate Symmetric Eigenvalue Decomposition nd Hyperbolic SVD. submitted to Linear Algebra Appl.

[45] Slapničar I. and Truhar N.: Relative perturbation theory for hyperbolic eigenvalue problem. Linear Algebra Appl. 309 (2000) 57–72.

[46] Slapničar I. and Veselić K.: Perturbations of the eigenprojections of a factorised Hermitian matrix. Linear Algebra Appl. 218 (1995) 273–280.

[47] Slapničar I. and Veselić K.: A bound for the condition of a hyperbolic eigenvector matrix. Linear Algebra Appl. 290 (1999) 247–255.

[48] Stewart G. W. and Sun J.-G.: Matrix Perturbation Theory, Academic Press, Boston, 1990.

[49] Truhar N. and Slapničar I.: Relative perturbation bound for invariant subspaces of graded indefinite Hermitian matrices. Linear Algebra Appl. 301 (1999) 171–185.

[50] Van der Sluis A.: Condition numbers and equilibration of matrices. Numer. Math. 14 (1969) 14–23.

[51] Veselić K.: An Eigenreduction Algorithm for Definite Matrix Pairs and its Applications to Overdamped Linear Systems. Num. Math. 64 (1992), 241-269.

[52] Veselić K. and Hari V.: A note on a one-sided Jacobi algorithm. Numer. Math. 56 (1989) 627-633.

[53] Veselić K. and Slapničar I.: Floating-point Perturbations of Hermitian Matrices. Linear Algbra Appl. 195 (1993) 81–116.

[54] Voevodin, V.: Cislennye metody linejnoj algebry. Nauka, Moscow 1966.

[55] Wilkinson J.: Note on the Quadratic Convergence of the Cyclic Jacobi Process. Numer. Math. 4 (1962) 296–300.

[56] Wilkinson J.: Almost Diagonal Matrices with Multiple or Close Eigenvalues. Linear Algebra Appl. 1 (1968) 1–12.

[57] Wilkinson J. and Reinsch C.: Handbook for Automatic Computation, Vol. 2, Linear Algebra. Springer-Verlag, New York 1971.

[58] Zha H.: A Note on The Existence of the Hyperbolic Singular Value Decomposition. Linear Algbra Appl. 240 (1996) 199–205.

MODELLING OF CURVED RODS

Mladen Jurak
Department of Mathematics, University of Zagreb
Bijenička cesta 30, 10000 Zagreb, Croatia
jurak@math.hr

Josip Tambača
Department of Mathematics, University of Zagreb
Bijenička cesta 30, 10000 Zagreb, Croatia
tambaca@math.hr

Zvonimir Tutek
Department of Mathematics, University of Zagreb
Bijenička cesta 30, 10000 Zagreb, Croatia
tutek@math.hr

Abstract Recent results on mathematical modelling of curved rods are presented. More precisely, elastic behavior of a curved rod-like 3-D body is approximated by elastic behavior of its middle curve. The method of approximation is asymptotic expansion with respect to the small parameter (diameter of the cross section of the rod). Certain convergence results are proved and the obtained 1-D approximation is compared with the Cosserat model and arch model.

Keywords: linearized elasticity, curved rod, evolution model, eigenvalue problem, equilibrium model

1. Introduction

Subject of this paper is mathematical modelling of elastic curved rods. Geometrically curved rod is a 3-D domain with one dimension, the length, much greater than the other two, so curved rod can be well approximated by a curve. The aim of rod modelling is to approximate elastic behavior of 3-D curved rod-like body by elastic behavior of the approximating curve. This problem is a part of lower dimensional modelling in elasticity theory; modelling of plates, shells and straight and curved rods. Resulting initial-boundary value problems are simpler from both, theoretical and numerical point of view.

The oldest approach in lower dimensional modelling is based on certain a priori assumptions on the approximating displacement and constitutive equations, see references in Trabucho and Viaño (1996) in case of straight rods. Second approach is foundation of 1-D and 2-D theories of elasticity, see Antman (1995), Germain (1962), Landau and Lifshitz (1970) and references cited there. Third method, the asymptotic expansion with respect to the small parameter, we call mathematical modelling because certain convergence results are usually proved. The approach was successful in case of plates Ciarlet and Destuynder (1979), Ciarlet and Kesavan (1981) and Raoult (1985), shells Ciarlet, Lods and Miara (1996) and Li–ming (1998), straight rods Bermudez and Viaño (1984), Tutek and Aganović (1986), Irago and Viaño (1998) and Kerdid (1993), shallow arch Álvarez–Dios and Viaño (1998) and curved rods Hay (1942), Jamal and Sanchez–Palencia (1996), Jurak and Tambača (1999), Jurak and Tambača (2001), Tambača (2001), Tambača and Tutek (2000b) and Tambača (2002). Here we briefly describe the rigorous derivation of a curved rod model. A posteriori Kirchhoff's hypotheses of approximate inextensibility, unshearability and rigidity of cross–sections of elastic rods are verified.

The rigorous derivation of curved rod model consists of five steps. First we write the variational formulation of the problem under consideration. The problem is posed on ε dependent domain, where the diameter of the cross section of the curved rod ε, is considered as a small parameter. This is done in Sections 2 and 3. Second, by the change of variables we rewrite the weak formulation on the canonical domain which is independent of ε. As a consequence, dependence on ε becomes explicit. This is content of Section 4. Third step consists of formal expansion of solution in terms of small parameter ε; lower order terms define the formal limit. The weak formulation of the problem satisfied by the formal limit, so called curved rod model, is identified and analyzed in Section 5. The problem is still posed on the canonical domain. Fourth step is justification of the formal limit procedure from the third step; it is based on uniform (with respect to ε) a priori estimates of the family of solutions. Sketch of the proof is given in Sections 6 and 7. Fifth step is return to the original $\varepsilon-$dependent domain; the real curved rod model is then written in more convenient differential form. This is done in Section 8. Finally, in Section 9 we show that our model coincides with the already known linearized Cosserat model and arch model.

2. Curved rod

In this section we introduce the curved rod as certain three-dimensional domain via its middle curve and cross-sections. Assumptions on the middle curve of curved rod are natural, i.e. it is a sufficiently smooth simple and regular curve. Generic curve is then a special case.

Let $\mathcal{C}$ be a simple and regular curve in $\mathbb{R}^3$ defined by its natural parametrization $\boldsymbol{\Phi} \in C^3([0,\ell];\mathbb{R}^3)$. The tangent on $\mathcal{C}$ at $\boldsymbol{\Phi}(y^1)$ is defined by $\boldsymbol{t}(y^1) = \boldsymbol{\Phi}'(y^1)$; obviously $\boldsymbol{t} \in C^2([0,\ell];\mathbb{R}^3)$. Then, see Jurak and Tambača (2001) or Tambača (1999), there exists a function $\mathbf{Q} \in C^2([0,\ell];\mathrm{SO}(3))$ such that

$$\mathbf{Q}\boldsymbol{e}_1 = \boldsymbol{t}.$$

The local basis $\boldsymbol{t}(y^1), \tilde{\boldsymbol{n}}(y^1), \tilde{\boldsymbol{b}}(y^1)$ on $\mathcal{C}$ at $\boldsymbol{\Phi}(y^1)$ can then be defined by

$$\boldsymbol{t}(y^1) = \mathbf{Q}(y^1)\boldsymbol{e}_1, \quad \tilde{\boldsymbol{n}}(y^1) = \mathbf{Q}(y^1)\boldsymbol{e}_2, \quad \tilde{\boldsymbol{b}}(y^1) = \mathbf{Q}(y^1)\boldsymbol{e}_3, \quad y^1 \in [0,\ell].$$

Smoothness of $\mathbf{Q}$ implies that $\boldsymbol{t}, \tilde{\boldsymbol{n}}, \tilde{\boldsymbol{b}} \in C^2([0,\ell];\mathbb{R}^3)$. Note that this basis exists even though the Frenet frame does not exist globally. In the case of generic curve (i.e. $|\boldsymbol{\Phi}''(y^1)| > 0$, $y^1 \in [0,\ell]$), $\mathbf{Q}$ can be chosen so that $\tilde{\boldsymbol{n}}$ and $\tilde{\boldsymbol{b}}$ be principal normal and binormal. Let

$$\mathbf{R}(y^1) = \left(\frac{d}{dy^1}\mathbf{Q}^T(y^1)\right)\mathbf{Q}(y^1), \quad y^1 \in [0,\ell]. \tag{2.1}$$

Matrix $\mathbf{R}(y^1)$ is skewsymmetric for all $y^1 \in [0,\ell]$. In the generic case one has

$$\mathbf{R} = \begin{pmatrix} 0 & \kappa & 0 \\ -\kappa & 0 & \tau \\ 0 & -\tau & 0 \end{pmatrix},$$

where κ and τ are the curvature and the torsion of $\mathcal{C}$, respectively.

Now we define the domain in $\mathbb{R}^3$ which represents the curved rod. Let $\varepsilon > 0$ be a small parameter and let $S \subset \mathbb{R}^2$ be a simply connected bounded domain with the Lipschitz boundary; points in S are denoted by (z^2, z^3). The coordinates z^2 and z^3 are chosen so that $(0,0)$ is the center of mass of S, i.e.

$$\int_S z^2 dz^2 dz^3 = \int_S z^3 dz^2 dz^3 = 0. \tag{2.2}$$

Let

$$\Omega^\varepsilon = (0,\ell) \times \varepsilon S, \quad \Gamma^\varepsilon = (0,\ell) \times \varepsilon\partial S, \quad B^\varepsilon_{y^1} = \{y^1\} \times \varepsilon S, \quad y^1 \in [0,\ell].$$

Γ^ε is the lateral boundary, ℓ is the length, $B^\varepsilon_{y^1}$ is the cross–section at $y^1 \in [0,\ell]$, while ε describes the thickness of the cylinder Ω^ε. Let $\boldsymbol{P} : \overline{\Omega}^\varepsilon \to \mathbb{R}^3$ be defined by

$$\boldsymbol{P}(y) = \boldsymbol{\Phi}(y^1) + y^2\,\tilde{\boldsymbol{n}}(y^1) + y^3\,\tilde{\boldsymbol{b}}(y^1). \tag{2.3}$$

Since $\mathcal{C}$ is simple and regular curve, there exists $\varepsilon_0 > 0$ such that the function $\boldsymbol{P}$ is injective (see Tambača (2000)), especially one has

$$g(y) = (\det \nabla \boldsymbol{P}(y))^2 = \left(1 - R_{12}y^2 - R_{13}y^3\right)^2 \geq m_g^2 > 0, \; y \in \Omega^{\varepsilon_0}, \quad (2.4)$$

for some constant $m_g > 0$. In the sequel we restrict ourselves to $\varepsilon \in (0, \varepsilon_0)$. The domain

$$\tilde{\Omega}^\varepsilon = \boldsymbol{P}(\Omega^\varepsilon)$$

is called the curved rod; $\tilde{\Gamma}^\varepsilon = \boldsymbol{P}(\Gamma^\varepsilon)$ is the lateral boundary, $\tilde{B}^\varepsilon_{y^1} = \boldsymbol{P}(B^\varepsilon_{y^1})$ is the cross–section at $\boldsymbol{\Phi}(y^1)$, while $\tilde{B}^\varepsilon_0$ and $\tilde{B}^\varepsilon_\ell$ are the bases of $\tilde{\Omega}^\varepsilon$. Because of (2.2) the curve $\mathcal{C}$ passes through the centers of mass of $\tilde{B}^\varepsilon_{y^1}, y^1 \in [0, \ell]$, thus we say that $\mathcal{C}$ is the middle curve of the curved rod $\tilde{\Omega}^\varepsilon$.

3. Three–dimensional equations of linearized elasticity

We assume that $\overline{\tilde{\Omega}}^\varepsilon$ is the natural state of a linearized isotropic homogenous elastic body with the Lamé constants λ and μ and constant mass density $\tilde{\varrho}^\varepsilon$. Let the rod be clamped on its bases $\tilde{B}^\varepsilon_0$ and $\tilde{B}^\varepsilon_\ell$. By $\tilde{\boldsymbol{U}}^\varepsilon$ we denote the displacement of the rod $\tilde{\Omega}^\varepsilon$. Then in the scope of linearized elasticity the stress tensor is given by

$$\tilde{\boldsymbol{\sigma}}^\varepsilon = \lambda \operatorname{tr}\left(\mathbf{e}(\tilde{\boldsymbol{U}}^\varepsilon)\right)\mathbf{I} + 2\mu \mathbf{e}(\tilde{\boldsymbol{U}}^\varepsilon),$$

where $\mathbf{e}(\tilde{\boldsymbol{V}})$ denotes the symmetrized gradient of the function $\tilde{\boldsymbol{V}}$, i.e. $\mathbf{e}(\tilde{\boldsymbol{V}}) = \frac{1}{2}(\nabla\tilde{\boldsymbol{V}} + (\nabla\tilde{\boldsymbol{V}})^T)$. Let us introduce the function space of admissible displacements

$$\mathcal{V}(\tilde{\Omega}^\varepsilon) = \{\tilde{\boldsymbol{V}} \in H^1(\tilde{\Omega}^\varepsilon)^3 \; : \; \tilde{\boldsymbol{V}}|_{\tilde{B}^\varepsilon_0} = \tilde{\boldsymbol{V}}|_{\tilde{B}^\varepsilon_\ell} = 0\},$$

which is a Hilbert space for the scalar product of $H^1(\tilde{\Omega}^\varepsilon)^3$. The unit outer normal on the boundary of the curved rod will be denoted $\tilde{\boldsymbol{\nu}}$.

3.1 Evolution problem

Let the rod be subject to the force with volume density $\tilde{\boldsymbol{F}}^\varepsilon$ and force free on the lateral boundary $\tilde{\Gamma}^\varepsilon$. Let $\tilde{\boldsymbol{U}}^\varepsilon_0$ and $\tilde{\boldsymbol{U}}^\varepsilon_1$ be the initial displacement and the initial velocity of the rod, respectively and let $T > 0$. Then the displacement of the curved rod is described by the function $\tilde{\boldsymbol{U}}^\varepsilon$ which is formally a solution of

$$
\begin{aligned}
&\tilde{\varrho}^\varepsilon \tilde{\boldsymbol{U}}^\varepsilon_{tt} - \operatorname{div} \tilde{\boldsymbol{\sigma}}^\varepsilon = \tilde{\boldsymbol{F}}^\varepsilon \quad \text{in } (0,T) \times \tilde{\Omega}^\varepsilon, \\
&\tilde{\boldsymbol{\sigma}}^\varepsilon = \lambda \operatorname{tr}\left(\mathbf{e}(\tilde{\boldsymbol{U}}^\varepsilon)\right)\mathbf{I} + 2\mu \mathbf{e}(\tilde{\boldsymbol{U}}^\varepsilon) \quad \text{in } (0,T) \times \tilde{\Omega}^\varepsilon, \\
&\tilde{\boldsymbol{U}}^\varepsilon = 0 \quad \text{on } (0,T) \times \left(\tilde{B}^\varepsilon_0 \cup \tilde{B}^\varepsilon_\ell\right), \\
&\tilde{\boldsymbol{\sigma}}^\varepsilon \tilde{\boldsymbol{\nu}} = \lambda \operatorname{tr}\left(\mathbf{e}(\tilde{\boldsymbol{U}}^\varepsilon)\right)\tilde{\boldsymbol{\nu}} + 2\mu \mathbf{e}(\tilde{\boldsymbol{U}}^\varepsilon)\tilde{\boldsymbol{\nu}} = 0 \quad \text{on } (0,T) \times \tilde{\Gamma}^\varepsilon, \\
&\tilde{\boldsymbol{U}}^\varepsilon|_{t=0} = \tilde{\boldsymbol{U}}^\varepsilon_0, \quad \tilde{\boldsymbol{U}}^\varepsilon_t|_{t=0} = \tilde{\boldsymbol{U}}^\varepsilon_1 \quad \text{in } \tilde{\Omega}^\varepsilon.
\end{aligned} \tag{3.1}
$$

The problem (3.1) can be stated in the variational form:
find $\tilde{\boldsymbol{U}}^\varepsilon$ such that

$$
\begin{aligned}
&\int_{\tilde{\Omega}^\varepsilon} \tilde{\varrho}^\varepsilon \tilde{\boldsymbol{U}}^\varepsilon_{tt} \cdot \tilde{\boldsymbol{V}} + \int_{\tilde{\Omega}^\varepsilon} \left(\lambda \operatorname{tr} \mathbf{e}(\tilde{\boldsymbol{U}}^\varepsilon) \operatorname{tr} \mathbf{e}(\tilde{\boldsymbol{V}}) + 2\mu \mathbf{e}(\tilde{\boldsymbol{U}}^\varepsilon) \cdot \mathbf{e}(\tilde{\boldsymbol{V}})\right) \\
&= \int_{\tilde{\Omega}^\varepsilon} \tilde{\boldsymbol{F}}^\varepsilon \cdot \tilde{\boldsymbol{V}}, \qquad \tilde{\boldsymbol{V}} \in \mathcal{V}(\tilde{\Omega}^\varepsilon), \quad 0 < t < T,
\end{aligned} \tag{3.2}
$$

$$
\tilde{\boldsymbol{U}}^\varepsilon|_{t=0} = \tilde{\boldsymbol{U}}^\varepsilon_0, \quad \tilde{\boldsymbol{U}}^\varepsilon_t|_{t=0} = \tilde{\boldsymbol{U}}^\varepsilon_1. \tag{3.3}
$$

Let

$$
\tilde{\boldsymbol{F}}^\varepsilon \in L^2(0,T; L^2(\tilde{\Omega}^\varepsilon)^3), \quad \tilde{\boldsymbol{U}}^\varepsilon_0 \in \mathcal{V}(\tilde{\Omega}^\varepsilon), \quad \tilde{\boldsymbol{U}}^\varepsilon_1 \in L^2(\tilde{\Omega}^\varepsilon)^3.
$$

The existence and uniqueness of the weak solution of (3.2) is mainly a consequence of the Korn inequality (see Nečas and Hlavaček (1981) or Ciarlet (1988)): there is a constant $C > 0$ such that

$$
\|\tilde{\boldsymbol{V}}\|_{H^1(\Omega^\varepsilon)^3} \le C \|\mathbf{e}(\tilde{\boldsymbol{V}})\|_{L^2(\Omega^\varepsilon)^9}, \quad \tilde{\boldsymbol{V}} \in \mathcal{V}(\tilde{\Omega}^\varepsilon). \tag{3.4}
$$

The constant C depends in general on the domain $\tilde{\Omega}^\varepsilon$ and for some special cases of domains the best constant can be found. Using the Korn inequality and the classical results on evolution equations (see Dautray and Lions (1992)) on can prove that for each $\varepsilon \in (0, \varepsilon_0)$ there exists a unique solution $\tilde{\boldsymbol{U}}^\varepsilon$ of the problem (3.2), (3.3) such that

$$
\tilde{\boldsymbol{U}}^\varepsilon \in C([0,T]; \mathcal{V}(\tilde{\Omega}^\varepsilon)), \ \tilde{\boldsymbol{U}}^\varepsilon_t \in C([0,T]; L^2(\tilde{\Omega}^\varepsilon)), \ \tilde{\boldsymbol{U}}^\varepsilon_{tt} \in L^2(0,T; \mathcal{V}(\tilde{\Omega}^\varepsilon)').
$$

3.2 Equilibrium problem

The equilibrium equations follow immediately from (3.1); for the force with volume density $\tilde{\boldsymbol{F}}^\varepsilon \in L^2(\tilde{\Omega}^\varepsilon)^3$, the equilibrium displacement $\tilde{\boldsymbol{U}}^\varepsilon$ is formally a solution of

$$\begin{aligned}
-\operatorname{div}\tilde{\boldsymbol{\sigma}}^{\varepsilon} &= \tilde{\boldsymbol{F}}^{\varepsilon} \quad \text{in } \tilde{\Omega}^{\varepsilon},\\
\tilde{\boldsymbol{\sigma}}^{\varepsilon} &= \lambda \operatorname{tr}\left(\mathbf{e}(\tilde{\boldsymbol{U}}^{\varepsilon})\right)\mathbf{I} + 2\mu\mathbf{e}(\tilde{\boldsymbol{U}}^{\varepsilon}) \quad \text{in } \tilde{\Omega}^{\varepsilon},\\
\tilde{\boldsymbol{U}}^{\varepsilon} &= 0 \quad \text{on } \tilde{B}_0^{\varepsilon} \cup \tilde{B}_\ell^{\varepsilon},\\
\tilde{\boldsymbol{\sigma}}^{\varepsilon}\tilde{\boldsymbol{\nu}} &= \lambda \operatorname{tr}\left(\mathbf{e}(\tilde{\boldsymbol{U}}^{\varepsilon})\right)\tilde{\boldsymbol{\nu}} + 2\mu\mathbf{e}(\tilde{\boldsymbol{U}}^{\varepsilon})\tilde{\boldsymbol{\nu}} = 0 \quad \text{on } \tilde{\Gamma}^{\varepsilon}.
\end{aligned} \tag{3.5}$$

The problem (3.5) can be stated in the variational form:
find $\tilde{\boldsymbol{U}}^{\varepsilon} \in \mathcal{V}(\tilde{\Omega}^{\varepsilon})$ such that

$$\int_{\tilde{\Omega}^{\varepsilon}} \left(\lambda \operatorname{tr}\mathbf{e}(\tilde{\boldsymbol{U}}^{\varepsilon}) \operatorname{tr}\mathbf{e}(\tilde{\boldsymbol{V}}) + 2\mu\mathbf{e}(\tilde{\boldsymbol{U}}^{\varepsilon}) \cdot \mathbf{e}(\tilde{\boldsymbol{V}})\right) = \int_{\tilde{\Omega}^{\varepsilon}} \tilde{\boldsymbol{F}}^{\varepsilon} \cdot \tilde{\boldsymbol{V}}, \tag{3.6}$$

for all $\tilde{\boldsymbol{V}} \in \mathcal{V}(\tilde{\Omega}^{\varepsilon})$. The Korn inequality (3.4) implies the $\mathcal{V}(\tilde{\Omega}^{\varepsilon})$–ellipticity of the form on the left hand side of (3.6) and the Lax–Milgram lemma implies the existence of a unique solution $\tilde{\boldsymbol{U}}^{\varepsilon} \in \mathcal{V}(\tilde{\Omega}^{\varepsilon})$ of (3.6) for each $\varepsilon \in (0, \varepsilon_0)$.

3.3 Eigenvalue problem

Harmonic oscillations, i.e. nontrivial solutions of the evolution equation (3.2) of the form $\tilde{\boldsymbol{U}}^{\varepsilon}(t, x) = e^{it\sqrt{\alpha^{\varepsilon}}}\mathbf{X}^{\varepsilon}(x)$, for some real number α^{ε}, are important in stability analysis. For such solutions problem (3.2) reduces to:
find all displacements $\mathbf{X}^{\varepsilon} \neq 0$ and real numbers α^{ε} satisfying

$$\begin{aligned}
-\operatorname{div}\tilde{\boldsymbol{\sigma}}^{\varepsilon} &= \alpha^{\varepsilon}\tilde{\varrho}^{\varepsilon}\mathbf{X}^{\varepsilon} \quad \text{in } \tilde{\Omega}^{\varepsilon},\\
\tilde{\boldsymbol{\sigma}}^{\varepsilon} &= \lambda \operatorname{tr}(\mathbf{e}(\mathbf{X}^{\varepsilon}))\mathbf{I} + 2\mu\mathbf{e}(\mathbf{X}^{\varepsilon}) \quad \text{in } \tilde{\Omega}^{\varepsilon},\\
\mathbf{X}^{\varepsilon} &= 0 \quad \text{on } \tilde{B}_0^{\varepsilon} \cup \tilde{B}_\ell^{\varepsilon},\\
\tilde{\boldsymbol{\sigma}}^{\varepsilon}\tilde{\boldsymbol{\nu}} &= \lambda (\operatorname{tr}\mathbf{e}(\mathbf{X}^{\varepsilon}))\,\tilde{\boldsymbol{\nu}} + 2\mu\mathbf{e}(\mathbf{X}^{\varepsilon})\tilde{\boldsymbol{\nu}} = 0 \quad \text{on } \tilde{\Gamma}^{\varepsilon}.
\end{aligned}$$

The above differential equations are then formally equivalent to:
find all $\mathbf{X}^{\varepsilon} \in \mathcal{V}(\tilde{\Omega}^{\varepsilon})\backslash\{0\}$ and $\alpha^{\varepsilon} \in \mathbb{R}$ such that for all $\tilde{\boldsymbol{V}} \in \mathcal{V}(\tilde{\Omega}^{\varepsilon})$

$$\int_{\tilde{\Omega}^{\varepsilon}} \left(\lambda \operatorname{tr}\mathbf{e}(\mathbf{X}^{\varepsilon}) \operatorname{tr}\mathbf{e}(\tilde{\boldsymbol{V}}) + 2\mu\mathbf{e}(\mathbf{X}^{\varepsilon}) \cdot \mathbf{e}(\tilde{\boldsymbol{V}})\right) = \alpha^{\varepsilon} \int_{\tilde{\Omega}^{\varepsilon}} \tilde{\varrho}^{\varepsilon}\mathbf{X}^{\varepsilon} \cdot \tilde{\boldsymbol{V}}. \tag{3.7}$$

The bilinear forms in the equation (3.7) are continuous and symmetric. By the Korn inequality the form on the left hand side is $\mathcal{V}(\tilde{\Omega}^{\varepsilon})$–elliptic. The space $\mathcal{V}(\tilde{\Omega}^{\varepsilon})$ is compactly embedded in $L^2(\tilde{\Omega}^{\varepsilon})^3$ so the spectral properties of (3.7) follow from the classical result for self–adjoint compact operators (see Tretter (2000), Weinberger (1974) or Davies (1995)). In particular, for each $\varepsilon \in (0, \varepsilon_0)$ there is an increasing sequence of eigenvalues of the spectral problem (3.7), each with finite multiplicity. Associated eigenfunctions are orthogonal in the scalar product defined by either the form on the left hand side or the right hand side of (3.7). The eigenfunctions form a complete set in $\mathcal{V}(\tilde{\Omega}^{\varepsilon})$ and $L^2(\tilde{\Omega}^{\varepsilon})^3$.

4. Curvilinear coordinates

Our goal is to find the limits of the problems in the previous sections when ε tends to zero. It is known from similar considerations made for straight rods that the asymptotic behavior of the longitudinal and transverse displacements of the rod is different; the transverse displacements satisfy a fourth order differential equation, while longitudinal displacement satisfies a second order differential equation. The same effect is expected in present setting. Therefore we need to consider asymptotic behavior of the local components of the displacement $\tilde{\boldsymbol{U}}^{\varepsilon}$. This can be done in many ways, but in order to preserve some important properties of bilinear forms such as positive definiteness and symmetry, we rewrite the equations in curvilinear coordinates defined by $\boldsymbol{P}$. Then we formulate an equivalent problem posed on domain independent of ε.

The covariant basis of the curved rod is defined by

$$\boldsymbol{g}_i = \partial_i \boldsymbol{P} : \Omega^{\varepsilon} \to \mathbb{R}^3, \quad i = 1, 2, 3.$$

Vectors $(\boldsymbol{g}^1, \boldsymbol{g}^2, \boldsymbol{g}^3)$ satisfying

$$\boldsymbol{g}^j \cdot \boldsymbol{g}_i = \delta_i^j \text{ on } \Omega^{\varepsilon}, \quad i, j = 1, 2, 3,$$

where δ_i^j is the Kronecker symbol, form the contravariant basis on $\tilde{\Omega}^{\varepsilon}$. The contravariant metric tensor $\mathbf{G} = (g^{ij})$ and the Christoffel symbols Γ_{jk}^i of the curved rod $\tilde{\Omega}^{\varepsilon}$ are defined by

$$g^{ij} = \boldsymbol{g}^i \cdot \boldsymbol{g}^j, \quad \Gamma_{jk}^i = \boldsymbol{g}^i \cdot \partial_j \boldsymbol{g}_k \text{ on } \Omega^{\varepsilon}, \quad i, j, k = 1, 2, 3.$$

The corresponding function space to $\mathcal{V}(\tilde{\Omega}^{\varepsilon})$ is the space

$$\mathcal{V}(\Omega^{\varepsilon}) = \{\boldsymbol{V} \in H^1(\Omega^{\varepsilon})^3 \ : \ \boldsymbol{V}|_{B_0^{\varepsilon}} = \boldsymbol{V}|_{B_\ell^{\varepsilon}} = 0\}.$$

The displacements and velocities are rewritten in the contravariant basis,

$$\tilde{\boldsymbol{U}}^{\varepsilon} \circ \boldsymbol{P} = U_i^{\varepsilon} \boldsymbol{g}^i, \ \tilde{\boldsymbol{U}}_0^{\varepsilon} \circ \boldsymbol{P} = (U_0^{\varepsilon})_i \boldsymbol{g}^i, \ \tilde{\boldsymbol{U}}_1^{\varepsilon} \circ \boldsymbol{P} = (U_1^{\varepsilon})_i \boldsymbol{g}^i, \ \tilde{\boldsymbol{V}} \circ \boldsymbol{P} = V_i \boldsymbol{g}^i,$$

while the force density is rewritten in the covariant basis of the curved rod, $\tilde{\boldsymbol{F}}^{\varepsilon} \circ \boldsymbol{P} = F^{\varepsilon i} \boldsymbol{g}_i$ on Ω^{ε}. New vector functions are defined by

$$\boldsymbol{U}^{\varepsilon} = U_i^{\varepsilon} \boldsymbol{e}_i, \ \boldsymbol{U}_0^{\varepsilon} = (U_0^{\varepsilon})_i \boldsymbol{e}_i, \ \boldsymbol{U}_1^{\varepsilon} = (U_1^{\varepsilon})_i \boldsymbol{e}_i, \ \boldsymbol{V} = V_i \boldsymbol{e}_i, \ \boldsymbol{F}^{\varepsilon} = F^{\varepsilon i} \boldsymbol{e}_i.$$

Let $\varrho^{\varepsilon} = \tilde{\varrho}^{\varepsilon} \circ \boldsymbol{P}$, $\boldsymbol{\Gamma}^i = (\Gamma_{jk}^i)$, $\gamma(\boldsymbol{V}) = \mathbf{e}(\boldsymbol{V}) - V_i \boldsymbol{\Gamma}^i$ for $\boldsymbol{V} \in \mathcal{V}(\Omega^{\varepsilon})$ and

$$\mathbf{A}\boldsymbol{\sigma} = \lambda\,(\operatorname{tr}\boldsymbol{\sigma})\,\mathbf{I} + 2\mu\boldsymbol{\sigma}, \quad \boldsymbol{\sigma} \in \operatorname{Sym}(\mathbb{R}^3, \mathbb{R}^3).$$

The equation of motion in the curvilinear coordinates $y \in \Omega^\varepsilon$ of the reference configuration $\tilde{\Omega}^\varepsilon$ reads as follows: find $\boldsymbol{U}^\varepsilon$ such that

$$\int_{\Omega^\varepsilon} \varrho^\varepsilon \mathbf{G} \boldsymbol{U}^\varepsilon_{tt} \cdot \boldsymbol{V} \sqrt{g}\, dy + \int_{\Omega^\varepsilon} \mathbf{A} \boldsymbol{\gamma}(\boldsymbol{U}^\varepsilon) \cdot \boldsymbol{\gamma}(\boldsymbol{V}) \sqrt{g}\, dy$$
$$= \int_{\Omega^\varepsilon} \boldsymbol{F}^\varepsilon \cdot \boldsymbol{V} \sqrt{g}\, dy, \quad \boldsymbol{V} \in \mathcal{V}(\Omega^\varepsilon), \quad 0 < t < T, \tag{4.1}$$
$$\boldsymbol{U}^\varepsilon|_{t=0} = \boldsymbol{U}^\varepsilon_0, \quad \boldsymbol{U}^\varepsilon_t|_{t=0} = \boldsymbol{U}^\varepsilon_1. \tag{4.2}$$

Problems for both $\tilde{\boldsymbol{U}}^\varepsilon$ and $\boldsymbol{U}^\varepsilon$ are posed on ε–dependent domains. Now we transform the problem (4.1), (4.2) to ε–independent domain. As a consequence, the coefficients of the resulting weak formulation will depend on ε explicitly.

Let $\Omega = (0, \ell) \times S$ and let $\boldsymbol{R}^\varepsilon : \overline{\Omega} \to \overline{\Omega^\varepsilon}$ be defined by

$$\boldsymbol{R}^\varepsilon(z) = (z^1, \varepsilon z^2, \varepsilon z^3), \quad z \in \Omega, \ \varepsilon \in (0, \varepsilon_0).$$

By $\Gamma = (0, \ell) \times \partial S$ and $B_{z^1} = \{z^1\} \times S$ we denote the lateral surface of Ω and the cross section at $z^1 \in [0, \ell]$, respectively. To the functions $\boldsymbol{U}^\varepsilon$, $\boldsymbol{U}^\varepsilon_0$, $\boldsymbol{U}^\varepsilon_1$, $\boldsymbol{F}^\varepsilon$, g, $\boldsymbol{g}_i$, $\boldsymbol{g}^i$, ϱ^ε, $\mathbf{G}$, Γ^i_{jk}, $i, j, k = 1, 2, 3$ defined on Ω^ε we associate the functions $\boldsymbol{u}(\varepsilon)$, $\boldsymbol{u}_0(\varepsilon)$, $\boldsymbol{u}_1(\varepsilon)$, $\boldsymbol{f}(\varepsilon)$, $g(\varepsilon)$, $\boldsymbol{g}_i(\varepsilon)$, $\boldsymbol{g}^i(\varepsilon)$, $\varrho(\varepsilon)$, $\mathbf{G}(\varepsilon)$, $\Gamma^i_{ij}(\varepsilon)$, $i, j, k = 1, 2, 3$ defined on Ω by composition with $\boldsymbol{R}^\varepsilon$. Let

$$\mathcal{V}(\Omega) = \{\boldsymbol{v} = (v_1, v_2, v_3) \in H^1(\Omega)^3 \ : \ \boldsymbol{v}|_{B_0} = \boldsymbol{v}|_{B_\ell} = 0\},$$
$$a(\varepsilon) : \mathcal{V}(\Omega) \times \mathcal{V}(\Omega) \to \mathbb{R}, \ a(\varepsilon)(\boldsymbol{v}, \boldsymbol{w}) = \frac{1}{\varepsilon^2} \int_\Omega \mathbf{A}(\varepsilon) \boldsymbol{\gamma}^\varepsilon(\boldsymbol{v}) \cdot \boldsymbol{\gamma}^\varepsilon(\boldsymbol{w}) \sqrt{g(\varepsilon)} dz,$$
$$b(\varepsilon) : L^2(\Omega)^3 \times L^2(\Omega)^3 \to \mathbb{R}, \quad b(\varepsilon)(\boldsymbol{v}, \boldsymbol{w}) = \frac{\varrho(\varepsilon)}{\varepsilon^2} \int_\Omega \mathbf{G}(\varepsilon) \boldsymbol{v} \cdot \boldsymbol{w} \sqrt{g(\varepsilon)} dz,$$
$$\boldsymbol{\gamma}^\varepsilon(\boldsymbol{v}) = \frac{1}{\varepsilon} \boldsymbol{\gamma}_z(\boldsymbol{v}) + \boldsymbol{\gamma}_y(\boldsymbol{v}) - v_i \boldsymbol{\Gamma}^i(\varepsilon),$$
$$\boldsymbol{\gamma}_y(\boldsymbol{v}) = \begin{pmatrix} \partial_1 v_1 & \frac{1}{2}\partial_1 v_2 & \frac{1}{2}\partial_1 v_3 \\ \frac{1}{2}\partial_1 v_2 & 0 & 0 \\ \frac{1}{2}\partial_1 v_3 & 0 & 0 \end{pmatrix},$$
$$\boldsymbol{\gamma}_z(\boldsymbol{v}) = \begin{pmatrix} 0 & \frac{1}{2}\partial_2 v_1 & \frac{1}{2}\partial_3 v_1 \\ \frac{1}{2}\partial_2 v_1 & \partial_2 v_2 & \frac{1}{2}(\partial_2 v_3 + \partial_3 v_2) \\ \frac{1}{2}\partial_3 v_1 & \frac{1}{2}(\partial_2 v_3 + \partial_3 v_2) & \partial_3 v_3 \end{pmatrix},$$
$$\mathbf{A}(\varepsilon)\boldsymbol{\sigma} = \lambda \operatorname{tr}(\mathbf{G}(\varepsilon)\boldsymbol{\sigma})\mathbf{G}(\varepsilon) + 2\mu \mathbf{G}(\varepsilon)\boldsymbol{\sigma}\mathbf{G}(\varepsilon), \quad \boldsymbol{\sigma} \in \operatorname{Sym}(\mathbb{R}^3).$$

Now we formulate the equilibrium, evolution and associated eigenvalue problem of three–dimensional linearized elasticity for curved rods in the curvilinear coordinates on the canonical domain Ω (independent of ε). As we only changed the variables and coordinate system in the three-dimensional problems, the existence and uniqueness results of weak solutions still hold.

4.1 Evolution problem

The problem (4.1), (4.2) is equivalent to the following problem: find $\boldsymbol{u}(\varepsilon)$ such that

$$\frac{d}{dt}b(\varepsilon)(\boldsymbol{u}_t(\varepsilon),\boldsymbol{v}) + a(\varepsilon)(\boldsymbol{u}(\varepsilon),\boldsymbol{v})$$
$$= (\frac{1}{\varepsilon^2}\boldsymbol{f}(\varepsilon)\sqrt{g(\varepsilon)}|\boldsymbol{v})_{L^2(\Omega)^3}, \qquad \boldsymbol{v} \in \mathcal{V}(\Omega),\ 0 < t < T, \tag{4.3}$$
$$\boldsymbol{u}(\varepsilon)|_{t=0} = \boldsymbol{u}_0(\varepsilon), \quad \boldsymbol{u}_t(\varepsilon)|_{t=0} = \boldsymbol{u}_1(\varepsilon). \tag{4.4}$$

Theorem 1. *Let*

$$\boldsymbol{f}(\varepsilon) \in L^2(0,T;L^2(\Omega))^3,\ \boldsymbol{u}_0(\varepsilon) \in \mathcal{V}(\Omega),\ \boldsymbol{u}_1(\varepsilon) \in L^2(\Omega)^3.$$

Then for $\varepsilon \in (0,\varepsilon_0)$ there exists a unique solution $\boldsymbol{u}(\varepsilon)$ of (4.3), (4.4) such that

$$\boldsymbol{u}(\varepsilon) \in C([0,T];\mathcal{V}(\Omega)), \boldsymbol{u}_t(\varepsilon) \in C([0,T];L^2(\Omega)^3), \boldsymbol{u}_{tt}(\varepsilon) \in L^2(0,T;\mathcal{V}(\Omega)').$$

4.2 Equilibrium problem

The equilibrium problem follows from (4.3), (4.4). It consists of: find $\boldsymbol{u}(\varepsilon) \in \mathcal{V}(\Omega)$ such that

$$a(\varepsilon)(\boldsymbol{u}(\varepsilon),\boldsymbol{v}) = (\frac{1}{\varepsilon^2}\boldsymbol{f}(\varepsilon)\sqrt{g(\varepsilon)}|\boldsymbol{v})_{L^2(\Omega)^3}, \quad \boldsymbol{v} \in \mathcal{V}(\Omega). \tag{4.5}$$

Theorem 2. *Let $\boldsymbol{f}(\varepsilon) \in L^2(\Omega)^3$. Then for each $\varepsilon \in (0,\varepsilon_0)$ the solution $\boldsymbol{u}(\varepsilon) \in \mathcal{V}(\Omega)$ of (4.5) exists and it is unique.*

4.3 Eigenvalue problem

The eigenvalue problem associated with (4.3), (4.4) is: find all $\boldsymbol{u}(\varepsilon) \in \mathcal{V}(\Omega)\backslash\{0\}$ and $\alpha(\varepsilon) = \alpha^\varepsilon \in \mathbb{R}$ such that

$$a(\varepsilon)(\boldsymbol{u}(\varepsilon),\boldsymbol{v}) = \alpha(\varepsilon)b(\varepsilon)(\boldsymbol{u}(\varepsilon),\boldsymbol{v}), \quad \boldsymbol{v} \in \mathcal{V}(\Omega). \tag{4.6}$$

Let us introduce the Rayleigh quotient

$$R(\varepsilon)(\boldsymbol{v}) = \frac{a(\varepsilon)(\boldsymbol{v},\boldsymbol{v})}{b(\varepsilon)(\boldsymbol{v},\boldsymbol{v})}.$$

Theorem 3. *For $\varepsilon \in (0,\varepsilon_0)$ there is an increasing sequence of eigenvalues of the spectral problem (4.6), each with finite multiplicity,*

$$0 < \alpha^1(\varepsilon) \le \alpha^2(\varepsilon) \le \alpha^3(\varepsilon) \le \cdots, \quad \lim_{n\to\infty}\alpha^n(\varepsilon) = \infty,$$

with associated eigenfunctions $\boldsymbol{u}^n(\varepsilon), n \in \mathbb{N}$. The eigenfunctions are orthogonal in scalar products defined by $a(\varepsilon)$ and $b(\varepsilon)$. If we normalize them in $b(\varepsilon)-$scalar product then

$$b(\varepsilon)(\boldsymbol{u}^n(\varepsilon), \boldsymbol{u}^k(\varepsilon)) = \delta^{nk}, \quad a(\varepsilon)(\boldsymbol{u}^n(\varepsilon), \boldsymbol{u}^k(\varepsilon)) = \alpha^n(\varepsilon)\delta^{nk}, \quad n, k \in \mathbb{N}.$$

The eigenfunctions form a complete set in $\mathcal{V}(\Omega)$ and $L^2(\Omega)^3$. The eigenvalues are characterized by

$$\alpha^n(\varepsilon) = \min\left\{R(\varepsilon)(\boldsymbol{v}) : b(\varepsilon)(\boldsymbol{v}, \boldsymbol{u}^k(\varepsilon)) = 0, k = 1, \ldots, n-1\right\}, \tag{4.7}$$

$$\alpha^n(\varepsilon) = \min_{L \in \mathcal{V}^n(\Omega)} \max_{\boldsymbol{v} \in L} R(\varepsilon)(\boldsymbol{v}), \tag{4.8}$$

where $\mathcal{V}^n(\Omega)$, $n \in \mathbb{N}$, is the family of all n–dimensional subspaces of $\mathcal{V}(\Omega)$.

5. One–dimensional models of curved rods

In this section we formulate and discuss the one–dimensional models of curved rods which are the limit problems of evolution problem (4.3), (4.4), equilibrium problem (4.5) and eigenvalue problem (4.6). We first introduce some notation. The Young modulus $E = \mu\frac{3\lambda+2\mu}{\lambda+\mu}$ and the Poisson ratio $\nu = \frac{1}{2}\frac{\lambda}{\lambda+\mu}$ are relevant characteristics of the material, while the geometry of the cross–section of the curved rod is described by constants

$$|S| = \int_S dz^2dz^3, \quad I_2 = \int_S (z^2)^2 dz^2dz^3, \quad I_3 = \int_S (z^3)^2 dz^2dz^3,$$

$$I_{23} = -\int_S z^2z^3dz^2dz^3, \quad K = \int_S (\partial_2 p - z^3)^2 + (\partial_3 p + z^2)^2 dz^2dz^3,$$

where $p \in H^1(S)$ is so called warping function, a unique solution of the variational equation

$$\int_S (\partial_2 p - z^3)\partial_2 r + (\partial_3 p + z^2)\partial_3 r dz^2dz^3 = 0, \; r \in H^1(S), \; \int_S p = 0. \tag{5.1}$$

$|S|$ is the area of the cross–section, I_2 and I_3 are the inertia moments, I_{23} is the inertia product of the cross–section of the curved rod, while the constant μK is the torsion rigidity.

Relevant function spaces are as follows:

$$\begin{aligned}
V_0 &= H_0^1(0,\ell) \times H_0^2(0,\ell) \times H_0^2(0,\ell),\\
\mathcal{V}_0 &= \{\boldsymbol{v} \in V_0 : v_1' - R_{12}v_2 - R_{13}v_3 = 0\},\\
H_0 &= L^2(0,\ell)^3,\\
\mathcal{H}_0 &= \{\boldsymbol{v} \in H_0^1(0,\ell) \times L^2(0,\ell) \times L^2(0,\ell) : v_1' - R_{12}v_2 - R_{13}v_3 = 0\},\\
W_0 &= V_0 \times H_0^1(0,\ell),\\
\mathcal{W}_0 &= \mathcal{V}_0 \times H_0^1(0,\ell).
\end{aligned}$$

Recall that functions R_{12} and R_{13} are elements of the skewsymmetric matrix $\mathbf{R}$ defined in (2.1); they depend on the geometry of the middle curve of the rod. The function space V_0 is usual in rod modelling. Its elements are admissible equilibrium displacements of the rod in present setting. The condition $v_1' - R_{12}v_2 - R_{13}v_3 = 0$ describes the approximate inextensibility of the rod. Therefore $\mathcal{V}_0$ is usually referred to as the space of inextensible displacements of the curved rod. Note that $\mathcal{H}_0$ is the closure of $\mathcal{V}_0$ in $L^2(0,\ell)^3$ topology. Besides of the displacements in the variational equations of the model an additional function appears which describes the angle of torsion of the cross–section. Therefore spaces W_0 and $\mathcal{W}_0$ are necessary for description of the model. Functions in V_0, $\mathcal{V}_0$, H_0 and $\mathcal{H}_0$ we denote by $\boldsymbol{v} = (v_1, v_2, v_3)$, $\boldsymbol{u} = (u_1, u_2, u_3)$, etc, and for elements of spaces W_0 and $\mathcal{W}_0$ we use the notations

$$\mathbf{v} = (\boldsymbol{v}, \psi), \quad \mathbf{u} = (\boldsymbol{u}, \phi), \quad \text{etc.}$$

The spaces V_0 and $\mathcal{V}_0$, W_0 and $\mathcal{W}_0$, H_0 and $\mathcal{H}_0$ are Hilbert spaces with the scalar products derived from the norms, respectively,

$$\begin{aligned}
\|\boldsymbol{v}\|_{V_0}^2 &= \int_0^\ell v_1'(z^1)^2 dz^1 + \int_0^\ell v_2''(z^1)^2 dz^1 + \int_0^\ell v_3''(z^1)^2 dz^1, \\
\|\mathbf{v}\|_{W_0}^2 &= \|\boldsymbol{v}\|_{V_0}^2 + \int_0^\ell \psi'(z^1)^2 dz^1, \\
\|\boldsymbol{v}\|_{H_0}^2 &= \int_0^\ell v_1(z^1)^2 dz^1 + \int_0^\ell v_2(z^1)^2 dz^1 + \int_0^\ell v_3(z^1)^2 dz^1.
\end{aligned}$$

We also define the linear forms $Q_i : W_0 \to \mathbb{R}$, $i = 1, \ldots, 9$ by

$$\begin{aligned}
Q_1(\mathbf{v}) &= v_1' - R_{12}v_2 - R_{13}v_3, \\
Q_2(\mathbf{v}) &= -(v_2' + R_{12}v_1 - R_{23}v_3)' - R_{13}\psi + R_{23}(v_3' + R_{13}v_1 + R_{23}v_2), \\
Q_3(\mathbf{v}) &= -(v_3' + R_{13}v_1 + R_{23}v_2)' - R_{23}(v_2' + R_{12}v_1 - R_{23}v_3) + R_{12}\psi, \\
Q_4(\mathbf{v}) &= \psi' + R_{12}(v_3' + R_{13}v_1 + R_{23}v_2) - R_{13}(v_2' + R_{12}v_1 - R_{23}v_3), \\
Q_5(\mathbf{v}) &= R_{12}(v_3' + R_{13}v_1 + R_{23}v_2) + R_{13}(v_2' + R_{12}v_1 - R_{23}v_3), \qquad (5.2) \\
Q_6(\mathbf{v}) &= -2R_{12}(v_2' + R_{12}v_1 - R_{23}v_3) - 2R_{23}\psi, \\
Q_7(\mathbf{v}) &= v_2' + 2R_{12}v_1 - 2R_{23}v_3, \\
Q_8(\mathbf{v}) &= v_3' + 2R_{13}v_1 + 2R_{23}v_2, \\
Q_9(\mathbf{v}) &= -2R_{13}(v_3' + R_{13}v_1 + R_{23}v_2) - 2R_{23}\psi.
\end{aligned}$$

The matrix

$$\mathbf{H} = \begin{pmatrix} \mu K & 0 & 0 \\ 0 & EI_3 & EI_{23} \\ 0 & EI_{23} & EI_2 \end{pmatrix} \qquad (5.3)$$

is symmetric and it satisfies

$$\mathbf{H}\boldsymbol{x} \cdot \boldsymbol{x} = \mu K x_1^2 + E \int_S (x_3 z^2 - x_2 z^3)^2 dz^2 dz^3 > 0, \quad \boldsymbol{x} \in \mathbb{R}\backslash\{0\}.$$

Therefore $\mathbf{H}$ is positive definite. Let us define the operator $\overline{\cdot}$ by

$$\overline{\boldsymbol{f}} = \int_S \boldsymbol{f}.$$

Forms $a_0 : W_0 \times W_0 \longrightarrow \mathbb{R}$ and $b_0 : H_0 \times H_0 \longrightarrow \mathbb{R}$,

$$a_0(\mathbf{u}, \mathbf{v}) = \int_0^\ell \mathbf{H} \begin{pmatrix} Q_4(\mathbf{u}) \\ Q_3(\mathbf{u}) \\ -Q_2(\mathbf{u}) \end{pmatrix} \cdot \begin{pmatrix} Q_4(\mathbf{v}) \\ Q_3(\mathbf{v}) \\ -Q_2(\mathbf{v}) \end{pmatrix} dz^1,$$

$$b_0(\boldsymbol{u}, \boldsymbol{v}) = \varrho |S| \int_0^\ell \boldsymbol{u} \cdot \boldsymbol{v} dz^1,$$

are bilinear, symmetric and continuous; positive number ϱ is considered as known. The first statement of the following lemma is obvious, while the proof of the second statement is analogous to the one in Jurak and Tambača (1999).

Lemma 1. *The bilinear form b_0 is H_0–elliptic. The bilinear form a_0 is $\mathcal{W}_0$–elliptic.*

Lemma 2. *For given $\boldsymbol{v} \in \mathcal{V}_0$ the solution $\psi_{\boldsymbol{v}} \in H_0^1(0, \ell)$ of the equation*

$$a_0((0, \psi_{\boldsymbol{v}}), (0, \eta)) = -a_0((\boldsymbol{v}, 0), (0, \eta)), \quad \eta \in H_0^1(0, \ell),$$

exists and it is unique. Moreover, the operator

$$\mathcal{D} : \mathcal{V}_0 \to H_0^1(0, \ell), \quad \mathcal{D}\boldsymbol{v} = \psi_{\boldsymbol{v}},$$

is continuous and injective.

Proof. Since the bilinear form a_0 is $\mathcal{W}_0$–elliptic it follows that the bilinear form $a_{\mathcal{D}} : H_0^1(0, \ell) \times H_0^1(0, \ell) \to \mathbb{R}$ defined by

$$a_{\mathcal{D}}(\psi, \eta) = a_0((0, \psi), (0, \eta))$$

is $H_0^1(0, \ell)$–elliptic. The linear functional $\eta \mapsto a_0((\boldsymbol{v}, 0), (0, \eta))$ is continuous and therefore, the Lax–Milgram lemma implies the statement of the lemma. □

Let us define the form $d_0 : \mathcal{V}_0 \times \mathcal{V}_0 \to \mathbb{R}$ by

$$d_0(\boldsymbol{u}, \boldsymbol{v}) := a_0((\boldsymbol{u}, 0), (\boldsymbol{v}, 0)) - a_0((0, \mathcal{D}\boldsymbol{u}), (0, \mathcal{D}\boldsymbol{v})).$$

The form d_0 is symmetric, bilinear and continuous. Using the definition of the operator $\mathcal{D}$ and $\mathcal{W}_0$–ellipticity of the form a_0 it follows for all $\mathbf{v} \in \mathcal{V}_0$,

$$d_0(\boldsymbol{v}, \boldsymbol{v}) = a_0((\boldsymbol{v}, \mathcal{D}\boldsymbol{v}), (\boldsymbol{v}, \mathcal{D}\boldsymbol{v})) \geq m_a \left(\|\boldsymbol{v}\|_{\mathcal{V}_0}^2 + \|\mathcal{D}\boldsymbol{v}\|_{H_0^1(0,\ell)}^2 \right),$$

and therefore the form d_0 is $\mathcal{V}_0$–elliptic; here m_a is the ellipticity constant of the form a_0.

5.1 Evolution model

For $\boldsymbol{u}_0 \in \mathcal{V}_0$, $\boldsymbol{u}_1 \in \mathcal{H}_0$ we consider the problem:
find $\mathbf{u} = (\boldsymbol{u}, \phi)$ such that

$$\frac{d}{dt} b_0(\boldsymbol{u}_t, \boldsymbol{v})_{H_0} + a_0(\mathbf{u}, \mathbf{v}) = (\overline{\boldsymbol{f}}|\boldsymbol{v})_{H_0}, \quad \mathbf{v} \in \mathcal{W}_0, \ 0 < t < T, \tag{5.4}$$

$$\boldsymbol{u}(0) = \boldsymbol{u}_0, \quad \boldsymbol{u}_t(0) = \boldsymbol{u}_1. \tag{5.5}$$

Function $\boldsymbol{u}$ describes the oscillations of the middle curve $\mathcal{C}$ and as it belongs to $\mathcal{V}_0$ it satisfies inextensibility condition. Moreover, the cross–sections remain normal to the deformed middle curve. In this model the flexion and the torsion effects are of the same order. Additional function ϕ that appears on the left hand side of the equation is the rotation angle of the cross–sections along the longitudinal axis.

Variable ϕ can be expressed in terms of variable $\boldsymbol{u}$ as follows. With the choice of test function $(0, \psi) \in \mathcal{W}_0$ we obtain from (5.4):

$$a_0((\boldsymbol{u}, \phi), (0, \psi)) = 0, \quad \psi \in H_0^1(0, \ell).$$

From Lemma 2 we get $\phi = \mathcal{D}\boldsymbol{u}$ and since we have

$$a_0((\boldsymbol{u}, \mathcal{D}\boldsymbol{u}), \mathbf{v}) = d_0(\boldsymbol{u}, \boldsymbol{v}), \quad \mathbf{v} \in \mathcal{W}_0,$$

it follows from (5.4), (5.5):

$$\frac{d}{dt} b_0(\boldsymbol{u}_t, \boldsymbol{v})_{H_0} + d_0(\boldsymbol{u}, \boldsymbol{v}) = (\overline{\boldsymbol{f}}|\boldsymbol{v})_{H_0}, \quad \boldsymbol{v} \in \mathcal{V}_0, \ 0 < t < T, \tag{5.6}$$

$$\boldsymbol{u}(0) = \boldsymbol{u}_0, \quad \boldsymbol{u}_t(0) = \boldsymbol{u}_1. \tag{5.7}$$

The problem (5.6), (5.7) is non–local evolution problem. Moreover, if $\boldsymbol{u}$ is a solution of (5.6), (5.7), then $\mathbf{u} = (\boldsymbol{u}, \mathcal{D}\boldsymbol{u})$ is a solution of (5.4), (5.5), hence problems (5.4), (5.5) and (5.6), (5.7) are equivalent.

The space $\mathcal{H}_0$ is a closed subspace of H_0 and therefore H_0 has the following decomposition:

$$H_0 = \mathcal{H}_0 \oplus \mathcal{H}_0^{\perp}, \quad \mathcal{H}_0^{\perp} = \{\boldsymbol{w} \in H_0 : (\boldsymbol{w}|\boldsymbol{v})_{H_0} = 0, \boldsymbol{v} \in \mathcal{H}_0\}.$$

Let us denote the orthogonal projector on $\mathcal{H}_0$ by $P_{\mathcal{H}_0}$. Then for $\boldsymbol{v} \in \mathcal{V}_0$ it holds

$$(\overline{\boldsymbol{f}}|\boldsymbol{v})_{H_0} = (P_{\mathcal{H}_0}\overline{\boldsymbol{f}}|\boldsymbol{v})_{H_0},$$

and the component of the linear force density $\overline{\boldsymbol{f}}$ in $\mathcal{H}_0^{\perp}$ has no influence on the solution of (5.4), (5.5).

For fixed $T > 0$ we define appropriate function spaces for problem (5.4), (5.5):

$$\mathcal{V}_0(0,T) = \left\{ \boldsymbol{u} \in L^\infty(0,T;\mathcal{V}_0) : \frac{d\boldsymbol{u}}{dt} \in L^\infty(0,T;\mathcal{H}_0), \frac{d^2\boldsymbol{u}}{dt^2} \in L^2(0,T;\mathcal{V}_0') \right\},$$

$$\mathcal{W}_0(0,T) = \mathcal{V}_0(0,T) \times L^\infty(0,T;H_0^1(0,\ell)),$$

$$\mathcal{V}_c(0,T) = \left\{ \boldsymbol{u} \in C([0,T];\mathcal{V}_0) : \frac{d\boldsymbol{u}}{dt} \in C([0,T];\mathcal{H}_0), \frac{d^2\boldsymbol{u}}{dt^2} \in L^2(0,T;\mathcal{V}_0') \right\},$$

$$\mathcal{W}_c(0,T) = \mathcal{V}_c(0,T) \times C([0,T];H_0^1(0,T)).$$

Obviously $\mathcal{W}_c(0,T) \subset \mathcal{W}_0(0,T)$. The following theorem is a consequence of the classical result for evolution problems (see Dautray and Lions (1992)) applied to problem (5.6), (5.7).

Theorem 4. *The solution of (5.4), (5.5) is unique in $\mathcal{W}_0(0,T)$ and it exists in $\mathcal{W}_c(0,T)$.*

Note that the uniqueness of the solution of (5.4), (5.5) in bigger space $\mathcal{W}_0(0,T)$ will be necessary in order to prove the convergence of the whole family of solutions $\boldsymbol{u}(\varepsilon)$.

The main convergence result is stated in the following theorem.

Theorem 5. *Let the assumption of Theorem 1 be fulfilled and let there exist $\boldsymbol{f}$ and ϱ, independent of ε, such that*

$$\boldsymbol{f}(\varepsilon) = \varepsilon^2 \boldsymbol{f}, \quad \varrho(\varepsilon) = \varepsilon^2 \varrho, \tag{5.8}$$

Moreover, let initial data satisfy

$$\|\frac{1}{\varepsilon}\gamma^\varepsilon(\boldsymbol{u}_0(\varepsilon))\|_{L^2(\Omega)^9} \leq C, \quad \boldsymbol{u}_0(\varepsilon) \rightharpoonup \boldsymbol{u}_0 \text{ weakly in } \mathcal{V}(\Omega), \tag{5.9}$$

where C is a constant independent of ε and

$$\boldsymbol{u}_1 \in \mathcal{H}_0, \qquad \boldsymbol{u}_1(\varepsilon) \rightharpoonup \boldsymbol{u}_1 \text{ weakly in } L^2(\Omega)^3. \tag{5.10}$$

Then

$$\boldsymbol{u}(\varepsilon) \overset{*}{\rightharpoonup} \boldsymbol{u} \text{ weak } * \text{ in } L^\infty(0,T;\mathcal{V}(\Omega)) \text{ as } \varepsilon \to 0,$$

$$\boldsymbol{u}_t(\varepsilon) \overset{*}{\rightharpoonup} \boldsymbol{u}_t \text{ weak } * \text{ in } L^\infty(0,T;L^2(\Omega)^3) \text{ as } \varepsilon \to 0,$$

$$\frac{1}{\varepsilon}\gamma^\varepsilon(\boldsymbol{u}(\varepsilon)) \overset{*}{\rightharpoonup} \gamma \text{ weak } * \text{ in } L^\infty(0,T;L^2(\Omega)^9) \text{ as } \varepsilon \to 0,$$

where $\mathbf{u} = (\boldsymbol{u}, \phi) \in \mathcal{W}_c(0,T)$ *is the unique solution of (5.4), (5.5), and symmetric matrix* $\boldsymbol{\gamma}$ *is given by*

$$\boldsymbol{\gamma} = \begin{pmatrix} Q_2(\mathbf{u})z^2 + Q_3(\mathbf{u})z^3 & \cdot & \cdot \\ \frac{1}{2}Q_4(\mathbf{u})(\partial_2 p - z^3) & -\nu\gamma_{11} & \cdot \\ \frac{1}{2}Q_4(\mathbf{u})(\partial_3 p + z^2) & 0 & -\nu\gamma_{11} \end{pmatrix}. \tag{5.11}$$

5.2 Equilibrium model

The equilibrium problem follows from the evolution problem (5.4), (5.5): find $\mathbf{u} = (\boldsymbol{u}, \phi) \in \mathcal{W}_0$ such that

$$a_0(\mathbf{u}, \mathbf{v}) = (\overline{\boldsymbol{f}}|\boldsymbol{v})_{H_0}, \quad \mathbf{v} = (\boldsymbol{v}, \psi) \in \mathcal{W}_0. \tag{5.12}$$

The consequence of the Lax–Milgram lemma and Lemma 1 is the following result.

Theorem 6. *The variational equation (5.12) has a unique solution.*

Our main result is the following theorem.

Theorem 7. *Let there exist* $\boldsymbol{f} \in L^2(\Omega)^3$ *such that*

$$\boldsymbol{f}(\varepsilon) = \varepsilon^2 \boldsymbol{f} \tag{5.13}$$

and let $\boldsymbol{u}(\varepsilon)$ *for* $\varepsilon \in (0, \varepsilon_0)$ *be a unique solution of the variational equation (4.3). Then*

$$\begin{aligned} &\boldsymbol{u}(\varepsilon) \to \boldsymbol{u}, \ \textit{in } V(\Omega) \textit{ strongly}, \\ &\frac{1}{\varepsilon}\boldsymbol{\gamma}^{\varepsilon}(\boldsymbol{u}(\varepsilon)) \to \boldsymbol{\gamma}, \ \textit{in } L^2(\Omega)^9 \textit{ strongly}, \end{aligned}$$

where $\mathbf{u} = (\boldsymbol{u}, \phi) \in \mathcal{W}_0$ *is the solution of (5.12) and* $\boldsymbol{\gamma}$ *is given by formula (5.11).*

5.3 Eigenvalue model

We consider the following problem: find all $\alpha_0 \in \mathbb{R}$ and $\mathbf{u} = (\boldsymbol{u}, \phi) \in \mathcal{W}_0 \backslash \{0\}$ such that

$$a_0(\mathbf{u}, \mathbf{v}) = \alpha_0 b_0(\boldsymbol{u}, \boldsymbol{v}), \qquad \mathbf{v} = (\boldsymbol{v}, \psi) \in \mathcal{W}_0. \tag{5.14}$$

If we denote

$$R_0(\boldsymbol{v}) = \frac{d_0(\boldsymbol{v}, \boldsymbol{v})}{b_0(\boldsymbol{v}, \boldsymbol{v})}, \quad \boldsymbol{v} \in \mathcal{V}_0,$$

then the properties of the spectral problem (5.14) are given in the following theorem (see Tambača (2000)).

Theorem 8. *There is an increasing sequence of eigenvalues, each of finite multiplicity,*

$$0 < \alpha_0^1 \le \alpha_0^2 \le \alpha_0^3 \le \cdots, \quad \lim_{n\to\infty} \alpha_0^n = \infty.$$

The associated eigenfunctions $\boldsymbol{u}^n \in \mathcal{V}_0$ are orthonormal in the scalar product defined by b_0 and form a complete set in both $\mathcal{V}_0$ and $\mathcal{H}_0$. Moreover, the following characterizations of the eigenvalues hold

$$\alpha^n = \min\{R_0(\boldsymbol{v}) : b_0(\boldsymbol{v}, \boldsymbol{u}_k) = 0, k = 1, \ldots, n-1\},$$
$$\alpha^n = \min_{L\in\mathcal{V}_0^n} \max_{\boldsymbol{v}\in L} R_0(\boldsymbol{v}),$$

where $\mathcal{V}_0^n$ denotes the family of all n–dimensional subspaces of $\mathcal{V}_0$.

Our main convergence result is stated in the following theorem.

Theorem 9. *Let $\varrho(\varepsilon) = \varepsilon^2\varrho$, where ϱ is a constant from the definition of the form b_0. Then*

a) for every $n \in \mathbb{N}$ and any sequence of $(\varepsilon, \varepsilon > 0)$ there is a subsequence, still denoted by ε, a real number $\alpha^n \in \mathbb{R}$, functions $\boldsymbol{u}^n \in \mathcal{V}_0$ and $\boldsymbol{\gamma}^n \in L^2(\Omega)^9$ such that, as ε tends to zero, the following convergence hold

$$\alpha^n(\varepsilon) \to \alpha^n, \tag{5.15}$$
$$\boldsymbol{u}^n(\varepsilon) \to \boldsymbol{u}^n \text{ strongly in } H^1(\Omega)^3, \tag{5.16}$$
$$\frac{1}{\varepsilon}\gamma^\varepsilon(\boldsymbol{u}^n(\varepsilon)) \to \gamma^n \text{ strongly in } L^2(\Omega)^9, \tag{5.17}$$

where the pair $(\alpha^n, (\boldsymbol{u}^n, \phi^n))$, $\phi^n = \mathcal{D}\boldsymbol{u}^n$ satisfies

$$a_0((\boldsymbol{u}^n, \phi^n), (\boldsymbol{v}, \psi)) = \alpha^n b_0(\boldsymbol{u}^n, \boldsymbol{v}), \quad (\boldsymbol{v}, \psi) \in \mathcal{W}_0. \tag{5.18}$$

The symmetric matrix function $\boldsymbol{\gamma}^n$ is given by

$$\boldsymbol{\gamma}^n = \begin{pmatrix} Q_2(\boldsymbol{u}^n, \phi^n)z^2 + Q_3(\boldsymbol{u}^n, \phi^n)z^3 & \cdot & \cdot \\ \frac{1}{2}Q_4(\boldsymbol{u}^n, \phi^n)(\partial_2 p - z^3) & -\nu\gamma_{11}^n & \cdot \\ \frac{1}{2}Q_4(\boldsymbol{u}^n, \phi^n)(\partial_3 p + z^2) & 0 & -\nu\gamma_{11}^n \end{pmatrix}. \tag{5.19}$$

b) *The sequence* $(\alpha^n, n \in \mathbb{N})$ *comprises all eigenvalues of the spectral problem (5.14), counting its multiplicities. The associated eigenfunctions* $(\boldsymbol{u}^n, n \in \mathbb{N})$ *form a complete set in* $\mathcal{V}_0$ *and* $\mathcal{H}_0$ *and satisfy*

$$|S|\varrho \int_0^\ell \boldsymbol{u}^n \cdot \boldsymbol{u}^k = \delta^{nk}, \quad n, k \in \mathbb{N}. \tag{5.20}$$

c) *For each* $n \in \mathbb{N}$ *the whole family* $\alpha^n(\varepsilon)$ *converges to* α^n. *If for* $n \in \mathbb{N}$, α^n *is a simple eigenvalue, then there is* $\varepsilon_0(n)$ *such that for all* $0 < \varepsilon \leq \varepsilon_0(n)$, $\alpha^n(\varepsilon)$ *is simple. The associated eigenfunctions* $\boldsymbol{u}^n(\varepsilon)$, $0 < \varepsilon \leq \varepsilon_0(n)$ *can be chosen such that the convergence (5.16) holds for the whole family.*

6. The Korn inequality

The Korn inequality already mentioned in (3.4) is essential for existence, uniqueness and properties of the three–dimensional equations of curved rods. The constant in the inequality is domain dependent so it depends in general on ε in an unknown way. Therefore the inequality (3.4) cannot be used for *a priori* estimates with respect to ε. The following inequality is designed specially for the curved rods and it is sharp with respect to ε.

Theorem 10 (generalized Korn's inequality). *There are positive constants* C_K *and* ε_K, *with* $\varepsilon_K \leq \varepsilon_0$, *such that for all* $\varepsilon \in (0, \varepsilon_K)$ *it holds*

$$\|\boldsymbol{v}\|_{H^1(\Omega)^3} \leq \frac{C_K}{\varepsilon} \|\boldsymbol{\gamma}^\varepsilon(\boldsymbol{v})\|_{L^2(\Omega)^9}, \quad \boldsymbol{v} \in \mathcal{V}(\Omega).$$

The above version of the Korn inequality is easy to prove by contradiction using the following Lemma, see Jurak and Tambača (2001), that will be used in identification of the limit of the solution of three–dimensional problems as well.

Lemma 3. *Let* $(\varepsilon_n)_{n\in\mathbb{N}}$ *be a sequence of positive real numbers such that* $\varepsilon_n \to 0$. *Let a sequence* $(\boldsymbol{v}(\varepsilon_n))_{n\in\mathbb{N}}$ *in* $\mathcal{V}(\Omega)$ *satisfies*

$$\boldsymbol{v}(\varepsilon_n) \rightharpoonup \boldsymbol{v} \qquad \text{weakly in } H^1(\Omega)^3, \tag{6.1}$$

$$\frac{1}{\varepsilon_n}\boldsymbol{\gamma}^{\varepsilon_n}(\boldsymbol{v}(\varepsilon_n)) \rightharpoonup \boldsymbol{\gamma} \qquad \text{weakly in } L^2(\Omega)^9, \tag{6.2}$$

as $\varepsilon_n \to 0$. *Then* $\boldsymbol{v} \in H_0^1(0,\ell) \times H_0^2(0,\ell) \times H_0^2(0,\ell)$ *and there is* $\psi \in H_0^1(0,\ell)$, $\mathrm{v} = (\boldsymbol{v}, \psi)$, *such that*

$$Q_1(\mathrm{v}) = 0, \; Q_2(\mathrm{v}) = \partial_2\gamma_{11}, \; Q_3(\mathrm{v}) = \partial_3\gamma_{11}, \; Q_4(\mathrm{v}) = \partial_2\gamma_{13} - \partial_3\gamma_{12}.$$

If the convergence in (6.2) is strong, then the convergence in (6.1) is also strong.

7. A priori estimates, convergence

We proceed with the proof of convergence Theorems 5, 7 and 9. Generalized Korn's inequality is the first step of the proof; from this inequality we will deduce a priori estimates, uniform with respect to ε, for families of the solutions of evolution problem (4.3), (4.4), equilibrium problem (4.5) and eigenvalue problem (4.6). Appropriate convergences are then a consequence of these estimates.

We begin with asymptotic properties of the coefficients of the evolution problem (4.3), (4.4). Direct calculation shows that there are constants $m_g, m_G, M_G, m_A, M_A > 0$, independent of $\varepsilon \in (0, \varepsilon_0)$, such that for all $z \in \Omega$,

$$m_g \le \sqrt{g(\varepsilon)} \le 2, \quad m_G\mathbf{I} \le \mathbf{G}(\varepsilon) \le M_G\mathbf{I}, \quad m_A\mathbf{I} \le \mathbf{A}(\varepsilon) \le M_A\mathbf{I}. \tag{7.1}$$

The functions $\boldsymbol{g}^i(\varepsilon), g^{ij}(\varepsilon), g(\varepsilon), A^{ijkl}(\varepsilon), \Gamma^i_{jk}(\varepsilon)$ are in $C(\overline{\Omega})$ by assumptions; moreover, simple calculation shows that there is a constant $C > 0$ such that for all $\varepsilon \in (0, \varepsilon_0)$,

$$\begin{aligned} &\|\boldsymbol{g}^i(\varepsilon) - \boldsymbol{g}^i(0)\|_\infty \le C\varepsilon, \ \|g^{ij}(\varepsilon) - g^{ij}(0)\|_\infty \le C\varepsilon, \\ &\|\sqrt{g(\varepsilon)} - 1\|_\infty \le C\varepsilon, \ \|\mathbf{A}(\varepsilon) - \mathbf{A}(0)\|_\infty \le C\varepsilon, \\ &\|\frac{1}{\varepsilon}\left(\Gamma^i_{jk}(\varepsilon) - \Gamma^i_{jk}(0)\right) - z^\alpha \left(\frac{\partial}{\partial y^\alpha}\Gamma^i_{jk}\right)(0)\|_\infty \le C\varepsilon, \end{aligned} \tag{7.2}$$

where $\|\cdot\|_\infty$ is the norm in $C(\overline{\Omega})$. From estimates (7.1) and the Korn inequality it follows

Lemma 4. *The bilinear forms $a(\varepsilon)$ and $b(\varepsilon)$ are continuous, symmetric and elliptic on their domains. One has:*

$$\begin{aligned} a(\varepsilon)(\boldsymbol{v}, \boldsymbol{v}) &\ge \frac{m_A m_g}{C_K^2}\|\boldsymbol{v}\|^2_{H^1(\Omega)^3}, \quad \boldsymbol{v} \in \mathcal{V}(\Omega), \\ b(\varepsilon)(\boldsymbol{v}, \boldsymbol{v}) &\ge \frac{\varrho(\varepsilon)}{\varepsilon^2} m_G m_g \|\boldsymbol{v}\|^2_{L^2(\Omega)^3}, \quad \boldsymbol{v} \in L^2(\Omega)^3. \end{aligned}$$

7.1 Equilibrium equation

In this subsection we will complete the proof of Theorem 7. Throughout this subsection we assume that (5.13) holds. The *a priori* estimates for the family of equilibrium displacements is easy to obtain in terms of the right hand side.

Lemma 5. *Let $(\boldsymbol{u}(\varepsilon), \varepsilon > 0)$ be a family of solutions of the equilibrium problem (4.5). Then for all $\varepsilon \in (0, \varepsilon_K)$ it holds*

$$\|\boldsymbol{u}(\varepsilon)\|_{\mathcal{V}(\Omega)} \le \frac{2C_K^2}{m_A m_g}\|\boldsymbol{f}\|_{L^2(\Omega)^3}, \ \|\frac{1}{\varepsilon}\boldsymbol{\gamma}^\varepsilon(\boldsymbol{u}(\varepsilon))\|_{L^2(\Omega)^9} \le \frac{4C_K^2}{m_A m_g}\|\boldsymbol{f}\|^2_{L^2(\Omega)^3}.$$

Proof. Inserting $\boldsymbol{u}(\varepsilon)$ in (4.5) and using (5.13) and Lemma 4 we obtain

$$\begin{aligned} \frac{m_A m_g}{C_K^2}\|\boldsymbol{u}(\varepsilon)\|_{H^1(\Omega)^3}^2 &\le a(\varepsilon)(\boldsymbol{u}(\varepsilon),\boldsymbol{u}(\varepsilon)) = \left(\frac{\boldsymbol{f}(\varepsilon)}{\varepsilon^2}\sqrt{g(\varepsilon)}|\boldsymbol{u}(\varepsilon)\right)_{L^2(\Omega)^3} \\ &\le \|\boldsymbol{f}\|_{L^2(\Omega)^3}\|\boldsymbol{u}(\varepsilon)\|_{H^1(\Omega)^3}. \end{aligned}$$

The statement of the lemma follows. □

These *a priori* estimates and the weak sequential compactness of the unit ball in $H^1(\Omega)^3$ and $L^2(\Omega)^9$ imply the convergence result.

Corollary 1. *There is a sequence in $(0,\varepsilon_K)$ converging toward zero, still denoted by ε, and functions $\boldsymbol{u}\in\mathcal{V}(\Omega)$ and $\boldsymbol{\gamma}\in L^2(\Omega)^9$, such that*

$$\boldsymbol{u}(\varepsilon) \rightharpoonup \boldsymbol{u} \text{ weakly in } \mathcal{V}(\Omega), \tag{7.3}$$

$$\frac{1}{\varepsilon}\boldsymbol{\gamma}^\varepsilon(\boldsymbol{u}(\varepsilon)) \rightharpoonup \boldsymbol{\gamma} \text{ weakly in } L^2(\Omega)^9, \tag{7.4}$$

when ε tends to zero.

Let the ε−sequence be as in Corollary 1. An application of Lemma 3 immediately implies that there is a function ϕ such that $\mathbf{u}=(\boldsymbol{u},\phi)\in\mathcal{W}_0$ and

$$Q_2(\mathbf{u}) = \partial_2\gamma_{11},\ Q_3(\mathbf{u}) = \partial_3\gamma_{11},\ Q_4(\mathbf{u}) = \partial_2\gamma_{13} - \partial_3\gamma_{12}.$$

We proceed with the proof of Theorem 7 by taking the limit in (4.5) for two special choices of the test functions.

The first choice is an arbitrary function $\boldsymbol{v}\in\mathcal{V}(\Omega)$ independent of ε. We insert such test function in (4.5) and take the limit when ε tends to zero. It follows (see Jurak and Tambača (1999)) that $\boldsymbol{\gamma}$ is related to $\mathbf{u}$ by

$$\gamma_{12} = \frac{1}{2}Q_4(\mathbf{u})(\partial_2 p - z^3),\ \gamma_{13} = \frac{1}{2}Q_4(\mathbf{u})(\partial_3 p + z^2), \tag{7.5}$$

$$\gamma_{22} = -\frac{1}{2}\frac{\lambda}{\lambda+\mu}\gamma_{11} + \gamma_{22}^H,\ \gamma_{33} = -\frac{1}{2}\frac{\lambda}{\lambda+\mu}\gamma_{11} + \gamma_{33}^H,\ \gamma_{23} = \gamma_{23}^H, \tag{7.6}$$

where $\gamma_{22}^H, \gamma_{23}^H, \gamma_{33}^H$ satisfy homogeneous system

$$\int_\Omega \left(\lambda(\gamma_{22}^H + \gamma_{33}^H) + 2\mu\gamma_{22}^H\right)\partial_2 w_2 + 2\mu\gamma_{23}^H\partial_3 w_2 = 0, \quad w_2\in W(\Omega), \tag{7.7}$$

$$\int_\Omega \left(\lambda(\gamma_{22}^H + \gamma_{33}^H) + 2\mu\gamma_{33}^H\right)\partial_3 w_3 + 2\mu\gamma_{23}^H\partial_2 w_3 = 0, \quad w_3\in W(\Omega). \tag{7.8}$$

The second choice of test function $\boldsymbol{v}(\varepsilon)\in\mathcal{V}(\Omega)$ is of the form $\boldsymbol{v}(\varepsilon)=\boldsymbol{v}+\varepsilon\boldsymbol{v}^1$ such that $\boldsymbol{\gamma}^\varepsilon(\boldsymbol{v}(\varepsilon)) = O(\varepsilon)$. It follows that $\boldsymbol{v}$ and $\boldsymbol{v}^1$ should satisfy

$$\boldsymbol{\gamma}_z(\boldsymbol{v}) = 0,\quad \boldsymbol{\gamma}_z(\boldsymbol{v}^1) + \boldsymbol{\gamma}_y(\boldsymbol{v}) - v_i\boldsymbol{\Gamma}^i(0) = 0, \tag{7.9}$$

so we can take $\boldsymbol{v} \in \mathcal{V}_0$ and

$$v_1^1 = -Q_7(\mathbf{v})z^2 - Q_8(\mathbf{v})z^3 + \hat{w}_1,\ v_2^1 = -\psi z^3 + \hat{w}_2,\ v_3^1 = \psi z^2 + \hat{w}_3,$$

where $\psi, \hat{w}_1, \hat{w}_2, \hat{w}_3 \in H_0^1(0,\ell)$. We insert such test function in (4.5) and take the limit when ε tends to zero. Using (7.5), (7.6) and (7.7), (7.8) it follows that $\mathbf{u} = (\boldsymbol{u}, \phi)$ satisfies (5.12). Now the uniqueness of solution of the problem (5.12) implies the convergence of the whole family $(\boldsymbol{u}(\varepsilon), \varepsilon \in (0, \varepsilon_K))$.

Then, considering the family

$$\Lambda^\varepsilon := \int_\Omega \mathbf{A}(\varepsilon)\left(\frac{1}{\varepsilon}\gamma^\varepsilon(\boldsymbol{u}^\varepsilon) - \boldsymbol{\gamma}\right) \cdot \left(\frac{1}{\varepsilon}\gamma^\varepsilon(\boldsymbol{u}^\varepsilon) - \boldsymbol{\gamma}\right)\sqrt{g(\varepsilon)}, \qquad \varepsilon > 0,$$

one is able to show that Λ^ε tends to zero (see Jurak and Tambača (1999)); this is equivalent to the strong convergence in (7.4) and $\gamma_{22}^H = \gamma_{23}^H = \gamma_{33}^H = 0$. Lemma 3 then implies the strong convergence in (7.3) and therefore Theorem 7 is proved.

7.2 Evolution equation

It is delicate to obtain the *a priori* estimates for the evolution problem (4.3). One would like to insert $\boldsymbol{u}_t(\varepsilon)$ as a test function in (4.3) and to integrate it over a finite time interval $(0,T)$ to obtain the inequality

$$\begin{aligned} &\varrho m_G m_g \|\boldsymbol{u}_t(\varepsilon)\|^2_{L^2(\Omega)^3} + a(\varepsilon)(\boldsymbol{u}(\varepsilon), \boldsymbol{u}(\varepsilon)) \\ &\quad \le \Big(\|\boldsymbol{f}\sqrt{g(\varepsilon)}\|^2_{L^2(0,T;L^2(\Omega)^3)} + b(\varepsilon)(\boldsymbol{u}_1(\varepsilon), \boldsymbol{u}_1(\varepsilon)) \\ &\qquad + a(\varepsilon)(\boldsymbol{u}_0(\varepsilon), \boldsymbol{u}_0(\varepsilon))\Big) e^{t/(\varrho m_G m_g)}, \quad t > 0. \end{aligned}$$

Unfortunately, $\boldsymbol{u}_t(\varepsilon)$ is not admissible test function. Nevertheless, this estimate holds and it is a by product of the proof of existence and uniqueness results for evolution equations by the Galerkin approximations, see Dautray and Lions (1992) or Tambača (2000) and Tambača (2002). From this estimate it is straightforward to obtain *a priori* estimates using (7.1) and Lemma 4.

Lemma 6. *Let us assume that hypotheses of Theorem 5 hold true. Then there exists a constant $C > 0$ such that for all $\varepsilon \in (0, \varepsilon_K)$*

$$\|\boldsymbol{u}(\varepsilon)\|_{\mathcal{V}(\Omega)} \le C, \quad \|\boldsymbol{u}_t(\varepsilon)\|_{L^2(\Omega)^3} \le C, \quad \|\frac{1}{\varepsilon}\gamma^\varepsilon(\boldsymbol{u}(\varepsilon))\|_{L^2(\Omega)^9} \le C, \quad (7.10)$$

for all $t \in [0,T]$.

A priori estimates (7.10) and the Banach–Alaoglu–Bourbaki theorem imply the convergence result.

Corollary 2. *Under the assumptions of Theorem 5 there is a sequence in* $(0, \varepsilon_K)$ *converging to zero, still denoted by* ε, *and functions* $\boldsymbol{u}$ *and* $\boldsymbol{\gamma}$ *such that* $\boldsymbol{u} \in L^\infty(0,T;\mathcal{V}(\Omega))$, $\boldsymbol{u}_t \in L^\infty(0,T;L^2(\Omega)^3)$, $\boldsymbol{\gamma} \in L^\infty(0,T;L^2(\Omega)^9)$ *and*

$$\boldsymbol{u}(\varepsilon) \overset{*}{\rightharpoonup} \boldsymbol{u} \text{ weak } * \text{ in } L^\infty(0,T;\mathcal{V}(\Omega)), \tag{7.11}$$

$$\boldsymbol{u}_t(\varepsilon) \overset{*}{\rightharpoonup} \boldsymbol{u}_t \text{ weak } * \text{ in } L^\infty(0,T;L^2(\Omega)^3), \tag{7.12}$$

$$\frac{1}{\varepsilon}\boldsymbol{\gamma}^\varepsilon(\boldsymbol{u}(\varepsilon)) \overset{*}{\rightharpoonup} \boldsymbol{\gamma} \text{ weak } * \text{ in } L^\infty(0,T;L^2(\Omega)^9), \tag{7.13}$$

when ε *tends to zero.*

Now we need to identify the equation satisfied by the limits. We do this in the sense of distributions by using Lemma 3. We fix $T > 0$, choose a function $\chi \in \mathcal{D}(0,T)$ and denote by $\bar{\cdot}^\chi$ the operator defined by

$$\overline{u}^\chi = \int_0^T u(t)\chi(t)\,dt.$$

Then, by continuity of the operator $\bar{\cdot}^\chi$ we obtain

$$\overline{\boldsymbol{u}}^\chi(\varepsilon) \rightharpoonup \overline{\boldsymbol{u}}^\chi \text{ weakly in } \mathcal{V}(\Omega), \tag{7.14}$$

$$\overline{\boldsymbol{u}}^\chi{}_t(\varepsilon) \rightharpoonup \overline{\boldsymbol{u}}^\chi{}_t \text{ weakly in } L^2(\Omega)^3, \tag{7.15}$$

$$\frac{1}{\varepsilon}\boldsymbol{\gamma}^\varepsilon(\overline{\boldsymbol{u}}^\chi(\varepsilon)) \rightharpoonup \overline{\boldsymbol{\gamma}}^\chi \text{ weakly in } L^2(\Omega)^9. \tag{7.16}$$

From Lemma 3 it follows easily that there is a function ϕ_χ such that $\mathbf{u}_\chi = (\overline{\boldsymbol{u}}^\chi, \phi_\chi) \in \mathcal{W}_0$ and

$$Q_2(\mathbf{u}_\chi) = \partial_2\overline{\gamma}^\chi{}_{11}, \; Q_3(\mathbf{u}_\chi) = \partial_3\overline{\gamma}^\chi{}_{11}, \; Q_4(\mathbf{u}_\chi) = \partial_2\overline{\gamma}^\chi{}_{13} - \partial_3\overline{\gamma}^\chi{}_{12}. \tag{7.17}$$

From the last equation it follows that there is a function ϕ belonging to the space $L^\infty(0,T;H_0^1(0,\ell))$ such that $\phi_\chi = \overline{\phi}^\chi$. The first two equations in (7.17) imply that $\boldsymbol{u} \in L^\infty(0,T;\mathcal{V}_0)$ and $\boldsymbol{u}_t \in L^\infty(0,T;\mathcal{H}_0)$. Then using the same reasoning and the same choices of test functions as in the equilibrium case one gets that the equation (5.4) is satisfied by the limit function. Moreover $\boldsymbol{u}_{tt} \in L^\infty(0,T;\mathcal{V}_0')$ and $\boldsymbol{\gamma}$ is given by (5.11) (for details see Tambača (2002) or Tambača (2000)). It is then rather standard to show that the limit function satisfies the initial conditions (5.5). Therefore the limit function $\mathbf{u} = (\boldsymbol{u}, \phi)$ belongs to the space $\mathcal{W}_0(0,T)$. Uniqueness of the solution of the problem (5.4), (5.5) in $\mathcal{W}_0(0,T)$ implies that $\mathbf{u} = (\boldsymbol{u}, \phi) \in \mathcal{W}_c(0,T)$ and that the whole families $(\boldsymbol{u}(\varepsilon), \varepsilon > 0)$ and $(\boldsymbol{\gamma}^\varepsilon(\boldsymbol{u}(\varepsilon)), \varepsilon > 0)$ converge. Therefore Theorem 5 is proved.

7.3 Eigenvalue equation

In this subsection we prove Theorem 9. Eigenpairs of the spectral problem (4.6) are $(\alpha^n(\varepsilon), \boldsymbol{u}^n(\varepsilon)) \in \mathbb{R} \times \mathcal{V}(\Omega)$, $n \in \mathbb{N}$, $\varepsilon \in (0, \varepsilon_0)$. Eigenfunctions are normalized as in Lemma 3. We begin with the *a priori* estimates for the eigenvalues.

Lemma 7. *For each $n \in \mathbb{N}$ there are constants $\varepsilon_n, \delta_n > 0$ independent of ε such that*

$$\forall\varepsilon, \quad 0 < \varepsilon \leq \varepsilon_n, \qquad \alpha^n(\varepsilon) \leq \delta_n. \tag{7.18}$$

Proof. Let $\boldsymbol{v}^0 \in \mathcal{V}_0$. We define $\psi = \mathcal{D}\boldsymbol{v}^0$ and $\boldsymbol{v}_{\boldsymbol{v}^0}(\varepsilon) = \boldsymbol{v}^0 + \varepsilon\tilde{\boldsymbol{v}}$, where

$$\begin{aligned}\tilde{v}_1(z) &= -Q_7(\boldsymbol{v},\psi)z^2 - Q_8(\boldsymbol{v},\psi)z^3,\\ \tilde{v}_2(z) &= -\psi(z^1)z^3,\\ \tilde{v}_3(z) &= \psi(z^1)z^2.\end{aligned}$$

From the definition of $\boldsymbol{\gamma}^\varepsilon(\boldsymbol{v})$ and the Christoffel symbols $\boldsymbol{\Gamma}^i(\varepsilon)$ we obtain

$$\frac{1}{\varepsilon}\boldsymbol{\gamma}^\varepsilon(\boldsymbol{v}_{\boldsymbol{v}^0}(\varepsilon)) = \boldsymbol{\Theta}^1(\boldsymbol{v}^0) - v_i^0\boldsymbol{\Sigma}^{0i}(\varepsilon) - \tilde{v}_i\boldsymbol{\Sigma}^{1i}(\varepsilon), \tag{7.19}$$

where

$$\begin{aligned}\boldsymbol{\Theta}^1(\boldsymbol{v}^0) &= \boldsymbol{\gamma}_y(\tilde{\boldsymbol{v}}) - \tilde{v}_i\boldsymbol{\Gamma}^i(0) - v_i^0 z^\alpha\left(\frac{\partial}{\partial y^\alpha}\boldsymbol{\Gamma}^i\right)(0),\\ \boldsymbol{\Sigma}^{0i}(\varepsilon) &= \frac{1}{\varepsilon}\boldsymbol{\Gamma}^i(\varepsilon) - \frac{1}{\varepsilon}\boldsymbol{\Gamma}^i(0) - z^\alpha\left(\frac{\partial}{\partial y^\alpha}\boldsymbol{\Gamma}^i\right)(0),\\ \boldsymbol{\Sigma}^{1i}(\varepsilon) &= \boldsymbol{\Gamma}^i(\varepsilon) - \boldsymbol{\Gamma}^i(0).\end{aligned}$$

Asymptotic result (7.2) implies the existence of a constant $C > 0$, independent of ε, such that

$$\|\boldsymbol{\Sigma}^{0i}(\varepsilon)\|_\infty \leq C\varepsilon, \quad \|\boldsymbol{\Sigma}^{1i}(\varepsilon)\|_\infty \leq C\varepsilon. \tag{7.20}$$

Let $n \in \mathbb{N}$ be fixed. By $\mathcal{V}_0^n$ we denote the family of all n–dimensional subspaces of $\mathcal{V}_0$. For $L_0 \in \mathcal{V}_0^n$ the space defined by

$$L = \left\{\boldsymbol{v}_{\boldsymbol{v}^0} \in \mathcal{V}(\Omega) : \boldsymbol{v}^0 \in L_0\right\}$$

belongs to the family $\mathcal{V}^n(\Omega)$ introduced in Lemma 3. Therefore (4.8) implies

$$\alpha^n(\varepsilon) \leq \min_{L_0\in\mathcal{V}_0^n}\max_{\boldsymbol{v}^0\in L_0} R(\varepsilon)(\boldsymbol{v}_{\boldsymbol{v}^0}(\varepsilon)).$$

The estimate

$$R(\varepsilon)(\boldsymbol{v}_{\boldsymbol{v}^0}(\varepsilon)) \leq \frac{6CM_A}{m_G m_g}\frac{d_0(\boldsymbol{v}^0,\boldsymbol{v}^0)}{b_0(\boldsymbol{v}^0,\boldsymbol{v}^0)} + \frac{6M_AC^2}{m_G m_g},$$

implies

$$\alpha^n(\varepsilon) \le \frac{6CM_A}{m_G m_g} \min_{L_0\in\mathcal{V}_0^n} \max_{v^0\in L_0} \frac{d_0(v^0, v^0)}{b_0(v^0, v^0)} + \frac{6M_A C^2}{m_G m_g}.$$

Using the characterization of the nth eigenvalue α_0^n from Lemma 8 the statement of the lemma follows with $\delta_n = \frac{6CM_A}{m_G m_g}\alpha_0^n + \frac{6M_A C^2}{m_G m_g}$. □

Now the application of the Korn inequality and Lemma 4 yield *a priori* estimates for the eigenfunctions.

Lemma 8. *For all $n \in \mathbb{N}$ there is a constant C such that for all $0 < \varepsilon \le \varepsilon_n$ holds true*

$$b(\varepsilon)(u^n(\varepsilon), u^n(\varepsilon)) = 1, \ \|u^n(\varepsilon)\|_{H^1(\Omega)^3} \le C, \ \|\frac{1}{\varepsilon}\gamma^\varepsilon(u^n(\varepsilon))\|_{L^2(\Omega)^9} \le C.$$

As before, the direct consequences of the preceding *a priori* estimates are appropriate convergences of the subsequences. Moreover, we are able to identify the limits of eigenpairs of (4.6) as the eigenpairs of (5.14) by the same reasoning as in the equilibrium case (limiting the equation (4.6) for the same choices of test functions). Therefore Theorem 9 a) follows. It is then rather technical to prove Theorem 9 b) and c) (for details see Tambača (2001)).

8. Differential form of curved rod model

In this section the variational problems (5.12) and (5.4), (5.5) and (5.14) are rewritten in differential form.

We begin with the equilibrium problem (5.12). In order to relax the inextensibility condition in the space $\mathcal{W}_0$ we formulate the mixed variational formulation. The first component of the contact force will then appear as a Lagrange multiplier, see also Sanchez–Hubert and Sanchez–Palencia (1999).

Let us define the bilinear functional $e_0(\cdot,\cdot)$ on $W_0 \times L^2(0,\ell)$ by

$$e_0(\mathbf{v},\eta) := \int_0^\ell (v_1' - R_{12}v_2 - R_{13}v_3)\eta dz^1.$$

We consider the following mixed variational formulation of problem (5.12): Find $(\mathbf{u},\xi) \in W_0 \times L^2(0,\ell)$ such that

$$\begin{aligned} a_0(\mathbf{u},\mathbf{v}) + e_0(\mathbf{v},\xi) &= (\boldsymbol{f}|\boldsymbol{v})_{H_0}, \qquad \mathbf{v}\in W_0, \\ e_0(\mathbf{u},\eta) &= 0, \qquad \eta \in L^2(0,\ell). \end{aligned} \tag{8.1}$$

Let $E_0\colon W_0 \to L^2(0,\ell)$ be the linear operator associated with the form $e_0(\cdot,\cdot)$. Obviously, the kernel of E_0 is $\mathcal{W}_0$. If the both coefficients R_{12} and R_{13} are zero (i.e. the rod is straight), then $\mathrm{Im}(E_0) = 1^\perp$ and therefore the image of E_0

is closed (in this case the multiplier is unique up to a constant); if at least one of the coefficients R_{12}, R_{13} is different from zero, then one can easily show that $\mathrm{Im}(E_0) = L^2(0,\ell)$. Since the form $a_0(\cdot,\cdot)$ is elliptic on $\mathcal{W}_0$ we have the following result (see Brezzi and Fortin (1991)):

Theorem 11. *The problem (8.1) has a unique solution*

$$(\mathbf{u},\xi) \in W_0 \times \left(L^2(0,\ell)/\operatorname{Ker}(E_0^t)\right).$$

The function **u** *is the solution of (5.12).*

Differential form of the equilibrium model (5.12) follows easily from the equation (8.1). Simple calculations leads to a system of differential equations with boundary conditions. We write it as the first order system. By using the notations

$$\boldsymbol{p} = \begin{bmatrix} \xi \\ E\left(I_2Q_2 - I_{23}Q_3\right)' - R_{13}\mu KQ_4 - R_{23}E\left(I_3Q_3 - I_{23}Q_2\right) \\ E\left(I_3Q_3 - I_{23}Q_2\right)' + R_{12}\mu KQ_4 + R_{23}E\left(I_2Q_2 - I_{23}Q_3\right) \end{bmatrix},$$

$$\boldsymbol{q} = \begin{bmatrix} \mu KQ_4 \\ E\left(I_3Q_3 - I_{23}Q_2\right) \\ -E\left(I_2Q_2 - I_{23}Q_3\right) \end{bmatrix}, \quad \boldsymbol{\omega} = \begin{bmatrix} \phi \\ -u_3' - R_{13}u_1 - R_{23}u_2 \\ u_2' + R_{12}u_1 - R_{23}u_3 \end{bmatrix},$$

we get from (8.1):

$$\boldsymbol{p}' - \mathbf{R}\boldsymbol{p} + \overline{\boldsymbol{f}} = 0, \tag{8.2}$$

$$\boldsymbol{q}' - \mathbf{R}\boldsymbol{q} + \boldsymbol{e}_1 \times \boldsymbol{p} = 0, \tag{8.3}$$

$$\boldsymbol{\omega}' - \mathbf{R}\boldsymbol{\omega} - \mathbf{H}^{-1}\boldsymbol{q} = 0, \quad \boldsymbol{\omega}(0) = \boldsymbol{\omega}(\ell) = 0, \tag{8.4}$$

$$\boldsymbol{u}' - \mathbf{R}\boldsymbol{u} + \boldsymbol{e}_1 \times \boldsymbol{\omega} = 0, \quad \boldsymbol{u}(0) = \boldsymbol{u}(\ell) = 0. \tag{8.5}$$

Since the displacement and the force were rewritten in local basis of the curved rod and on canonical domain Ω, we rewrite the above differential equations once more in terms of physical quantities. Let

$$\tilde{\boldsymbol{p}}^\varepsilon = \varepsilon^4\mathbf{Q}\boldsymbol{p},\ \tilde{\boldsymbol{q}}^\varepsilon = \varepsilon^4\mathbf{Q}\boldsymbol{q},\ \tilde{\boldsymbol{\omega}} = \mathbf{Q}\boldsymbol{\omega},\ \tilde{\boldsymbol{u}} = \mathbf{Q}\boldsymbol{u},\ \tilde{\boldsymbol{f}}^\varepsilon = \varepsilon^4\mathbf{Q}\overline{\boldsymbol{f}},$$

$$\tilde{\boldsymbol{u}}_0 = \mathbf{Q}\boldsymbol{u}_0,\ \tilde{\boldsymbol{u}}_1 = \mathbf{Q}\boldsymbol{u}_1,\ \alpha_0^\varepsilon = \varepsilon^2\alpha_0,$$

and let us define

$$|\varepsilon S| = \int_{\varepsilon S} dy^2dy^3,\ I_2^\varepsilon = \int_{\varepsilon S}(y^2)^2dy^2dy^3,\ I_3^\varepsilon = \int_{\varepsilon S}(y^3)^2dy^2dy^3,$$

$$I_{23}^\varepsilon = -\int_{\varepsilon S} y^2y^3dy^2dy^3,\ K^\varepsilon = \int_{\varepsilon S}(\partial_2p^\varepsilon - y^3)^2 + (\partial_3p^\varepsilon + y^2)^2dy^2dy^3.$$

Here $p^\varepsilon \in H^1(\varepsilon S)$ is the unique solution of the variational equation

$$\int_{\varepsilon S} (\partial_2 p^\varepsilon - y^3)\partial_2 r + (\partial_3 p^\varepsilon + y^2)\partial_3 r dy^2 dy^3 = 0, \quad r \in H^1(\varepsilon S),$$

such that $\int_{\varepsilon S} p^\varepsilon dy^2 dy^3 = 0$. These quantities describe the geometry of the cross–section of the curved rod. The elastic properties of the material the rod is made of appear in the matrix

$$\mathbf{H}^\varepsilon = \begin{pmatrix} \mu K^\varepsilon & 0 & 0 \\ 0 & EI_3^\varepsilon & EI_{23}^\varepsilon \\ 0 & EI_{23}^\varepsilon & EI_2^\varepsilon \end{pmatrix}.$$

Then it follows $|\varepsilon S| = \varepsilon^2 |S|$, $I_\alpha^\varepsilon = \varepsilon^4 I_\alpha$, $\alpha = 2, 3$, $I_{23}^\varepsilon = \varepsilon^4 I_{23}$, $K^\varepsilon = \varepsilon^4 K$, $p^\varepsilon \circ \boldsymbol{R}^\varepsilon = \varepsilon^2 p$, $\mathbf{H}^\varepsilon = \varepsilon^4 \mathbf{H}$ and $\varrho^\varepsilon = \varepsilon^2 \varrho$.

Now the equations (8.2) - (8.5) can be written as

$$(\tilde{\boldsymbol{p}}^\varepsilon)' + \tilde{\boldsymbol{f}}^\varepsilon = 0, \tag{8.6}$$
$$(\tilde{\boldsymbol{q}}^\varepsilon)' + \boldsymbol{t} \times \tilde{\boldsymbol{p}}^\varepsilon = 0, \tag{8.7}$$
$$\tilde{\boldsymbol{\omega}}' - \mathbf{Q}(\mathbf{H}^\varepsilon)^{-1}\mathbf{Q}^T \tilde{\boldsymbol{q}}^\varepsilon = 0, \quad \tilde{\boldsymbol{\omega}}(0) = \tilde{\boldsymbol{\omega}}(\ell) = 0, \tag{8.8}$$
$$\tilde{\boldsymbol{u}}' + \boldsymbol{t} \times \tilde{\boldsymbol{\omega}} = 0, \quad \tilde{\boldsymbol{u}}(0) = \tilde{\boldsymbol{u}}(\ell) = 0. \tag{8.9}$$

The equations (8.6) and (8.7) are one–dimensional equilibrium equations of an elastic rod. The function $\tilde{\boldsymbol{p}}^\varepsilon$ is the contact force, $\tilde{\boldsymbol{f}}^\varepsilon$ is the linear density of the body force while $\tilde{\boldsymbol{q}}^\varepsilon$ is the contact couple about the center of the mass of the cross-section. The linear density of the contact couple does not appear in the equation (8.7) since it is a higher order quantity with respect to ε.

It can be shown by asymptotic expansion method that the approximation of the displacement $\tilde{\boldsymbol{U}}^\varepsilon \circ \boldsymbol{P}$ is given by the function

$$\tilde{\boldsymbol{u}} + \tilde{\boldsymbol{\omega}} \times (y_2 \tilde{\boldsymbol{n}} + y_3 \tilde{\boldsymbol{b}}).$$

Therefore the function $\tilde{\boldsymbol{u}}$ is the displacement of the middle curve $\mathcal{C}$, while the function $\tilde{\boldsymbol{\omega}}$ is the infinitesimal rotation of the cross–section of the rod with respect to the center of mass. The components of the infinitesimal rotation orthogonal to the tangent vector (ω_2 and ω_3) are expressed through the displacement of the middle curve and its derivatives, while the tangential component ($\omega_1 = \phi$) appears as an independent variable in the variational equation. The necessity of the appearance of the new variable ϕ in the description of the equilibrium displacement was already recognized and discussed in Sanchez–Hubert and Sanchez–Palencia (1999) and also in Jamal and Sanchez–Palencia (1996).

The last two equations (8.8) and (8.9) are one-dimensional constitutive equations. The third equation (8.8) gives the connection between dynamical and kinematical quantities, while the equation (8.9) is usually referred to as a constitutive or material restriction. It describes the approximate unshearability and inextensibility of the rod, see Antman (1995), Sanchez–Hubert and Sanchez–Palencia (1999) or Jurak and Tambača (2001). Therefore the deformed cross–sections remain approximately perpendicular to the deformed middle curve and the middle curve is approximately inextensible.

8.1 Evolution model

Similarly we obtain differential formulation of the evolution model (5.4), (5.5):

$$(\tilde{\boldsymbol{p}}^\varepsilon)' + \tilde{\boldsymbol{f}}^\varepsilon = \varrho^\varepsilon |\varepsilon S| \frac{\partial^2 \tilde{\boldsymbol{u}}}{\partial t^2}, \tag{8.10}$$

$$(\tilde{\boldsymbol{q}}^\varepsilon)' + \boldsymbol{t} \times \tilde{\boldsymbol{p}}^\varepsilon = 0, \tag{8.11}$$

$$\tilde{\boldsymbol{\omega}}' - \mathbf{Q}(\mathbf{H}^\varepsilon)^{-1}\mathbf{Q}^T \tilde{\boldsymbol{q}}^\varepsilon = 0, \quad \tilde{\boldsymbol{\omega}}|_{z^1=0} = \tilde{\boldsymbol{\omega}}|_{z^1=\ell} = 0, \tag{8.12}$$

$$\tilde{\boldsymbol{u}}' + \boldsymbol{t} \times \tilde{\boldsymbol{\omega}} = 0, \quad \tilde{\boldsymbol{u}}|_{z^1=0} = \tilde{\boldsymbol{u}}|_{z^1=\ell} = 0, \tag{8.13}$$

$$\tilde{\boldsymbol{u}}|_{t=0} = \tilde{\boldsymbol{u}}_0, \quad \tilde{\boldsymbol{u}}_t|_{t=0} = \tilde{\boldsymbol{u}}_1. \tag{8.14}$$

The first two equations (8.10), (8.11) are well–known one–dimensional conservation laws. The first equation (8.10) is the one–dimensional linear impulse–momentum law, while the second equation (8.11) is the one–dimensional angular impulse–momentum law. The equations (8.10) and (8.11) are the classical forms of the equations of motion for rods, see Antman (1995). The function $\tilde{\boldsymbol{p}}^\varepsilon$ is the contact force, $\tilde{\boldsymbol{q}}^\varepsilon$ is the contact couple. The third and the fourth equation are constitutive equations.

8.2 Eigenvalue model

The differential form of the eigenvalue model (5.14) is given by: find $\alpha_0^\varepsilon \in \mathbb{R}$ and $(\tilde{\boldsymbol{p}}^\varepsilon, \tilde{\boldsymbol{q}}^\varepsilon, \tilde{\boldsymbol{\omega}}, \tilde{\boldsymbol{u}}) \neq 0$ such that

$$\begin{aligned}
&(\tilde{\boldsymbol{p}}^\varepsilon)' = -\alpha_0^\varepsilon \varrho^\varepsilon |\varepsilon S| \tilde{\boldsymbol{u}},\\
&(\tilde{\boldsymbol{q}}^\varepsilon)' + \boldsymbol{t} \times \tilde{\boldsymbol{p}}^\varepsilon = 0,\\
&\tilde{\boldsymbol{\omega}}' - \mathbf{Q}(\mathbf{H}^\varepsilon)^{-1}\mathbf{Q}^T \tilde{\boldsymbol{q}}^\varepsilon = 0, \quad \tilde{\boldsymbol{\omega}}(0) = \tilde{\boldsymbol{\omega}}(\ell) = 0,\\
&\tilde{\boldsymbol{u}}' + \boldsymbol{t} \times \tilde{\boldsymbol{\omega}} = 0, \quad \tilde{\boldsymbol{u}}(0) = \tilde{\boldsymbol{u}}(\ell) = 0.
\end{aligned}$$

9. Comparison of various curved rod models

In this section we describe briefly another two methods in modeling of curved rods. First method is so called special Cosserat theory, see Antman (1995). Second method uses well established shell models; assuming that the shell has constant cross-section, various rod models can be derived by singular perturbation. The resulting models are then compared with our model. For simplicity we consider the equilibrium problem only. Throughout this section parameter ε is fixed, thus we omit it for simplicity.

9.1 The Special Cosserat Theory of Rods

In this section we describe the nonlinear model of the special Cosserat rods, Antman (1995), and compare it with the obtained model (8.6)–(8.9).

Let the referent configuration of the curved rod be described by $\boldsymbol{\Phi}, \boldsymbol{n}, \boldsymbol{b}$ defined in Section 2 through the mapping $\boldsymbol{P}$ defined in (2.3). Under the applied loads we assume that the equilibrium configuration of the curved rod is described by

$$\boldsymbol{r}, \boldsymbol{d}_2, \boldsymbol{d}_3 : [0, \ell] \to \mathbb{R}^3, \tag{9.1}$$

where $\boldsymbol{d}_2$ and $\boldsymbol{d}_3$ are orthonormal vectors called directors. Let $\boldsymbol{d}_1 = \boldsymbol{d}_2 \times \boldsymbol{d}_3$, so $\mathbf{D} = \begin{bmatrix} \boldsymbol{d}_1 & \boldsymbol{d}_2 & \boldsymbol{d}_3 \end{bmatrix}$ is rotation. $\boldsymbol{r}$ defines the equilibrium deformation of the middle curve of the rod, while $\boldsymbol{d}_2$ and $\boldsymbol{d}_3$ describe the equilibrium position of the deformed cross section. Equilibrium displacement is

$$\tilde{\boldsymbol{z}}(y) - \boldsymbol{P}(y) = \boldsymbol{r}(y^1) - \boldsymbol{\Phi}(y^1) + y^2(\boldsymbol{d}_2(y^1) - \boldsymbol{n}(y^1)) + y^3(\boldsymbol{d}_3(y^1) - \boldsymbol{b}(y^1)). \tag{9.2}$$

More general models can be obtained by use of more than two directors and nonlinear terms of the cross–sectional variables in (9.2).

The axial vector of the skewsymmetric matrix $\mathbf{B} = (\mathbf{D}^T)'\mathbf{D}$ is denoted by $-\boldsymbol{v}$. Then

$$\mathbf{D}' = -\mathbf{D}\mathbf{B}. \tag{9.3}$$

We also define

$$\tilde{\boldsymbol{w}} = \boldsymbol{r}', \quad \boldsymbol{w} = \mathbf{D}^T \tilde{\boldsymbol{w}}.$$

$\boldsymbol{v}$ and $\boldsymbol{w}$ are strain variables; they determine the equilibrium configuration up to the rigid displacement. The components v_2 and v_3 measure flexure, v_1 measures torsion, w_2 and w_3 measure shear and w_1 measures dilatation.

Now we assume that the rod is clamped at both ends $y^1 = 0$, $y^1 = \ell$ and that it is submitted to the action of body force $\tilde{\boldsymbol{f}}$ and body couple $\tilde{\boldsymbol{l}}$ about 0 per unit reference length. Let $\tilde{\boldsymbol{p}}(y^1)$ and $\tilde{\boldsymbol{q}}(y^1)$ denote the contact force and the contact spin at $\boldsymbol{r}(y^1)$. Then the equilibrium equations for the special Cosserat theory of rods are given by

$$\tilde{\boldsymbol{p}}'(s) + \tilde{\boldsymbol{f}}(s) = 0, \tag{9.4}$$

$$\tilde{\boldsymbol{q}}'(s) + \boldsymbol{r}'(s) \times \tilde{\boldsymbol{p}}(s) + \tilde{\boldsymbol{l}}(s) = 0. \tag{9.5}$$

Let $\boldsymbol{p} = \mathbf{D}^T \tilde{\boldsymbol{p}}$, $\boldsymbol{q} = \mathbf{D}^T \tilde{\boldsymbol{q}}$, $\boldsymbol{f} = \mathbf{D}^T \tilde{\boldsymbol{f}}$, $\boldsymbol{l} = \mathbf{D}^T \tilde{\boldsymbol{l}}$. The functions p_2 and p_3 are the shear forces, $\tilde{\boldsymbol{p}} \cdot \boldsymbol{r}'/|\boldsymbol{r}'|$ is the tension, q_2 and q_3 are the bending couples and q_1 is the twisting couple.

The dynamical and kinematical quantities are connected by the constitutive equations. We consider a special case:

$$\tilde{\boldsymbol{w}} = \boldsymbol{d}_1, \quad \boldsymbol{q}(y^1) = \boldsymbol{q}^c(\boldsymbol{v}(y^1), y^1), \tag{9.6}$$

for a given $\boldsymbol{q}^c$. Equation of the form $(9.6)_1$ is sometimes called material restriction. This one is the simplest and the most important. It describes the fact that cross–sections are orthogonal to the middle curve in the equilibrium configuration and that the rod is inextensible. The function $\boldsymbol{p}$ is not prescribed in the constitutive equations, so it remains free to take any value to maintain the constraint $(9.6)_1$.

Now we linearize the equilibrium equations (9.3), (9.4), (9.5), (9.6) about the reference configuration which is assumed to be natural. We first rewrite (9.4) and (9.5) in terms of $\boldsymbol{p}, \boldsymbol{q}$ and $\overline{\boldsymbol{f}}$ and then (9.3) in terms of $\mathbf{E} = \mathbf{D}\mathbf{Q}^T$. We suppose that the equilibrium position of the rod is a small perturbation of the reference configuration which is assumed to be natural. It follows that the contact forces and couples vanish in referent configuration. Therefore unknowns in (9.3), ..., (9.6) in the reference configuration take the values

$$\boldsymbol{p}^0 = 0, \quad \boldsymbol{q}^0 = 0, \quad \boldsymbol{v}^0 = \begin{bmatrix} -R_{23} \\ R_{13} \\ -R_{12} \end{bmatrix}, \quad \mathbf{E}^0 = \mathbf{I}, \quad \boldsymbol{r}^0 = \boldsymbol{\Phi}, \quad \mathbf{B}^0 = \mathbf{R}.$$

Calculating the first order approximation we obtain

$$\tilde{\boldsymbol{p}}' + \tilde{\boldsymbol{f}} = 0, \tag{9.7}$$

$$\tilde{\boldsymbol{q}}' + \boldsymbol{t} \times \tilde{\boldsymbol{p}} + \tilde{\boldsymbol{l}} = 0, \tag{9.8}$$

$$\tilde{\boldsymbol{\omega}}' - \mathbf{Q}\mathbf{H}^{-1}\mathbf{Q}^T\tilde{\boldsymbol{q}} = 0, \tag{9.9}$$

$$\tilde{\boldsymbol{u}}' + \boldsymbol{t} \times \tilde{\boldsymbol{\omega}} = 0, \tag{9.10}$$

where $\boldsymbol{v}^0$ is the axial vector corresponding to the matrix $\mathbf{R}$; $\tilde{\boldsymbol{u}}$ is the infinitesimal displacement of the middle curve of the rod, i.e. the first order approximation to $\boldsymbol{r} - \boldsymbol{\Phi}$; $\tilde{\boldsymbol{\omega}}$ is the axial vector of the first order approximation of $\mathbf{E} - \mathbf{I}$ (hence $\tilde{\boldsymbol{\omega}}$ describes the infinitesimal rotation of the cross-sections); $\mathbf{H}(y^1) = \partial_v \boldsymbol{q}^c(\boldsymbol{v}^0(y^1), y^1)$. Then the displacement (9.2) is approximated by

$$\mathbf{U}(y) = \tilde{\boldsymbol{u}}(y^1) + \tilde{\boldsymbol{\omega}} \times \left(y^2\boldsymbol{n}(y^1) + y^3\boldsymbol{b}(y^1)\right). \tag{9.11}$$

It follows that the one–dimensional model (8.6)–(8.9) is equal to the linearized special Cosserat rod model for which it is assumed that the body couple is zero (the couple is force generated), the rod is inextensible and unshearable and the matrix $\mathbf{H}$ from the constitutive equation is of the form (5.3).

9.2 Arch model

Let now $\mathcal{C}$ be a plane curve such that $\tilde{\boldsymbol{b}}(y^1) = \mathbf{e}_3,\ y^1 \in (0,l)$. Curve $\mathcal{C}$ represents a shell with thickness h, finite or infinite in $\mathbf{e}_3$ direction. In present setting $\mathcal{C}$ is an arch. Under certain assumptions on the applied forces the equilibrium displacement takes the form, see Bernadou and Ducatel (1982),

$$\boldsymbol{u}(y^1) = u_1(y^1)\boldsymbol{t}(y^1) + u_2(y^1)\tilde{\boldsymbol{n}}(y^1),\ y^1 \in (0,l).$$

For a shell of thickness h in $\mathbf{e}_3$ direction as well, the Koiter shell model reduces then to the following variational equation, so called arch model,

$$\int_0^l \frac{Eh^2}{1-\nu^2}\left\{(u_1' - \kappa u_2)(v_1' - \kappa v_2) + \frac{h^2}{12}((u_2' + \kappa u_1)'(v_2' + \kappa v_1)'\right\}$$
$$= \int_0^l (\overline{f}_1 v_1 + \overline{f}_2 v_2), \quad (v_1, v_2) \in H_0^1(0,l) \times H_0^2(0,l);$$

here κ stands for R_{12}. As usual, ν^2 is neglected with respect to 1. For a thin arch, i.e. in the limit when $h \to 0$, it follows formally, see Zhang (1992), that (u_1, u_2) belongs to $\tilde{\mathcal{V}}_0$ and it satisfies the variational equation

$$EI\int_0^l (u_2' + \kappa u_1)'(v_2' + \kappa v_1)' = \int_0^l (\overline{f}_1 v_1 + \overline{f}_2 v_2), \qquad (v_1, v_2) \in \tilde{\mathcal{V}}_0$$

where $I = \frac{h^4}{12}$ is in present setting the moment of inertia of the cross section and $\tilde{\mathcal{V}}_0$ is the space of inextensible displacements, i. e.

$$\tilde{\mathcal{V}}_0 = \{(v_1, v_2) \in H_0^1(0,\ell) \times H_0^2(0,\ell) : v_1' - \kappa v_2 = 0\}.$$

The above equation coincides with the problem (5.12) for a plane curve $\mathcal{C}$.

References

Álvarez–Dios, J. A. and Viaño, J. M. (1995). Une théorie asymptotique de flexion–extension pour les poutres élastiques faiblement courbées. *C. R. Acad. Sci. Paris, série I*, 321: 1395–1400.

Álvarez–Dios, J. A. and Viaño, J. M. (1998). Mathematical justification of a one–dimensional model for general shallow arches. *Math. Meth. Appl. sci.*, 21: 281–325.

Antman, S. S. (1995). *Nonlinear Problems of Elasticity*. Springer–Verlag.

Bermúdez, A. and Viaño, J. M. (1984). Une justification des équations de la thermo–élasticité des poutres à section variable par des méthodes asymptotiques. *RAIRO Anal. Numér.*, 17: 121–136.

Bernadou, M. and Ducatel, Y. (1982). Approximation of General Arch Problems by Straight Beam Elements. *Numer. Math.*, 40: 1–29.

Brezzi, F. and Fortin, M. (1991). *Mixed and Hybrid Finite Element Methods*. Springer-Verlag.

Chatelin, F. (1983). *Spectral Approximation of Linear Operators*. Academic Press.

Ciarlet, P. G. (1988). *Mathematical Elasticity, Vol. I: Three–Dimensional Elasticity*. North–Holland.

Ciarlet, P. G. and Kesavan, S. (1981). Two–dimensional approximations of three–dimensional eigenvalue problems in plate theory. *Comp. Meth. Appl. Mech. Engrg.*, 26: 145–172.

Ciarlet, P. G. and Destuynder, P. (1979). A justification of the two dimensional linear plate model. *J. Mécanique*, 18: 315–344.

Ciarlet, P. G., Lods, V. and Miara, B. (1996). Asymptotic analysis of linearly elastic shells. II. Justification of flexular shell equations. *Arch. Rational Mech. Anal.*, 136: 163-190.

Cimetière, A., Geymonat, G., Le Dret, H., Raoult, A. and Tutek, Z. (1986). Une dérivation d'un modéle non linéaire de poutres a partir de l'élasticité tridimensionelle. *C. R. Acad. Sci. Paris, série I*, 302: 697–700.

Cimetière, A., Geymonat, G., Le Dret, H., Raoult, A. and Tutek, Z. (1988). Asymptotic theory and analysis for displacement and stress distributions in nonlinear elastic straight slender rods. *J. Elasticity*, 19: 111–161.

Dautray, R. and Lions, J. L. (1992). *Mathematical Analysis and Numerical Methods for Science and Tehnology. Volume 5, Evolution Problems I.* Springer–Verlag.

Davies, E. B. (1995). *Spectral Theory and Differential Operators*, University Press, Cambridge.

Diestel, J. and Uhl, J. J. (1977). *Vector measures.* AMS, Mathematical Surveys, No 15.

Figueiredo, I. N. and Trabucho, L. (1993). A Galerkin approximation for curved beams. *Comp. Meth. Appl. Mech. Engrg.* 102: 235–253.

Germain, P. (1962). Méchanique des Milieux Continus. Masson.

Giarult, V. and Raviart, P. (1986). *Finite Element Methods for Navier–Stokes Equations.* Springer–Verlag.

Hay, G. E. (1942). The finite displacement of thin rods. *Trans. Amer. Math. Soc.*, 51: 65–102.

Irago, H. and Kerdid, N. and Viaño, J. M. (1998). Analyse asymptotique des modes de hautes fréquences dans les poutres minces, *C. R. Acad. Sci. Paris, série I*, 326: 1255–1260.

Irago, H. and Viaño, J. M. (1998). Second–order asymptotic approximation of flexural vibrations in elastic rods, *Mat. Mod. Meth. Appl. Sci.*, 8: 1343–1362.

Jamal, R. (1998). *Modélisation asymptotique des comportements statique et vibratoire des tiges courbes élastiques*, Thése de doctorat, De L'Université Pierre et Marie Curie, Paris 6.

Jamal, R. and Sanchez–Palencia, É. (1996). Théorie asymptotique des tiges courbes anisotropes. *C. R. Acad. Sci. Paris, série I*, 322: 1099–1106.

Jurak, M. and Tambača, J. (1999). Derivation and justification of a curved rod model. *Math. Mod. Meth. in Appl. Sci.*, 9: 991-1014.

Jurak, M. and Tambača, J. (2001). Linear curved rod model. General curve. *Math. Mod. Meth. Appl. Sci.*, 11: 1237–1253.

Jurak, M., Tambača, J. and Tutek, Z. (1999). Derivation of a curved rod model by Kirchhoff assumptions. *ZAMM*, 79: 455-463.

Jurak, M. and Tutek, Z. (1999). Wrinkled rod. *Math. Mod. Meth. Appl. Sci.*, 9: 665–692.

Kerdid, N. (1993). Comportement asymptotique quand l'épaisseur tend vers zéro du problème de valeurs propres pour une poutre mince encastré linéare, *C. R. Acad. Sci. Paris, série I*, 316: 755–758.

Kikuchi, F. (1982). Accuracy of some finite element models for arch problems. *Comp. Meth. Appl. Mech. Engrg.*, 35: 315–345.

Landau, L. D. and Lifshitz, E. M. (1970). *Theory of Elasticity.* Pergamon Press.

Le Dret, H. (1989). Modeling of the junction between two rods. *J. Math. Pures. Appl.*,, 68: 365–397.

Le Dret, H. (1991). *Problémes variationneles dans les multi-domains.* Masson.

Li–ming, X. (1998). Asymptotic analysis of dynamic problems for linearly elastic shells – Justification of equations for dynamic membrane shells. *Asymptotic Analysis*, 17: 121–134.

Nečas, J. and Hlaváček, I. (1981) *Mathematical Theory of Elastic and Elasto–Plastic Bodies: An Introduction*. Elsevier.

Raoult, A. (1985). Construction d'un modèle d'évolution de plaques avec terme d'inerte de rotation. *Annali di Matematica Pura et Applicata, Serie Quarta*, 139: 361–400.

Saleeb, A. F. and Chang, T. Y. (1987). On the hybrid–mixed formulation of C^0 curved beam elements. *Comp. Meth. Appl. Mech. Engrg.*, 60: 95–121.

Sanchez–Hubert, J. and Sanchez–Palencia, É. (1999). Statics of curved rods on account of torsion and flexion. *Eur. J. Mech. A/Solids*, 18: 365–390.

Sanchez–Palencia, É. (1992). Asymptotic and spectral properties of a class of singular–stiff problems. *J. Math. Pures. Appl.*, 71: 379–406.

Tambača, J. (1999). *One–dimensional models in theory of elasticity*, master's thesis, Department of Mathematics, University of Zagreb. (In Croatian).

Tambača, J. (2000). *Evolution model of curved rods.* PhD thesis, Department of Mathematics, University of Zagreb. (in Croatian).

Tambača, J. and Tutek, Z. (2000). Dynamic Curved Rod Model. *Proceedings of the Conference on Applied Mathematics and Computation, Dubrovnik 1999, eds. V. Hari et al.*

Tambača, J. and Tutek, Z. (2000). Evolution model of curved rods. *Proceedings of the Fifth International Conference on Mathematical and Numerical Aspects of Wave Propagation (Santiago de Compostela, 2000)*, SIAM: 197–201.

Tambača, J. (2001). One–dimensional approximations of the eigenvalue problem of curved rods. *Math. Meth. Appl. Sci.*, 24: 927–948.

Tambača, J. (2002). Justification of the dynamic model of curved rods. Submitted to *Asymptotic Analysis.*

Trabucho, L. and Viaño, J. M. (1996). Mathematical Modelling of Rods in *Handbook of Numerical Analysis, Vol. IV*, Eds. P. G. Ciarlet, J. L. Lions. North–Holland.

Tretter, C. (2000). Linear operator pencils $A - \lambda B$ with discrete spectrum. *Integral Eq. Oper. Th.*, 37: 357–373.

Tutek, Z. and Aganović, I. (1986). A justification of the one–dimensional linear model of elastic beam, *Math. Meth. in the Appl. Sci.*, 8: 1–14.

Weinberger, H. F. (1974). *Variational Methods for Eigenvalue Approximation*, SIAM.

Zerner, M. (1994). An asymptotically optimal finite element scheme for the arch problem. *Numer. Math.*, 69: 117–123.

Zhang, Z. (1992). Arch beam models: Finite element analysis and superconvergence. *Numer. Math.*, 61: 117–143.

INCOMPRESSIBLE NEWTONIAN FLOW THROUGH THIN PIPES

Eduard Marušić-Paloka
Department of Mathematics, University of Zagreb
Bijenička 30, 10000 Zagreb, Croatia
emarusic@math.hr

Abstract The aim of this paper is to present some results about asymptotic approximations of the incompressible viscous flow through thin (or long) pipes. The ratio between the length and the cross-section is considered as the small parameter. Using the asymptotic analysis with respect to that small parameter, the effective behaviour of the flow is found. After a simple case of one straight pipe, we study the fluid flow through a network of intersected pipes with prescribed pressure at their ends. At each junction an explicit formula for computing the value of the pressure is found. The interior layer phenomenon in vicinity of the junction is studied. The effects of the curved pipes on the flow profile are considered. Correctors for the Poiseuille flow, due to the curvedness of the pipe are computed.

Keywords: Thin pipes, Navier-Stokes equations, Poiseuille flow, junction, Kirchoff law, Leray problem, curved pipes.

1. Introduction

It is well known that the stationary Navier-Stokes system describing the viscous flow in pipes with impermeable walls governed by the prescribed pressure drop has a solution in the form of the Poiseuille[1] flow, which in case of circular pipe reads

$$v(r) = \frac{\Delta p}{4\mu L}(R^2 - r^2), \qquad \left\{ \begin{array}{ll} \mu & - \text{ viscosity} \\ v & - \text{ velocity} \\ \Delta p & - \text{ pressure drop} \\ L, R & - \text{ pipe's length and radius} \end{array} \right. , \tag{1}$$

or, in general case (13), (14).

However, Poiseuille formula gives an exact solution only in case of time independent, laminar flow through one straight pipe with constant cross-section. If the flow is nonstationary or the pipe has a variable cross-section or it is curved, the appropriate version of the Poiseuille formula could be derived but it only gives an approximation for the solution, which is (nevertheless) frequently used in the engineering literature. That kind of approximations can be generalized even on networks of pipes, but, again, they only give an asymptotic approximation of the flow for the small pipe diameter.

There are several methods that allow to find the asymptotic behaviour of the flow. In this paper we intend to describe only some of them.

If we are only interested in one-dimensional approximation we can take an average over the cross section of the pipe, i.e., we could integrate the Navier-Stokes system with respect to the all but one variable (obviously, that one is directed along the pipe). Such simple one-dimensional approximation, frequently used in the engineering literature, can be derived based on an assumption that, in case of very thin or very long pipe the variations of the solution on the cross section are of no importance for the global flow. Obtained approximation, usually, has a low order of accuracy and does not describe the flow profile in the pipe.

Another, more sophisticated and more precise, approach is based on the fine asymptotic analysis with respect to the small parameter ε, being the ratio between the pipes length and its diameter. Such methods were applied in 50's for deriving, formally, lower dimensional approximations for fluid flow through thin domains (mostly in theory of lubrication and meteorology, see [26, 25, 27] and the references therein). Their method relies on two-scale asymptotic expansions in powers of the small parameter ε of the form

$$U^0(x_1, \frac{x'}{\varepsilon}) + \varepsilon U^1(x_1, \frac{x'}{\varepsilon}) + \cdots , \qquad \varepsilon^2 P^0(x_1) + \varepsilon^3 P^1(x_1, \frac{x'}{\varepsilon}) + \cdots .$$

The variable x_1 is along the pipe and $x' = (x_2, x_3)$ on the cross section. The role of dilated (so called) fast variable $y = \frac{x'}{\varepsilon}$ is to describe the fast changes of the solution on the cross-section of the thin domain. Substituting expansion in the Navier-Stokes system and collecting equal powers of ε leads to the recursive sequence of linear problems that can be solved (explicitely, if the geometry of the cross-section is simple, numerically, in general). However, in their work, no rigorous justification of such asymptotic expansion (error estimate, convergence...) was given.

Necessary mathematical tools for rigorous analysis were developed by Ciarlet and Destuynder [4] (see also Ciarlet [4], Aganović and Tutek [24], Le Dret [12] and the references therein) for study of plates and bars in elasticity theory. A direct application on fluid mechanics was done by Bayada and Chambat [2] (see also its improvement in [6]), again, in case of lubrication theory and Nazarov [21], Nazarov and Piletskas [22] for more general applications in fluid mechanics, including the pipe with variable cross-section.

2. Poiseuille approximation for pipe with variable cross-section

2.1 The problem

2.1.1 The geometry. To describe the thin pipe with a small parameter ε appearing explicitly, we start with the thick pipe. We suppose that the thick pipe has the form

$$\Omega = \{x = (x_1, y) \in \mathbf{R}^3 \; ; \; x_1 \in]0, \ell[\; , \; y = (y_2, y_3) \in \; S(x_1)\}$$

where the family of bounded locally Lipschitz domains $\{S(x_1)\}_{x_1 \in [0,\ell]}, S(\cdot) \subset \mathbf{R}^2$ is chosen such that Ω is locally Lipschitz. We now define the thin pipe (with variable cross section and the thickness ε) by

$$\begin{aligned}
&\Omega_\varepsilon = \{x = (x_1, x') \in \mathbf{R}^3 \; ; \; x_1 \in]0, \ell[\; , \; x' = (x_2, x_3) \in \; \varepsilon \, S(x_1)\} \\
&\Sigma_\varepsilon^i = \varepsilon \; S(i) \\
&\Gamma_\varepsilon = \{x = (x_1, x') \in \mathbf{R}^3 \; ; \; x_1 \in]0, \ell[\; , \; x' = (x_2, x_3) \in \; \varepsilon \, \partial \, S(x_1)\} \; .
\end{aligned}$$

2.1.2 The equations. For the sake of simplicity, we restrict ourselves to the stationary case. The flow is assumed to be governed by the pressure drop. In that case we have the boundary value problem for the Navier-Stokes system that reads:
For given pressures p_i, $i = 0, \ell$ find $(u^\varepsilon, p^\varepsilon)$ such that

$$-\mu \, \Delta u^\varepsilon + (u^\varepsilon \nabla) u^\varepsilon + \nabla p^\varepsilon = 0 \;\text{ in } \Omega_\varepsilon \tag{2}$$

$$\operatorname{div} u^\varepsilon = 0 \;\text{ in } \Omega_\varepsilon \tag{3}$$

$$u^\varepsilon = 0 \;\text{ on } \Gamma_\varepsilon \tag{4}$$

$$\mathbf{e}_1 \times u^\varepsilon = 0, \quad p^\varepsilon = p_i \;\text{ on } \Sigma_\varepsilon^i \, , \; i = 1, \ell \; , \tag{5}$$

where $\mathbf{e}_1 = (1, 0, 0)$.

Remark 1. *Imposing that the tangential component of the velocity $\mathbf{e}_1 \times u^\varepsilon$ equals zero on Σ_ε^i is not a serious restriction because the only part of the velocity on the boundary that counts is the normal part, due to the St-Venant's principle for thin domains. It is well-known that two flows with same normal velocities on Σ_ε^i and different tangential velocities, differ only in some small boundary layer in vicinity of Σ_ε^i (see [19, 15]).*

The existence and uniqueness of the solution for such problem was discussed in [17, 5, 9]. The solution exists as long as the Reynolds number is not too large. More precisely the nonexistence will not appear if p_i have the order of magnitude smaller than ε^{-2}. Such solution is unique in some ball around 0. In the stationary case we cannot prove that there are no some other, large, solutions. In the evolutional case the uniqueness can be proved if the tube is supposed to be axially symmetric (the same as for the Dirichlet problem). For larger Reynolds numbers the existence can be obtained if the dynamic pressure $p + \frac{1}{2}|u|^2$ instead of the pressure is prescribed (see [5]).

Our goal is to find the asymptotic behaviour of the flow, as the thickness $\varepsilon \to 0$. The classical approach is to rescale the domain and to write the problem on Ω instead of Ω_ε. Introducing

$$U^\varepsilon(x_1, y) = u^\varepsilon(x_1, \varepsilon\, y), \quad P^\varepsilon(x_1, y) = p^\varepsilon(x_1, \varepsilon\, y)$$

we can write our equations as

$$-\mu\left(\frac{1}{\varepsilon^2}\Delta_y U^\varepsilon + \frac{\partial^2 U^\varepsilon}{\partial x_1^2}\right) + U_1^\varepsilon \frac{\partial U^\varepsilon}{\partial x_1} + $$
$$+\frac{1}{\varepsilon}(U^\varepsilon \nabla_y)U^\varepsilon + \frac{1}{\varepsilon}\nabla_y P^\varepsilon + \frac{\partial P^\varepsilon}{\partial x_1}\,\mathbf{e}_1 = 0 \ \text{ in } \Omega \tag{6}$$
$$\operatorname{div}_y U^\varepsilon + \varepsilon \frac{\partial U_1}{\partial x_1} = 0 \ \text{ in } \Omega\,, \tag{7}$$

where

$$\Delta_y U = \frac{\partial^2 U}{\partial y_2^2} + \frac{\partial^2 U}{\partial y_3^2}\,, \ \operatorname{div}_y U = \frac{\partial U_2}{\partial y_2} + \frac{\partial U_3}{\partial y_3} \text{ and } \nabla_y \phi = \frac{\partial \phi}{\partial y_2}\,\mathbf{e}_2 + \frac{\partial \phi}{\partial y_3}\,\mathbf{e}_3.$$

By taking U^ε as the test function in weak formulation of (6) we prove the a priori estimate for U^ε. By duality, we then prove the estimate for the pressure.

Lemma 1. *There exists a constant $C > 0$, independent from ε, such that*

$$|U^\varepsilon|_{L^2(\Omega)} \le C\,\varepsilon^2, \quad |P^\varepsilon|_{L^2(\Omega)} \le C\,.$$

We are now trying to find the asymptotic expansion of the solution in powers of ε. A priori estimates from lemma 1 suggest that the leading power in expansion for U^ε should be 2, while in the expansion for P^ε we start with ε^0. We look for an expansion in the form

$$U^\varepsilon = \varepsilon^2\, U^0 + \varepsilon^3\, U^1 + \cdots, \quad P^\varepsilon = P^0 + \varepsilon\, P^1 + \cdots.$$

Substituting such expansion in (6) and collecting equal powers of ε leads to the recursive sequence of equations that can be solved by partial separation of variables

$$
\begin{aligned}
\varepsilon^{-1} \quad &: \quad \nabla_y P^0 = 0 \;\Rightarrow\; P^0 = P^0(x_1) \\
\varepsilon^0 \quad &: \quad -\mu\Delta_y U^0 + \nabla y P^1 = -\frac{\partial P^0}{\partial x_1}\,\mathbf{e}_1 \;\Rightarrow \\
&\Rightarrow \begin{cases} U^0(x_1,y) = W(x_1,y)\,\frac{\partial P^0}{\partial x_1}(x_1)\,\mathbf{e}_1 \\ P^1 = 0 \\ \mu\Delta_y W(x_1,\cdot\,) = 1 \;\text{ in } S(x_1) \\ W(x_1,\cdot\,) = 0 \;\text{ on } \partial S(x_1) \end{cases} \\
\varepsilon^k \quad &: \quad -\mu\Delta_y U^k + \nabla y P^{k+1} = -\frac{\partial P^k}{\partial x_1}\mathbf{e}_1 + \mu\frac{\partial^2 U^{k-2}}{\partial x_1^2} - \\
&- \sum_{i+j=k-1} (U^i\nabla_y)U^j + \sum_{i+j=k-2} U_1^i\,\frac{\partial U^j}{\partial x_1} \\
&U^k, P^k = 0 \;\text{ for }\; k < 0\,.
\end{aligned}
$$

In addition (7) implies

$$
\operatorname{div}_y U^{k+1} + \frac{\partial U_1^k}{\partial x_1} = 0\,.
$$

Integration over $S(x_1)$, with respect to y gives

$$
\frac{\partial}{\partial x_1}\left(\int_{S(x_1)} U_1^k\right) = 0
$$

or

$$
\int_{S(x_1)} U_1^k = \text{const.}
$$

In particular, denoting

$$
\theta(x_1) = \int_{S(x_1)} W(x_1,y)\,dy,
$$

we get

$$
\theta\,\frac{\partial P^0}{\partial x_1} = C_1
$$

and

$$
P^0(x_1) = C_1 \int_0^{x_1} \frac{dt}{\theta(t)} + C_2\,.
$$

Using the boundary conditions (5) leads to

$$C_1 = \left(\int_0^\ell \frac{dt}{\theta(t)} \right)^{-1} (p_\ell - p_0), \quad C_2 = p_0$$

and finally

$$P^0(x_1) = p_0 + (p_\ell - p_0) \left(\int_0^\ell \frac{dt}{\theta(t)} \right)^{-1} \int_0^{x_1} \frac{dt}{\theta(t)}$$

$$U^0(x_1, y) = \frac{W(x_1, y)}{\theta(x_1)} \left(\int_0^\ell \frac{dt}{\theta(t)} \right)^{-1} (p_\ell - p_0) \,.$$

Remark 2. *In case of circular pipe* $S(x_1) = \{y \in \mathbf{R}^2 \ ; \ |y| < R(x_1)\}$, *where* $R : [0, \ell] \to]0, +\infty[$ *is strictly positive smooth function, we can compute* $W = (|y|^2 - R(x_1)^2)/4\mu$. *Therefore* $\theta = -\frac{\pi}{8} R^4$ *and*

$$P^0 = p_0 + (p_\ell - p_0) \left(\int_0^\ell \frac{dt}{R^4(t)} \right)^{-1} \int_0^{x_1} \frac{dt}{R^4(t)}$$

$$U^0 = \frac{(p_0 - p_\ell)}{4\mu \, R^4} (|y|^2 - R(x_1)^2) \int_0^{x_1} \frac{dt}{R^4(t)} \,.$$

Other terms in the expansion could be computed in the similar manner (see [8] for the 2D case).

For the approximation of k-th order

$$u(k) = \sum_{j=0}^{k} \varepsilon^{j+2} U^j(x_1, \frac{x'}{\varepsilon}), \quad p(k) = P^0(x_1) + \sum_{j=1}^{k} \varepsilon^j P^j(x_1, \frac{x'}{\varepsilon})$$

using the standard a priori estimates for the Navier-Stokes system, we can prove the error estimate (see e.g. [8]). It is important to notice that the domain Ω_ε is shrinking so that the integral $L^q(\Omega_\varepsilon)$ norms are not appropriate for the error estimate. Indeed, the convergence of $u^\varepsilon - u(k)$ to zero in $L^q(\Omega_\varepsilon)$ norm does not mean anything since any L^∞ - bounded function converges to zero in such norm. The appropriate convergence is that on the rescaled domain Ω. It is easy to see that $L^2(\Omega)$ norm for the rescaled function U^ε is the same as the rescaled norm

$$\| \cdot \|_\varepsilon = \varepsilon^{-1} \| \cdot \|_{L^2(\Omega_\varepsilon)}$$

of u^ε on Ω_ε.[2] Therefore we prove the error estimates in $\| \cdot \|_\varepsilon$ norm.

Theorem 1. *There exists a constant* $C > 0$ *such that*

$$\|u^\varepsilon - u(k)\|_\varepsilon \le C \|\nabla(u^\varepsilon - u(k))\|_\varepsilon \le C \, \varepsilon^{k+2}$$

$$\|p^\varepsilon - p(k)\|_\varepsilon \le C \, \varepsilon^{k+1}$$

3. Justification of the Kirchoff's law

Often, in real-life situations, two (or several) pipes are interconnected (for instance watering systems and water-works are networks of thin pipes). Also, at some point, a pipe can separate in two or several pipes, not necessarily having the same profile (as for example in case of blood vessels or, again, water-works). Such multiple pipes systems may be as small as two pipes separating or rejoining or as complex as several hundreds of interconnected pipes forming a massive network. In either case, the basic principles of analysis are the same (although the complexity of the computation depends on the complexity of the system).

Indeed, in each pipe of our system the flow basically remains Poiseuille flow (at least far from the junction). To know the form of that Poiseuille solution we need the pressure drop along the pipe. The value of the pressure on one end of the pipe is prescribed by our boundary condition, but to know the value of the pressure drop between the pipe's end and the junction we need to know the value of the pressure at the junction point as well. That value is given by an explicit formula (16), frequently refered as the Kirchoff's law, in analogy with electric circuits. Such behaviour is confirmed by our asymptotic study of the flow, as $\varepsilon \to 0$. In vicinity of the junction the behaviour is somewhat different and an interior layer appears. In that layer the flow behaves as the rescaled solution of the Leray's problem (18)-(21) chosen to match the exterior Poiseuille solution. With those two flows (Poiseuille's and Leray's) we construct a matched asymptotic expansion that approximates our flow with an error smaller than $C\varepsilon\sqrt{\varepsilon\log(1/\varepsilon)}$, for the velocity and for the pressure. For the velocity gradient the error is of order $\sqrt{\varepsilon\log(1/\varepsilon)}$. If we leave out the interior Leray's problem and take just the Poiseuille's solutions, we lose the accuracy of our approximation and the error estimate is only of order $\sqrt{\varepsilon\log(1/\varepsilon)}$. The estimate for the velocity gradient is, in that case, lost.

3.1 The junction

We begin by studying the flow through one junction and we get a construction applicable to more complex systems.

3.2 Position of the problem

3.2.1 The geometry. We suppose for the simplicity that the cross-sections of all pipes are circular even though our method can easily be generalized (see [17]). The fluid domain Ω_ε consists of m pipes $\mathcal{O}_i^\varepsilon$, $i = 1, 2, \dots, m$ with lengths L_i connected at junction point O, where we chose to put the origin. The cross-section of i-th pipe is the circle $\mathcal{C}_i^\varepsilon$ with radius $\varepsilon\, r_i$, where $\varepsilon \ll 1$ is our small parameter. Now our pipes, in (possibly) different coordinate systems,

$(\mathbf{e}_k^i)_{k=1,2,3}$, $i = 1, \ldots, m$ have the following form

$$\mathcal{O}_i^\varepsilon = \{x = (x_1^i, \tilde{x}^i) \in \mathbf{R}^3 \ , \ 0 < x_1^i < L_i \ , \ \tilde{x}^i = (x_2^i, x_3^i) \in \mathcal{C}_i^\varepsilon\} \ . \quad (8)$$

We denote by Γ_ε^i the pipe's lateral boundary, by $\Gamma_\varepsilon = \cup_{i=1}^m \Gamma_\varepsilon^i$ and by Σ_ε^i its exterior end.

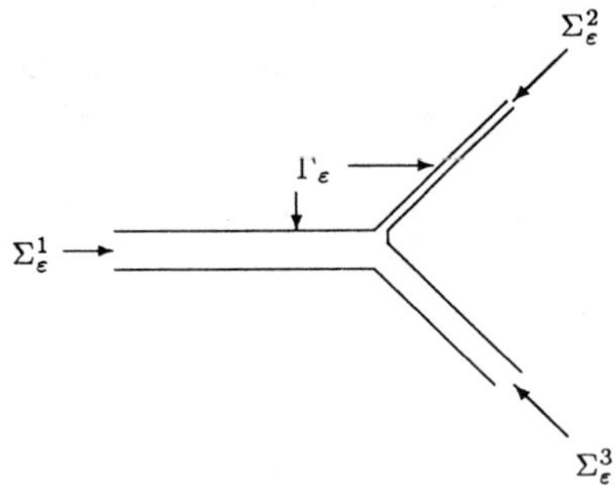

Figure 1. Three thin pipes in junction

3.2.2 The equations. Our flow can be described by the following boundary-value problem for the Navier-Stokes system:
For given pressures p_i, $i = 1, \ldots, m$ find $(u^\varepsilon, p^\varepsilon)$ such that

$$-\mu \ \Delta u^\varepsilon + (u^\varepsilon \nabla) u^\varepsilon + \nabla p^\varepsilon = 0 \ \text{ in } \Omega_\varepsilon \quad (9)$$

$$\operatorname{div} u^\varepsilon = 0 \ \text{ in } \Omega_\varepsilon \quad (10)$$

$$u^\varepsilon = 0 \ \text{ on } \Gamma_\varepsilon \quad (11)$$

$$\mathbf{e}_1^i \times u^\varepsilon = 0, \quad p^\varepsilon = p_i \ \text{ on } \Sigma_\varepsilon^i \ , \ i = 1, \ldots, m \ . \quad (12)$$

The same arguments about the existence and uniqueness of the solution, used in the previous section, can be used here. Again, we are interested in its asymptotic bahaviour as $\varepsilon \to 0$.

3.2.3 The matching procedure. In this section we derive the asymptotic behavior of the fluid far from the junction (Poiseuille region) and in vicinity of the junction (Leray's region) such that the two approximations are well matched between two regions. We use a version of the classical matching procedure (see [25, 27] as the general references on asymptotic methods as well as the matching techniques).

3.2.4 Poiseuille flow. Supposing that the pressure is continuous in the junction point O, the Poiseuille flow in i-th pipe $\mathcal{O}_i^\varepsilon$ reads

$$P_i^\varepsilon = \frac{1}{L_i}(p_i - q)\, x_1^i + q \tag{13}$$

$$U_i^\varepsilon = \varepsilon^2\, \mathcal{U}_i^\varepsilon(x) = \varepsilon^2\, \frac{w^i(\tilde{x}^i/\varepsilon)}{L_i}(q - p_i)\mathbf{e}_1^i \,, \tag{14}$$

where $w^i(y)$, $y = (y_1, y_2)$ is the solution of the i-th auxiliary problem on the cross-section

$$-\mu\, \Delta w^i = 1 \ \text{ for } |y| < r_i, \quad w^i = 0 \ \text{ for } |y| = r_i \tag{15}$$

i.e.

$$w^i(y) = \frac{1}{4\,\mu}(r_i^2 - |y|^2)\,.$$

and q is the value of the pressure at the junction point.

Now the incompressibility of our flow implies that

$$\sum_{i=1}^m \int_{\Sigma_\varepsilon^i} U_i^\varepsilon \cdot \mathbf{e}_1^i = 0\,,$$

leading to the junction condition

$$q = \frac{\sum_{i=1}^m p_i\, \frac{\langle w^i \rangle}{L_i}}{\sum_{i=1}^m \frac{\langle w^i \rangle}{L_i}}\,, \tag{16}$$

where $\langle w^i \rangle = \int_{|y|<r_i} w^i = \frac{\pi\, r_i^4}{8\mu}$. If all the pipes have the same length and the same cross-section, then q is simply the mean value of p_i-s, i.e. $q = \frac{1}{m}\sum_{i=1}^m p_i$. Otherwise we get the general expression

$$q = \frac{\sum_{i=1}^m L_i^{-1}\, r_i^4\, p_i}{\sum_{i=1}^m L_i^{-1}\, r_i^4}\,.$$

3.2.5 Leray's flow. We denote the rescaled junction by $\mathcal{G}$ and its unbounded outlets by $\mathcal{O}_i$. More precisely we define

$$\mathcal{O}_i = \{y = (y_1^i, \tilde{y}^i) \in \mathbf{R}^3 \,,\ 0 < y_1^i \,,\ \tilde{y}^i = (y_2^i, y_3^i)\,, |\tilde{y}^i| < r_i\} \tag{17}$$

and

$$\mathcal{G} = \cup_{i=1}^m \mathcal{O}_i\,.$$

In the vicinity of junction, in an interior layer

$$\mathcal{G}_\varepsilon = \cup_{i=1}^m \mathcal{G}_\varepsilon^i\,,$$

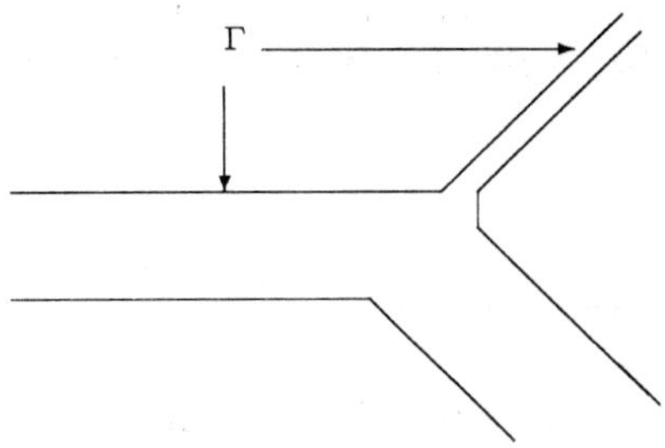

Figure 2. Interior layer $\mathcal{G}$ in vicinity of junction

where

$$\mathcal{G}_\varepsilon^i = \{x = (x_1^i, x_2^i, x_3^i) \in \mathcal{O}_i^\varepsilon \ ; \ 0 < x_1^i < \varepsilon \log(1/\varepsilon)\}$$
$$\omega_i^\varepsilon = \{x = (x_1^i, x_2^i, x_3^i) \in \mathcal{O}_i^\varepsilon \ ; \ \varepsilon \log(1/\varepsilon) < x_1^i < L_i\}$$

by passing to the fast variable $y = x/\varepsilon$ we get the Leray's problem:

$$-\mu\,\Delta V + \nabla Q = 0 \ \text{ in } \mathcal{G} \tag{18}$$
$$V = 0 \ \text{ on } \partial\mathcal{G} \tag{19}$$
$$\operatorname{div} V = 0 \ \text{ in } \mathcal{G} \tag{20}$$
$$\lim_{|y|\to+\infty} V = U^i \equiv \frac{w^i(\tilde{y}^i)}{L_i}(q - p_i)\mathbf{e}_1^i \ \text{ in } \mathcal{O}_i \ , \ i = 1, 2 \ldots, m \ , \tag{21}$$

Obviously the condition (21) is chosen to match the exterior Poiseuille approximation, far from the junction. The problem (18)-(21) is the linear Leray's problem and, unlike the nonlinear one, it has a unique solution for any choice of pressures p_i (see e.g. [1, 7]). Furthermore the solution (V, Q) decays exponentially as $|y| \to +\infty$ in the sense that

$$|V(y) - U^i(\tilde{y}^i)| + |\nabla V(y) - \nabla U^i(\tilde{y}^i)| \le C\, e^{-\tau\,|y|} \ \text{ in } \mathcal{G}_i \tag{22}$$

and Q (which is determined up to a constant) can be chosen such that

$$|Q(y) - \frac{1}{L_i}(p_i - q)\, y_1^i| \le C\, e^{-\tau\,|y|} \ \text{ in } \mathcal{G}_i \ , \tag{23}$$

(see [7], section VI.2). Now our approximation in $\mathcal{G}_\varepsilon$ reads

$$V^\varepsilon(x) = \varepsilon^2\, \mathcal{V}^\varepsilon(x) = \varepsilon^2\, V(x/\varepsilon), \quad Q^\varepsilon(x) = \varepsilon\, \mathcal{Q}^\varepsilon(x) = \varepsilon\, Q(x/\varepsilon) \ .$$

Remark 3. *In the Leray's problem (18)-(21) we could impose the condition at infinity only on the velocity (21) and not on the pressure. Fortunately we get, a posteriori, that the Leray's pressure also tends to the Poiseuille's pressure. This nice gift, offered to us by the theory of Leray's problems, will allow to match the velocity and the pressure at the same time.*

3.2.6 Matching. We can now define our matched asymptotic approximation as

$$\mathcal{W}^\varepsilon = \begin{cases} \mathcal{U}_i^\varepsilon & \text{in } \omega_i^\varepsilon \\ \mathcal{V}^\varepsilon & \text{in } \mathcal{G}_\varepsilon \end{cases}, \quad \Pi^\varepsilon = \begin{cases} P_i^\varepsilon & \text{in } \omega_i^\varepsilon \\ Q^\varepsilon & \text{in } \mathcal{G}_\varepsilon \end{cases} \tag{24}$$

3.2.7 The error estimate. As explained in the previous section, we express the error estimate in the rescaled norm

$$\|f\|_\varepsilon = \varepsilon^{-1} \|f\|_{L^2(\Omega_\varepsilon)} ,$$

which is an appropriate L^2 norm for shrinking domains. It has been proved in [18] (see also [17]) that the following estimates hold:

Theorem 2. *Let $\mathcal{M}_\varepsilon = \cup_{j=1}^m \omega_j^\varepsilon \cup \mathcal{G}_\varepsilon$. Then there exists a constant $C > 0$ such that*

$$\|\varepsilon^{-2} u^\varepsilon - \mathcal{W}^\varepsilon\|_\varepsilon \leq C\varepsilon \sqrt{\varepsilon \, log(1/\varepsilon)} \tag{25}$$

$$\varepsilon^{-1} \|\nabla(\varepsilon^{-2} u^\varepsilon - \mathcal{W}^\varepsilon)\|_{L^2(\mathcal{M}_\varepsilon)} \leq C \sqrt{\varepsilon \, log(1/\varepsilon)} \tag{26}$$

$$\|p^\varepsilon - \Pi^\varepsilon\|_\varepsilon \leq C\varepsilon \sqrt{\varepsilon \, log(1/\varepsilon)} \, . \tag{27}$$

For the pure Poiseuille solution, without the interior layer corrector $(V^\varepsilon, Q^\varepsilon)$, we still have the convergence but the error estimate is worst. We can prove that the interior layer corrector improves the approximation in the L^2 norm with an order ε :

$$\varepsilon^{-1} \|\frac{u^\varepsilon}{\varepsilon^2} - \mathcal{U}_i^\varepsilon\|_{L^2(\mathcal{O}_\varepsilon^i)} \leq C\sqrt{\varepsilon} \log(1/\varepsilon)$$

$$\varepsilon^{-1} \|p^\varepsilon - P_i^\varepsilon\|_{L^2(\mathcal{O}_\varepsilon^i)} \leq C\sqrt{\varepsilon} \log(1/\varepsilon), \quad i = 1, \ldots, m \, .$$

The estimate for the velocity gradient is left out in case of the pure Poiseuille approximation. An easy computation gives that

$$\varepsilon^{-1} \|\nabla \mathcal{W}^\varepsilon\|_{L^2(\mathcal{G}_\varepsilon)} = O(\varepsilon^{-1/2})$$

proving that the interior layer corrector is important for the gradient estimate and there is no hope to get one without it.

Remark 4. *The generalization of the Poiseuille flow and the Kirchoff law on compressible flow could be found in [17]. Its rigorous justification is still an open question. In fact, there is no rigorous justifications of any lower-dimensional model for compressible flows.*

3.2.8 An application. Let us apply the Poiseuille-Kirchoff method to a network with two junctions (and four elbows) as on figure 3. We have four pipes enumerated by $0, \ldots, 3$ having radii $\varepsilon \, r_i$, $i = 0, \ldots, 3$ and lengths L_i.

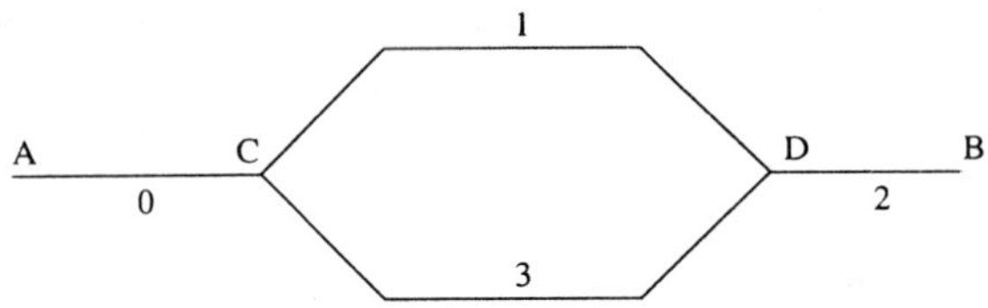

Figure 3.

We denote by $k_i = L_i^{-1}\ r_i^4$ and by p_α , $\alpha = A, \ldots, D$ the pressure at point $\alpha \in \{A, B, C, D\}$. We suppose that the pressures p_A and p_B are known. Now at first junction C, using (16) we obtain

$$p_C = \frac{k_0\, p_A + (k_1 + k_3)\, p_D}{k_0 + k_1 + k_3}\ .$$

For the second junction we analogously have

$$p_D = \frac{k_2\, p_B + (k_1 + k_3)\, p_C}{k_1 + k_2 + k_3}\ .$$

Combining those two equations we arrive at

$$\begin{aligned} p_C &= \frac{k_0\,(k_1 + k_2 + k_3)\, p_A + k_2\,(k_1 + k_3)\, p_B}{k_0\, k_2 + (k_0 + k_2)\,(k_1 + k_3)} \\ p_D &= \frac{k_2\,(k_0 + k_1 + k_3)\, p_B + k_0\,(k_1 + k_3)\, p_A}{k_0\, k_2 + (k_0 + k_2)\,(k_1 + k_3)}\ . \end{aligned}$$

Now the pressure drops between each two points are easy to compute:

$$\begin{aligned} p_C - p_D &= \frac{k_0\ k_2}{k_0\, k_2 + (k_0 + k_2)(k_1 + k_3)}\,(p_A - p_B) \\ p_C - p_A &= \frac{k_2\ (k_1 + k_3)}{k_0\, k_2 + (k_0 + k_2)(k_1 + k_3)}\,(p_B - p_A) \\ p_D - p_B &= \frac{k_0(k_1 + k_3)}{k_0\, k_2 + (k_0 + k_2)(k_1 + k_3)}\,(p_A - p_B)\ . \end{aligned}$$

Poiseuille solutions in each pipe can now be directly written as in (13)-(14).

4. Curved pipes

In our network some pipes can be curved. It turns out that curvedness of the pipe affects only the flow profile but not the mean flow. The main part of the flow remains in the Poiseuille form, but the flow profile is not perfectly parabolic any more and it should be corrected. In each point of the central curve of the pipe we define the local coordinate system consisting of the tangent, the normal and the binormal. Now for the Poiseuille velocity in the pipe (which, of course, has the direction of the tangent) we compute the tangential, the normal and the binormal corrector. The tangential corrector is proportional to the curvature of the central curve of the pipe while the normal and the binormal ones are proportional to its torsion. Those correctors are of lower order and they affect only the flow profile and not the mean flow. Therefore they should be used only for the precise analysis of the pipes flow.

4.1 Position of the problem

We define the geometry of the curved pipe and we write Navier-Stokes system (9) -(12) in a form appropriate for our asymptotic analysis. To do so we use the local coordinate system attached to the central curve of the pipe.

4.1.1 The Geometry. We suppose that γ, the central curve of the pipe, is the smooth generic curve of class C^5. For simplicity we assume that γ is parameterized by its arc length $y_1 \in [0, \ell]$. We denote by $\mathbf{x} : [0, \ell] \to \mathbf{R}^3$ its natural parametrization. In each point $\mathbf{x}(y_1)$, $y_1 \in [0, \ell]$ of the curve γ we define its curvature as $\kappa(y_1) = |\mathbf{x}''(y_1)|$ and the Frenet's basis, $\mathbf{t}(y_1) = \mathbf{x}'(y_1)$ (the tangent), $\mathbf{n}(y_1) = \frac{\mathbf{x}''(y_1)}{\kappa(y_1)}$ (the normal)[3] and $\mathbf{b}(y_1) = \mathbf{t}(y_1) \times \mathbf{n}(y_1)$ (the binormal). As usual, we denote by $\tau(y_1) = -|\mathbf{b}'(y_1)|$ the torsion.

We define an undeformed pipe with cross-section $\mathcal{C}^\varepsilon$ as

$$\mathcal{P}_\varepsilon = \{y = (y_1, y_2, y_3) \in \mathbf{R}^3 \ ; \ y_1 \in [0, \ell] \ , \ (y_2, y_3) \in \mathcal{C}^\varepsilon\} \ .$$

As this is the case of single tube we do not need r_i any more and we can denote by $\mathcal{C}^\varepsilon$ the circle with radius ε, where $\varepsilon \ll 1$. We define the mapping $\Phi : \mathcal{P}_\varepsilon \to \mathbf{R}^3$ by

$$\Phi(y) = \mathbf{x}(y_1) + y_2\mathbf{n}(y_1) + y_3\mathbf{b}(y_1) \ .$$

In order to have the local injectivity of Φ, we suppose that ε is sufficiently small, more precisely, we assume that

$$\varepsilon\,|\kappa(y_1)| < 1/2, \quad y_1 \in [0, \ell] \ . \tag{28}$$

Indeed, we will prove in section 2 that $\det \nabla\Phi = 1 - y_2\kappa(y_1)$ so that the supposition (28) assures the local injectivity of Φ. We are now ready to define the curved pipe with the central curve γ and the cross section $\mathcal{C}^\varepsilon$ by

$$P_\varepsilon = \Phi(\mathcal{P}_\varepsilon) \ .$$

Obviously, the curve γ is passing the centroid of each cross section of the pipe. We use the symbols Σ_0^ε and Σ_1^ε to denote the ends of the pipe and Γ_ε to denote its side.

4.1.2 Geometric tools. In this section we study the curve γ and we develop some geometric tools that are used in our method. The idea is to define the appropriate curvilinear coordinate system in P_ε and to write the Navier-Stokes system in such coordinates. For more details about geometric tools and an interesting generalization to nongeneric curves see [10].

- Covariant basis is, in fact, the gradient of the mapping Φ, i.e. it consists of the vectors $\mathbf{a}_i(y) = \frac{\partial \Phi}{\partial y_i}(y)$. In our case, we have

$$\mathbf{a}_1(y) = [1 - y_2\kappa(y_1)]\mathbf{t}(y_1) - y_3\tau(y_1)\mathbf{n}(y_1) + y_2\tau(y_1)\mathbf{b}(y_1) , \quad (29)$$

while $\mathbf{a}_2(y) = \mathbf{n}(y_1)$, $\mathbf{a}_3(y) = \mathbf{b}(y_1)$. The covariant metric tensor $[\underline{\mathbf{M}}]_{ij} = g_{ij} = \mathbf{a}_i \cdot \mathbf{a}_j$ is now easily computed as $\underline{\mathbf{M}}(y) = \mathbf{I} + \mathbf{H}$, where

$$\mathbf{H} = \begin{bmatrix} -2y_2\kappa(y_1) + y_2^2\kappa(y_1)^2 + y_3^2\tau(y_1)^2 + y_2^2\tau(y_1)^2 & -y_3\tau(y_1) & y_2\tau(y_1) \\ -y_3\tau(y_1) & 0 & 0 \\ y_2\tau(y_1) & 0 & 0 \end{bmatrix}$$

and its determinant equals

$$g_\varepsilon(y) = \det \underline{\mathbf{M}}(y) = [1 - y_2\kappa(y_1)]^2 .$$

- The contravariant basis is the dual basis to $\mathbf{a}_i$ basis, i.e. it is defined by the relation $\mathbf{a}^i \cdot \mathbf{a}_j = \delta_{ij}$. It is easily computed as

$$\mathbf{a}^1(y) = \frac{\mathbf{t}(y_1)}{\sqrt{g_\varepsilon(y)}} , \quad \mathbf{a}^2(y) = \mathbf{n}(y_1) + \frac{y_3\tau(y_1)}{\sqrt{g_\varepsilon(y)}}\mathbf{t}(y_1) ,$$

$$\mathbf{a}^3(y) = \mathbf{b}(y_1) - \frac{y_2\tau(y_1)}{\sqrt{g_\varepsilon(y)}}\mathbf{t}(y_1) . \quad (30)$$

The contravariant metric tensor $[\overline{\mathbf{M}}]_{ij} = g_\varepsilon^{ij} = \mathbf{a}^i \cdot \mathbf{a}^j$ now has the form

$$\overline{\mathbf{M}}(y) = \frac{1}{g_\varepsilon(y)} \begin{bmatrix} 1 & y_3\tau(y_1) & -y_2\tau(y_1) \\ y_3\tau(y_1) & g_\varepsilon(y) + y_3^2\tau(y_1)^2 & -y_2y_3\tau(y_1)^2 \\ -y_2\tau(y_1) & -y_2y_3\tau(y_1)^2 & g(y)_\varepsilon + y_2^2\tau(y_1)^2 \end{bmatrix} .$$

Obviously

$$\det \overline{\mathbf{M}}(y) = \frac{1}{g_\varepsilon(y)} .$$

- The asymptotic behaviour of the above basis is

$$\mathbf{a}_1 = \mathbf{t} + O(\varepsilon), \quad \mathbf{a}^1 = \mathbf{t} + O(\varepsilon) \quad (31)$$

$$\mathbf{a}_2 = \mathbf{n} + O(\varepsilon), \quad \mathbf{a}^2 = \mathbf{n} + O(\varepsilon) \quad (32)$$

$$\mathbf{a}_3 = \mathbf{b} + O(\varepsilon), \quad \mathbf{a}^3 = \mathbf{b} + O(\varepsilon) . \quad (33)$$

Therefore we may think about our covariant and contravariant basis as of $(\mathbf{t}, \mathbf{n}, \mathbf{b})$ basis. We also find the asymptotic behaviour of $\overline{\mathbf{M}}$ and Christoffel symbols Γ^{ℓ}_{ij} that will be important in the sequel. Using the notations $z_\alpha = y_\alpha/\varepsilon$, $\alpha = 2, 3$ we get

$$\overline{\mathbf{M}} = \mathbf{I} + \varepsilon \begin{bmatrix} 2z_2\kappa & z_3\tau & -z_2\tau \\ z_3\tau & 0 & 0 \\ -z_2\tau & 0 & 0 \end{bmatrix} + O(\varepsilon^2) . \tag{34}$$

- Christoffel symbols, defined by $\Gamma^{\ell}_{ij} = \mathbf{a}^{\ell} \frac{\partial \mathbf{a}_j}{\partial y_i}$, can be computed as

$$\Gamma^1_{11} = \frac{1}{\sqrt{g_\varepsilon}}(\kappa\tau y_3 - \kappa' y_2), \quad \Gamma^1_{12} = \Gamma^1_{21} = -\frac{\kappa}{\sqrt{g_\varepsilon}}$$

$$\Gamma^2_{11} = (1 - \kappa y_2)\kappa - \tau^2 y_2 - \tau' y_3 + \frac{1}{\sqrt{g_\varepsilon}}(\tau\kappa' y_2 y_3 + \kappa\tau^2 y_3^2)$$

$$\Gamma^2_{12} = \Gamma^2_{21} = -\kappa\tau \frac{y_3}{\sqrt{g_\varepsilon}}, \quad \Gamma^2_{13} = \Gamma^2_{13} = -\tau$$

$$\Gamma^3_{11} = y_2\tau' - y_3\tau^2 + \frac{1}{\sqrt{g_\varepsilon}}(\tau\kappa' y_2^2 - y_2 y_3 \tau^2 \kappa)$$

$$\Gamma^3_{12} = \Gamma^3_{21} = \tau + \kappa\tau \frac{y_2}{\sqrt{g_\varepsilon}} ,$$

while all the other Γ^k_{ij} are equal to zero. For nonzero Christoffel's symbols we obtain

$$\begin{array}{ll} \Gamma^2_{13} = -\tau, & \Gamma^1_{11} = O(\varepsilon) \\ \Gamma^1_{12} = -\kappa + O(\varepsilon), & \Gamma^3_{11} = O(\varepsilon) \\ \Gamma^3_{12} = \tau + O(\varepsilon), & \Gamma^2_{12} = O(\varepsilon) \\ \Gamma^2_{11} = \kappa + O(\varepsilon) . & \end{array} \tag{35}$$

Let $\mathcal{U}_\varepsilon(y) = u^\varepsilon(\mathbf{x})$ and $Q^\varepsilon(y) = p^\varepsilon(\mathbf{x})$. Denoting by $\mathcal{U}^\ell_\varepsilon = \mathcal{U}_\varepsilon \cdot \mathbf{a}^\ell$, the Navier-Stokes system can now be written in the form (see Michal [20]):

$$\begin{aligned} -\mu \sum_{i,j=1}^{3} g^{ij} \Bigg\{ \frac{\partial^2 \mathcal{U}^n_\varepsilon}{\partial y_i \partial y_j} + \sum_{m=1}^{3} \Bigg[\Gamma^n_{mj} \frac{\partial \mathcal{U}^m_\varepsilon}{\partial y_i} + \Gamma^n_{mi} \frac{\partial \mathcal{U}^m_\varepsilon}{\partial y_j} - \Gamma^m_{ij} \frac{\partial \mathcal{U}^n_\varepsilon}{\partial y_m} + \\ + \Bigg(\frac{\partial \Gamma^n_{mi}}{\partial y_j} + \sum_{k=1}^{3} \Gamma^n_{kj}\Gamma^k_{mi} - \Gamma^n_{mk}\Gamma^k_{ij} \Bigg) \mathcal{U}^m_\varepsilon \Bigg\} + \\ + \sum_{s=1}^{3} \Bigg(\mathcal{U}^s_\varepsilon \frac{\partial \mathcal{U}^n_\varepsilon}{\partial y_s} + \sum_{h=1}^{3} \Gamma^n_{hs} \mathcal{U}^h_\varepsilon \mathcal{U}^s_\varepsilon \Bigg) + \sum_{k=1}^{3} g^{nk} \frac{\partial Q_\varepsilon}{\partial y_k} = 0 \end{aligned} \tag{36}$$

$$\sum_{i=1}^{3} \frac{\partial}{\partial y_i} (\sqrt{g}\, \mathcal{U}^i_\varepsilon) = 0, \quad \text{in } \mathcal{P}_\varepsilon, \quad n = 1, 2, 3 . \tag{37}$$

4.2 Asymptotic expansion

We seek an asymptotic expansion for $(\mathcal{U}_\varepsilon, \mathcal{Q}_\varepsilon)$ in the form:

$$\begin{cases} \mathcal{U}_\varepsilon(y) = \varepsilon^2\, \mathcal{U}_0(y_1, \frac{y_2}{\varepsilon}, \frac{y_3}{\varepsilon}) + \varepsilon^3\, \mathcal{U}_1(y_1, \frac{y_2}{\varepsilon}, \frac{y_3}{\varepsilon}) + \cdots \\ \mathcal{Q}_\varepsilon(y) = \mathcal{Q}_0(y_1) + \varepsilon\, \mathcal{Q}_1(y_1, \frac{y_2}{\varepsilon}, \frac{y_3}{\varepsilon}) + \cdots \end{cases} \tag{38}$$

After noticing that for $z_\alpha = \frac{y_\alpha}{\varepsilon}$, $\alpha = 2, 3$ a simple rule for differentiation holds

$$\frac{\partial}{\partial y_\alpha} = \frac{1}{\varepsilon}\frac{\partial}{\partial z_\alpha} .$$

By substitution of (38) in (36), after collecting equal powers of ε, we obtain

$$\begin{cases} 1: & -\mu\Delta_{z'}\mathcal{U}_0 + \left(\frac{\partial \mathcal{Q}_0}{\partial y_1}, \frac{\partial \mathcal{Q}_1}{\partial z_2}, \frac{\partial \mathcal{Q}_1}{\partial z_3}\right) = 0 \\ \varepsilon: & \operatorname{div}_{z'} \mathcal{U}_0 = 0 \\ \varepsilon^2: & \frac{\partial \mathcal{U}_0^1}{\partial y_1} + \operatorname{div}_{z'} \mathcal{U}^1 = 0 , \end{cases} \tag{39}$$

where

$$z' = (z_2, z_3), \quad \operatorname{div}_{z'} V = \frac{\partial V^2}{\partial z_2} + \frac{\partial V^3}{\partial z_3}, \quad \Delta_{z'} V = \frac{\partial^2 V}{\partial z_2^2} + \frac{\partial^2 V}{\partial z_3^2} ,$$

for a vector function $V = (V^1, V^2, V^3)$.

By simple integration of the last equation in (39) we get

$$\frac{\partial}{\partial y_1}\left(\int_{|z'|<1} \mathcal{U}_0^1\right) = 0 . \tag{40}$$

The system (39)-(40) (with corresponding boundary conditions) can be solved by taking

$$\begin{aligned} \mathcal{U}_0^\alpha &= 0, \quad \alpha = 2, 3 \\ \mathcal{Q}_1 &= \mathcal{Q}_1(y_1) = (p_1 - p_0)\,\tfrac{y_1}{\ell} + p_0 \\ \mathcal{U}_0^1 &= -w(z')\tfrac{\partial \mathcal{Q}_0}{\partial y_1} = \tfrac{w(z')}{\ell}\,(p_1 - p_0) , \end{aligned} \tag{41}$$

where $w(z')$ denotes the solution of the auxiliary problem (15) posed on the cross section $|z'| < 1$, i.e.

$$w(z') = \frac{1}{4\,\mu}(1 - |z'|^2) .$$

The second term is

$$\varepsilon : \quad -\mu\Delta\mathcal{U}_1 - \mu\Big(-3\kappa\frac{\partial\mathcal{U}_0^1}{\partial z_2}, -2\tau\frac{\partial\mathcal{U}_0^1}{\partial z_3}, 2\tau\frac{\partial\mathcal{U}_0^1}{\partial z_2}\Big) + \tag{42}$$

$$+\Big(\frac{\partial\mathcal{Q}_1}{\partial y_1} + 2\kappa z_2\frac{\partial\mathcal{Q}_0}{\partial y_1}, \frac{\partial\mathcal{Q}_2}{\partial z_2} + \tau z_3\frac{\partial\mathcal{Q}_0}{\partial y_1}, \frac{\partial\mathcal{Q}_2}{\partial z_3} - \tau z_2\frac{\partial\mathcal{Q}_0}{\partial y_1}\Big) = 0$$

$$\varepsilon^3 : \quad \mathrm{div}_{z'}\,\mathcal{U}_2 + \frac{\partial}{\partial y_1}(\mathcal{U}_1^1 - \kappa\, z_2\,\mathcal{U}_0^1) = 0 \tag{43}$$

and, integrating the incompressibility equation (43) over $S = \{z' \in \mathbf{R}^2 : |z'| < 1\}$

$$\frac{\partial}{\partial y_1}\left(\int_{|z'|<1}\mathcal{U}_1^1 - \kappa\frac{p_0 - p_1}{\ell}\int_{|z'|<1} z_2 w(z')dz'\right) = 0\,. \tag{44}$$

We seek $\mathcal{U}_1^1$ in the form

$$\mathcal{U}_1^1(y_1, z') = \chi(z')\frac{p_0 - p_1}{\ell}\kappa(y_1)\,. \tag{45}$$

where χ is the solution of the second auxiliary problem on S

$$-\mu\Delta\chi = 2z_2 - 3\mu\frac{\partial w}{\partial z_2} \ \text{ for } |z'| < 1, \quad \chi = 0 \ \text{ for } |z'| = 1\,. \tag{46}$$

We notice that

$$\int_S \chi = \mu\int_S \nabla w\nabla\chi = 2\int_S z_2 w - 3\mu\int_S w\frac{\partial w}{\partial z_2} = 2\int_S z_2 w = 0$$

so that the mean value $\int_S \mathcal{U}_1^1 = 0$, i.e. it does not affect the mean velocity. For $\mathcal{U}_1^\alpha$, $\alpha = 2, 3$ and $\mathcal{Q}_2$ we put

$$\mathcal{U}_1^\alpha(y_1, z') = v_\alpha(z')\frac{p_0 - p_1}{\ell}\tau(y_1) \tag{47}$$

$$\mathcal{Q}_2(y_1, z') = q(z')\frac{p_0 - p_1}{\ell}\tau(y_1)\,, \tag{48}$$

where (v, q) is the solution of the Stokes auxiliary problem posed on the cross-section S:

$$-\mu\Delta_{z'}\, v + \nabla_{z'}\, q = 2\mu\Big(-\frac{\partial(w - \frac{z_3^2}{4\mu})}{\partial z_3}, \frac{\partial(w - \frac{z_2^2}{4\mu})}{\partial z_2}\Big) = \tag{49}$$

$$= 2\mu\,\mathrm{curl}(w - \frac{|z'|^2}{4\mu}) \ \text{ for } |z'| < 1$$

$$\mathrm{div}_{z'}\, v = 0 \ \text{ for } |z'| < 1, \quad v = 0 \ \text{ for } |z'| = 1\,. \tag{50}$$

Those two auxiliary problems can be solved explicitely by putting

$$\chi(\rho,\varphi) = -\frac{7}{16\mu}(\rho^3-1)\cos\varphi$$

and

$$\begin{aligned} v_1(\rho,\varphi) &= -\frac{1}{8\mu}(\rho^3-\rho)\sin\varphi \\ v_2(\rho,\varphi) &= \frac{1}{8\mu}(\rho^3-\rho)\cos\varphi \\ q &= 0\,, \end{aligned}$$

where (ρ,φ) are polar coordinates on the cross-section $|z'|<1$.

For the mean value we get, because of the symmetry, that $\int_{|z'|<1} v_\alpha = 0$ so that for circular tube in the second order approximation the flexion and the torsion affect only the flow profile. We can now write our approximation in the form

$$\begin{aligned} \mathcal{U}_\varepsilon(y) &= \frac{p_0-p_1}{\ell\,\mu}\{\varepsilon^2\,\frac{1}{4}(1-\rho^2)\mathbf{t}(y_1) + \\ &+\varepsilon^3\,[(\frac{7}{16}+\frac{\rho}{4}-\frac{11\,\rho^3}{16})\cos\varphi\,\kappa(y_1)\,\mathbf{t}(y_1) + \\ &+\tau(y_1)\,\frac{\rho-\rho^3}{8}\,(-\sin\varphi\,\mathbf{n}(y_1)+\cos\varphi\,\mathbf{b}(y_1)\,)]\} \\ \mathcal{Q}_\varepsilon(y) &= \frac{p_1-p_0}{\ell}\,y_1+p_0 \end{aligned}$$

The error estimate in $\|\cdot\|_\varepsilon$ norm is of order ε for the pure Poiseuille solution and of order ε^2 if we take into account the asymptotic effects of flexion and torsion of the pipe, i.e. with corrector $\mathcal{U}^1$ (see [15] for details). Part of those results were generalized to the case of non-Newtonian fluids in [13], using the two-scale convergence introduced in [14].

5. Conclusion

The engineering approach to the flow through pipes (including junctions and elbows) is based, either on the Poiseuille-Kirchoff approximation described in our paper, or on the empirical Darcy-Weisbach formula for computing the energy loss due to the friction (see e.g. [23]). Both methods, widely used in applications, seam to be giving results fairly consistent with the experience. Unlike the Poiseuille-Kirchoff method, the order of accuracy of the Darcy-Weisbach approach is not clear, since it is not derived from the basic conservation laws (i.e. the Navier-Stokes system). Furthermore, it allows to compute only the mean velocity (the mass flux) and not the velocity profile in the pipe. Even though the pipes are thin, for more precise computations (as, for instance, in hemodynamics) the velocity profile is needed. In this paper we have presented more accurate asymptotic approximations. Such approach has two levels of accuracy. If we are interested only in the global flow and we do not have to analyze precisely the local flow in vicinity of junction, we only have to take the Poiseuille-Kirchoff approximation (13), (14) with the junction pressure computed from (16). If we want to study more precisely the local flow around the junction, where the peculiar behaviour of the fluid, with eddies and rapid changes of the velocity as well as of the pressure, has been experimentally observed, we need to add the interior layer corrector of the Leray's type by solving (18)-(21).

In case of curved pipe the flow profile is not perfectly parabolic any more. The Poiseuille flow has to be corrected by a lower order term containing the effects of flexion and torsion of the pipe. As presented, those correctors can be explicitely computed. Such, more accurate approximation, is particularly interesting if ε (the ratio between the length and the thickness of the pipe) is not too small (e.g. $\varepsilon = 10^{-1}$).

Notes

1. The formula (1) is sometimes refered as the Hagen-Poiseuille flow. According to Landau and Lifchitz [11] only the average $\overline{v} = \pi R^4 \Delta p / 8\mu L$ of that formula was empirically found by G.Hagen in 1839 and, independently, by J.L.M.Poiseuille in 1840, while its theoretical derivation via (1) is due to G.G.Stokes (1845).
2. Convergence in such norm is also called the strong two-scale convergence on Ω_ε. Weak convergence on Ω is called the two-scale convergence on Ω_ε (see [14]).
3. Assuming that $\mathbf{n}$ is extended by continuity in points where curvature is 0.

References

[1] Amick C.J., Steady Solutions of the Navier-Stokes Equations in Unbounded Channels and Pipes, Ann.Sc.Norm.Sup.Pisa, IV (3) (1977), 473-513.

[2] Bayada G., Chambat M., The Transition Between the Stokes Equation and the Reynolds Equation: a Mathematical Proof, Appl.Math.Optim., 14 (1986), 73-93.

[3] Ciarlet P., Plates and Junctions in Elastic Multi-structures, An Asymptotic Analysis, Masson, 1990.

[4] Ciarlet P., Destuynder P., A Justification of the Two-Dimensional Linear Plate Model, J. de Mécanique, 18 (1979), 315-344.

[5] Conca C., Murat F., Pironneau O., The Stokes and Navier-Stokes equations with boundary conditions involving the pressure, Japan J.Math., 20 (1994), 263-318.

[6] Duvnjak A., Marušić-Paloka E., Derivation of the Reynolds equation for lubrication of the rotating shaft, Arch. Math., t 36, No 4 (2000), 239-253.

[7] Galdi G.P., An Introduction to the Mathematical Theory of the Navier–Stokes Equations, I, II, Springer–Verlag, 1994.

[8] Gipouloux O., Marušić-Paloka E., Asymptotic behaviour of the incompressible Newtonian flow through thin constricted fracture, to appear in proceedings of *Multiscale problems in science and technology*, Dubrovnik 2000, Springer–Verlag.

[9] Heywood J.G., Rannacher R., Turek S., Artificial boundaries and flux and pressure conditions for the incompressible Navier-Stokes equations, Int.J.Numer.Methods Fluids, 22 (1996), 325-352.

[10] Jurak M., Tambača J., Linear curved rod model. General curve. Math.Models Methods Appl.Sci., Vol 11, No 7 (2001), 1237–1253.

[11] Landau L., Lifchitz E., Physique théorique: Mécanique des fluides, 3ème édition, Ellipses, 1994.

[12] Le Dret H., Problèmes variationnels dans les multi-domaines, Masson, Paris, 1991.

[13] Marušić S., The asymptotic behaviour of quasi-Newtonian flow through a very thin or a very long curved pipe, Asymptotic Anal. 26 (2001), 73-89.

[14] Marušić S., Marušić-Paloka E., Two scale convergence for thin domains and its applications to some lower-dimensional models in fluid mechanics, Asymptotic Anal., 23 (2000), 23-58.

[15] Marušić-Paloka E., The effects of flexion and torsion for the fluid flow through a curved pipe, Appl.Math.Optim., 44 (2001), 245-272.

[16] Marušić-Paloka E., Non-Newtonian flow through unbounded channels and pipes, Mathematical Models and Methods in Applied Sciences M3AS, Vol 10, No 9 (2000), 1425-1445.

[17] Marušić-Paloka E., Fluid flow through a network of thin pipes, C.R.Acad.Sci. Paris, Série II b, t 329 (2001),103-108.

[18] Marušić-Paloka E., Junction of pipes filled with viscous fluid, preprint, Department of Mathematics, University of Zagreb, submitted for publication

[19] Marušić-Paloka E., Marušić S., Decay of a fluid flow in a weakly permeable domain, ZAMM, 79 (1999) 6, 427-432.

[20] Michal A.D., Matrix and Tensor Calculus, John Wiley and Sons, New York, 1947.

[21] Nazarov S.A., Asymptotic solution of the Navier-Stokes problem on the flow of a thin layer of fluid, Sib.Math.J.,31(1990), 296-307.

[22] Nazarov S.A., Piletskas K.I., The Reynolds Flow in a Thin Three-Dimensional Channel, Lithuanian Math.J., 30(1991), 366-375.

[23] Street R.L., Watters G.Z., Vennard J.K., Elementary fluid mechanics, seventh edition, Wiley, 1996.

[24] Tutek Z., Aganović I., A Justification of the One-Dimensional Linear Model of Elastic Beam, Math. Methods Appl. Sci., 8 (1986), 502-515.

[25] Van Dyke M., Perturbation methods in fluid mechanics, The Parabolic Press, Stanford, 1975.

[26] Wanier G.H., A contribution to the hydrodynamics of lubrication, Quart.Appl.Math. 8 (1950), 1-32.

[27] Zeytounian R.Kh., Modélisation asymptotique en mécanique des fluides newtoniens, Collection SMAI, Springer Verlag, Berlin Heidelberg, 1994.

FIRST ORDER EIGENVALUE PERTURBATION THEORY AND THE NEWTON DIAGRAM

Julio Moro
Departamento de Matemáticas
Universidad Carlos III de Madrid, Spain
jmoro@math.uc3m.es

Froilán M. Dopico
Departamento de Matemáticas
Universidad Carlos III de Madrid, Spain
dopico@math.uc3m.es

Abstract First order perturbation theory for eigenvalues of arbitrary matrices is systematically developed in all its generality with the aid of the Newton diagram, an elementary geometric construction first proposed by Isaac Newton. In its simplest form, a square matrix A with known Jordan canonical form is linearly perturbed to $A(\varepsilon) = A + \varepsilon B$ for an arbitrary perturbation matrix B, and one is interested in the leading term in the ε-expansion of the eigenvalues of $A(\varepsilon)$. The perturbation of singular values and of generalized eigenvalues is also covered.

Keywords: eigenvalues, singular values, generalized eigenvalue problems, perturbation theory, asymptotic expansions, Newton diagram, Jordan canonical form.

1. Introduction

Eigenvalue perturbation theory has been an issue in applied mathematics since Lord Rayleigh laid its foundations in [25]. One of his calculations aimed at determining both the eigenfrequencies and eigenmodes of an oscillatory string with constant elasticity modulus and whose mass density was a small deviation of a constant value. This particular problem illustrates perfectly the typical setting of eigenvalue perturbation theory: the matrix or operator under study is assumed to be a slight deviation from some close, simpler matrix or operator for which the spectral problem is completely (and, in most cases, easily) solved. The given operator $\widetilde{A}$ is replaced by a neighboring operator

A whose eigenvalues and eigenvectors are known. Then, the influence of the difference $B = \widetilde{A} - A$ on the spectral objects is analyzed using an appropriate perturbation theory. The usual approach is introducing a perturbation parameter ε and considering the (matrix or) operator

$$A(\varepsilon) = A + \varepsilon B. \tag{1}$$

Its eigenvalues and eigenvectors depend on ε and *are assumed* to converge to the corresponding eigenvalues and eigenvectors as ε goes to zero. In a first stage of the analysis the leading terms of the ε-expansions of the spectral objects of $A(\varepsilon)$ are determined (hence the name of *first order* perturbation theory). In a second stage the convergence of these expansions is justified up to $\varepsilon = 1$. This paper shall be mainly focused on the first stage, although some mention will be made of the appropriate convergence results.

This general framework owes much to the formalism proposed by Schrödinger [33] in his approach to quantum mechanics. He described any observable of a quantum mechanical system as a selfadjoint operator H (the Hamiltonian in the case of the energy) defined in a certain Hilbert space. An isolated eigenvalue λ_0 and its corresponding normalized eigenvector x_0 are interpreted as a bound state x_0 with energy level λ_0. If the system is influenced by some external field, or if some previously neglected interaction has to be taken into account, a perturbation H_1 must be added to the operator H. The question arises of whether there is at least one bound state in the neighborhood of the unperturbed one. One also wants to calculate the bound states and energy levels of $H + H_1$ in the neighborhood of x_0, λ_0. Schrödinger was one of the first to take the approach (1), considering $H + \varepsilon H_1$ and *postulating* the analytic dependence of the perturbed spectral objects on the perturbation parameter ε. The explicit formulas he obtained for eigenvalues and eigenvectors are known as perturbation series in quantum mechanics. Even the splitting of multiple eigenvalues was studied in the first order approximation.

Schrödinger, however, did not prove the convergence of such expansions. That issue was finally settled for isolated eigenvalues of selfadjoint operators in Hilbert space by Rellich in a series of papers [26, 27, 28, 29, 30] which stimulated further advances in similar problems. On one hand, Sz.-Nagy [38] translated Rellich's arguments into the complex domain using the Cauchy-Riesz integral method. This led, in particular, to the extension of Rellich's results to the case of nonselfadjoint operators. On the other hand, Friedrichs [4] developed the perturbation theory of continuous spectra, which turned out to be extremely useful in scattering theory and quantum field theory. Meanwhile, the perturbation theory for one-parameter semigroups was developed by Hille and Phillips [23, 6], and a general framework for the perturbation theory of linear operators was presented by Kato in his well-known treatise [9].

As can be seen from this discussion of the early history of the subject, a great deal of effort was invested in developing eigenvalue perturbation theory for infinite-dimensional operators. In other words, first order eigenvalue perturbation theory began with a strong flavor of functional analysis. The essential ingredients, however, are purely finite-dimensional and can be presented in a strictly finite-dimensional setting without losing any of the mathematical subtleties of the original problem[1]. This is precisely the main goal of the present paper: presenting, in an accessible way and without much technical apparatus, the essentials of first order eigenvalue perturbation theory. It is also our intention to show in §3.2 that the general problem of determining the leading term of the ε-expansion of the eigenvalues of (1) as a function of B is far from being completely solved, even in the finite-dimensional case.

We choose not to present eigenvector perturbation results for two main reasons: the first one is to keep the presentation relatively concise. The second, and more important one, is that the techniques we will employ are able to produce such kind of results only indirectly, and under quite restrictive assumptions.

Making use of a powerful algebraic tool like the Newton diagram method will allow us to present a systematic development of the relevant results by means of purely matrix-analytic techniques. The results we will present are chiefly those obtained by Vishik and Lyusternik [41] and Lidskii [14], together with some consequences and extensions which, to our knowledge, are new (see §3.2.1 and §4 below). The results in [41], intended to be applied on differential operators, were generalized by Lidskii [14] for the finite-dimensional case. He obtained simple explicit formulas for the perturbation coefficients and provided, at the same time, a much more elementary proof. The results in both [41] and [14] were later refined by Baumgärtel (see [1] § 7.4), in the sense of dealing not only with perturbation series for eigenvalues and eigenvectors, but also with the corresponding eigenprojections as functions of ε. Vainberg and Trenogin ([39] § 32), on the other hand, offer a fairly thorough account of similar results, obtained for Fredholm operators by applying the techniques of branching theory. Langer and Najman [10, 11, 12] generalized Lidskii's results to analytic matrix functions $A(\lambda) + B(\lambda, \varepsilon)$, using the local Smith normal form of parameter-dependent matrices (see section 5 for more references).

From the functional analytic bias mentioned above one might get the impression that first order eigenvalue perturbation theory is suited mostly for dealing with infinite-dimensional operators, the finite-dimensional results being just a simple by-product of the former. Also, numerical analysts have not been very keen on first order eigenvalue perturbation results or, as they are also known, on *perturbation expansions*. That is partly true, mainly because perturbation

[1] Actually, the original problem can be reduced to a purely finite-dimensional one via the resolvent integration method (see [1, chapter 10])

bounds, as opposed to perturbation expansions, are more amenable to backward rounding-error analysis, the most widespread tool for analyzing the numerical stability of algorithms for spectral problems. However, first order eigenvalue perturbation results are still extremely useful for the numerical analyst whenever some qualitative information is needed, e.g. on the splitting directions of a multiple eigenvalue, even if perturbation bounds are available (which is not always the case). Borrowing the very words of Stewart and Sun in their book [35, p. 292],

> Although these [perturbation] expansions are usually corollaries of more general results, in many cases the general results themselves were conjectured by looking at first order expansions. The reason is that the expansions often tell ninety percent of the story and yet are free of the clutter that accompanies rigorous upper bounds.

Thus, first order eigenvalue perturbation theory has a role to play in applied mathematics, as well as a definite place among the many different tools-of-the-trade of the applied mathematician.

The paper is organized as follows: the Newton diagram technique is presented in detail in section 2, together with some elementary examples. Section 3 contains the results on perturbation of the standard eigenvalue problem: after setting the stage for using the Newton diagram in this context, the main result (Theorem 2) is proved and discussed in § 3.1. The cases outside the scope of Theorem 2 (what we call the *nongeneric* case) are analyzed in § 3.2. Particular attention is given in § 3.2.1 to the case when the perturbation B has small rank. Section 4 presents first order perturbation results for singular values, which can be easily recovered from the results in section 3. The final section 5 is devoted to a brief review of some of the available results for the perturbation of generalized eigenvalue problems.

2. The Newton Diagram

Although the Newton diagram (also called Newton *polygon*, or Newton-Puiseux diagram) can be applied in much more general settings (including analytic perturbation of infinite-dimensional operators, see [2, § III.8.3], [39, § 1.2] or [1, Appendix A7]), we strictly confine our presentation to the particular context we are dealing with, namely eigenvalues as roots of the characteristic polynomial of a parameter-dependent matrix $A(\varepsilon) = A + \varepsilon B$. Therefore, consider a complex polynomial equation

$$P(\lambda, \varepsilon) = \lambda^n + \alpha_1(\varepsilon)\lambda^{n-1} + \ldots + \alpha_{n-1}(\varepsilon)\lambda + \alpha_n(\varepsilon) = 0 \qquad (2)$$

in λ, with analytic coefficients

$$\alpha_k(\varepsilon) = \hat{\alpha}_k \varepsilon^{a_k} + \ldots, \qquad k = 1, \ldots, n, \qquad (3)$$

where a_k is the *leading exponent* and $\hat{\alpha}_k$ the *leading coefficient* of $\alpha_k(\varepsilon)$ (i.e. $\hat{\alpha}_k \neq 0$ and no term of order lower than a_k appears in the expansion of $\alpha_k(\cdot)$). For our convenience we set $\alpha_0(\varepsilon) \equiv 1$, i.e. $\hat{\alpha}_0 = 1$ and $a_0 = 0$.

It is well known [1, 9] that the roots λ of (2) are given by expansions in fractional powers of ε. Our goal is to determine both the leading exponents and the leading coefficients of these ε-expansions. To do that, we make the *Ansatz*

$$\lambda(\varepsilon) = \mu\varepsilon^{\beta} + \ldots, \tag{4}$$

with μ, β to be determined. Substituting the Ansatz into (2), each $\alpha_k(\varepsilon)\lambda^{n-k}$ produces a term of order $\varepsilon^{a_k+(n-k)\beta}$ plus higher order terms. Hence, $P(\lambda, \varepsilon)$ is an infinite sum of powers of ε, each of them multiplied by an ε-independent coefficient. If, as we assume, $\lambda(\varepsilon)$ is a root of (2), then all the coefficients must be zero. In particular, the lowest order in ε must be present at least twice among the exponents $\{a_k + (n-k)\beta\}_{k=0}^{n}$, i.e. there exist at least two indices $i, j \in \{0, 1, \ldots, n\}$ such that

$$a_i + (n-i)\beta = a_j + (n-j)\beta \leq a_k + (n-k)\beta, \qquad k = 0, 1, \ldots, n.$$

In order to interpret geometrically this inequality, we plot the values a_k versus k for $k = 0, 1, \ldots, n$ on a cartesian grid. Then, the segment S joining (i, a_i) with (j, a_j) has slope β, and, since $\varepsilon^{a_i+(n-i)\beta}$ is the term with the lowest order, no other point (k, a_k) lies below the straight line containing S. Therefore, if we draw the lower boundary of the convex hull of all the points $\{(k, a_k)\}_{k=0}^{n}$, it is clear that S must be on that boundary. Thus, the slopes of the segments on the lower boundary are just the exponents β appearing in (4). By imposing for each β that the coefficient of $\varepsilon^{a_i+(n-i)\beta}$ be zero we obtain that the leading coefficients μ of the eigenvalues of order ε^{β} are the solutions of

$$\sum_{k\in I_s} \mu^{n-k}\hat{\alpha}_k = 0, \tag{5}$$

where the set $I_s = \{k : (k, a_k) \in S\}$ may contain indices other than i, j (e.g. see Fig. 1(a) below, where three points (k, a_k) lie on S). If we denote $k_{\max} = \max I_s$ and $k_{\min} = \min I_s$, then the polynomial equation (5) has $n - k_{\max}$ zero roots, corresponding to the roots of order higher than β, and $k_{\max} - k_{\min}$ nonzero roots which are the leading coefficients we are looking for. As a consequence, the number of roots of (2) with leading exponent β is given by the length of the projection on the horizontal axis of the segment S. Notice that the total sum of the lengths of these projections is n.

Thus, we have derived a quite simple geometrical construction, going back to Newton, which very simply provides us with all the leading powers and leading coefficients we were looking for. The crucial ingredient is the so-called *Newton diagram* or *Newton polygon*.

Definition of the Newton diagram: *Given a polynomial equation of the form (2) with analytic coefficients given by (3), plot* a_k *versus* k *for* $k = 0, 1, \ldots, n$ *(if* $\alpha_k(\cdot) \equiv 0$, *the corresponding point is disregarded). Denote each of these points by* $\pi_k = (k, a_k)$ *and let*

$$\Pi = \{\pi_k : \alpha_k \not\equiv 0\}$$

be the set of all plotted points. Then, the Newton diagram *associated with* $P(\lambda, \varepsilon)$ *is the lower boundary of the convex hull of the set* Π.

With this terminology, the procedure outlined above for determining the leading terms of the asymptotic expansions (4) goes as follows:

Newton diagram procedure: *Given a polynomial (2) with analytic coefficients given by (3),*

1) *draw the associated Newton diagram.*

2) *The leading exponents* β *of the roots of (2) are the different slopes of the segments forming the Newton diagram.*

3) *The number of roots of order* ε^β *is given by the length of the projection on the horizontal axis of the segment with slope* β.

4) *The leading coefficients* μ *for each root of order* ε^β *are the nonzero roots of equation (5), where* S *is the segment of the Newton diagram with slope* β.

We illustrate the procedure with two specific examples, depicted in Figure 1 below.

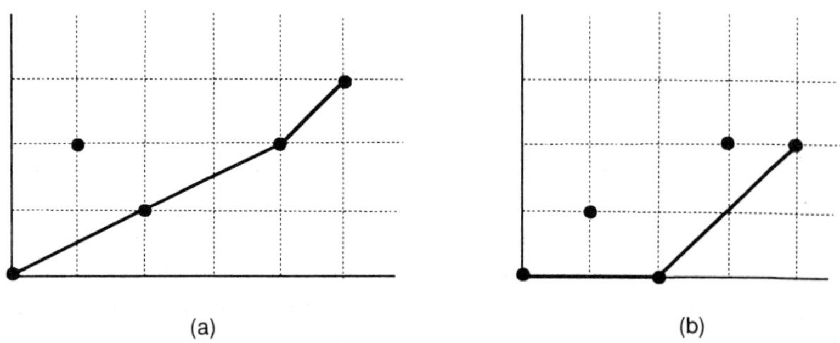

Figure 1. *Newton diagrams associated with the polynomials: a)* $\lambda^5 + (2\varepsilon^2 - \varepsilon^3)\lambda^4 - \varepsilon\lambda^3 + (-6\varepsilon^2 + 3\varepsilon^5)\lambda + \varepsilon^3 - \varepsilon^4$; *and b)* $\lambda^4 - (\varepsilon + 2\varepsilon^2)\lambda^3 + (\varepsilon^2 - 1)\lambda^2 + (\varepsilon^2 - \varepsilon^3)\lambda + 2\varepsilon^2$

Example 1. Let $P_1(\lambda, \varepsilon) = \lambda^5 + (2\varepsilon^2 - \varepsilon^3)\lambda^4 - \varepsilon\lambda^3 + (-6\varepsilon^2 + 3\varepsilon^5)\lambda + \varepsilon^3 - \varepsilon^4$. Then, $\Pi = \{\pi_0, \pi_1, \pi_2, \pi_4, \pi_5\}$ and the Newton diagram associated with P_1 is the one in Figure 1(a). It consists of two segments, one of slope $1/2$ connecting the points π_0, π_2, π_4, and one of slope 1 connecting π_4 with π_5. Therefore, P_1 has roots of orders $\sqrt{\varepsilon}$ and ε. More precisely, there are four roots of order $\sqrt{\varepsilon}$ whose leading coefficients are the nonzero roots of

$$\hat{\alpha}_0\mu^5 + \hat{\alpha}_2\mu^3 + \hat{\alpha}_4\mu = \mu^5 - \mu^3 - 6\mu = \mu(\mu^2 - 3)(\mu^2 + 2),$$

i.e $\pm\sqrt{3}$ and $\pm i\sqrt{2}$. Finally, there is one root of order ε whose leading coefficient is $1/6$, the only root of

$$\hat{\alpha}_4\mu + \hat{\alpha}_5 = -6\mu + 1.$$

Example 2: Let $P_2(\lambda, \varepsilon) = \lambda^4 - (\varepsilon + 2\varepsilon^2)\lambda^3 + (\varepsilon^2 - 1)\lambda^2 + (\varepsilon^2 - \varepsilon^3)\lambda + 2\varepsilon^2$. Now $\Pi = \{\pi_i\}_{i=0}^{4}$. The corresponding Newton diagram is drawn in Figure 1(b). Hence, there are two roots of $O(1)$ with leading coefficients ± 1, the nonzero roots of

$$\mu^4 + \hat{\alpha}_2\mu^2 = \mu^4 - \mu^2,$$

and two roots of $O(\varepsilon)$, whose leading coefficients $\pm\sqrt{2}$ solve

$$\hat{\alpha}_2\mu^2 + \hat{\alpha}_4 = -\mu^2 + 2 = 0.$$

We stress that the whole argument above, leading to the Newton diagram technique *relies completely on the initial assumption that a convergent asymptotic expansion* (4) *exists*. Otherwise, we might be computing the leading term of a nonexisting quantity. Of course, that was not a concern for Newton: he proposed this method in letters to Leibniz and Oldenburg (see [2, pp. 372-375], [21, pp. 20-42]), and developed it later in his treatises [22], handling infinite series but saying nothing as to their convergence. Only in the 19th century, Puiseux [24] proved, in the course of his investigations on singularities, that the expansions obtained through the Newton diagram converge in a neighborhood of $\varepsilon = 0$. Hence the name of *Newton-Puiseux* diagram.

Once this tool is at our disposal, we set to the task of obtaining first order results on perturbation of matrix eigenvalues

3. First order perturbation bounds for the standard eigenvalue problem

Let λ_0 be an eigenvalue of algebraic multiplicity a of the complex matrix $A \in \mathbb{C}^{n\times n}$, and consider a perturbation

$$A(\varepsilon) = A + \varepsilon B$$

for arbitrary $B \in \mathbb{C}^{n\times n}$. It is a well known fact [1, 9] that, for sufficiently small ε, the matrix $A(\varepsilon)$ has a eigenvalues $\lambda_j(\varepsilon)$ with $\lambda_j(0) = \lambda_0$, each of them admitting an expansion in fractional powers of ε. Our goal is to determine the leading term of each expansion applying the Newton diagram technique to the characteristic polynomial $P(\lambda, \varepsilon) = \det(\lambda I - A - \varepsilon B)$ of $A(\varepsilon)$.

In order to prove our main perturbation result (Theorem2 below), it is crucial to carefully determine which points (k, a_k) may appear on the Newton diagram *for a particular given Jordan structure* of the unperturbed matrix A: let

$$\left[\begin{array}{c|c} J & \\ \hline & \widehat{J} \end{array}\right] = \left[\begin{array}{c} Q \\ \hline \widehat{Q} \end{array}\right] A \left[\begin{array}{c|c} P & \widehat{P} \end{array}\right] \tag{6}$$

be a Jordan decomposition of A, i.e.

$$\left[\begin{array}{c} Q \\ \hline \widehat{Q} \end{array}\right] \left[\begin{array}{c|c} P & \widehat{P} \end{array}\right] = I. \tag{7}$$

The matrix J contains all Jordan blocks associated with the eigenvalue of interest λ_0, while $\widehat{J}$ is the part of the Jordan form containing the other eigenvalues. Let

$$J = \Gamma_1^1 \oplus \ldots \oplus \Gamma_1^{r_1} \oplus \ldots \oplus \Gamma_q^1 \oplus \ldots \oplus \Gamma_q^{r_q}, \tag{8}$$

where, for $j = 1, \ldots, q$,

$$\Gamma_j^1 = \ldots = \Gamma_j^{r_j} = \begin{bmatrix} \lambda_0 & 1 & & & \\ & \cdot & \cdot & & \\ & & \cdot & \cdot & \\ & & & \cdot & 1 \\ & & & & \lambda_0 \end{bmatrix}$$

is a Jordan block of dimension n_j, repeated r_j times, and ordered so that

$$n_1 > n_2 > \ldots > n_q.$$

The n_j are called the *partial multiplicities* for λ_0. The eigenvalue λ_0 is semisimple (nondefective) if $q = n_1 = 1$ and nonderogatory if $q = r_1 = 1$. The algebraic and geometric multiplicities of λ_0 are, respectively,

$$a = \sum_{j=1}^{q} r_j n_j \qquad \text{and} \qquad g = \sum_{j=1}^{q} r_j. \tag{9}$$

We further partition

$$P = \left[\begin{array}{c|c|c|c|c|c|c} P_1^1 & \ldots & P_1^{r_1} & \ldots & P_q^1 & \ldots & P_q^{r_q} \end{array}\right]$$

conformally with (8). The columns of each P_j^k form a right Jordan chain of A with length n_j corresponding to λ_0. If we denote by x_j^k the first column of P_j^k, each x_j^k is a right eigenvector of A associated with λ_0. Analogously, we split

$$Q = \begin{bmatrix} Q_1^1 \\ \hline \vdots \\ \hline Q_1^{r_1} \\ \hline \vdots \\ \hline Q_q^1 \\ \hline \vdots \\ \hline Q_q^{r_q} \end{bmatrix},$$

also conformally with (8). The rows of each Q_j^k form a left Jordan chain of A of length n_j corresponding to λ_0. Hence, if we denote by y_j^k the last (i.e. n_j–th) row of Q_j^k, each y_j^k is a left eigenvector corresponding to λ_0. With these eigenvectors we build up matrices

$$Y_j = \begin{bmatrix} y_j^1 \\ \vdots \\ y_j^{r_j} \end{bmatrix}, \qquad X_j = [x_j^1, \dots, x_j^{r_j}],$$

for $j = 1, \dots, q$,

$$W_s = \begin{bmatrix} Y_1 \\ \vdots \\ Y_s \end{bmatrix}, \qquad Z_s = [X_1, \dots, X_s],$$

for $s = 1, \dots, q$, and define square matrices Φ_s and E_s of dimension

$$f_s = \sum_{j=1}^{s} r_j \tag{10}$$

by

$$\Phi_s = W_s B Z_s, \qquad s = 1, \dots, q,$$

$$E_1 = I, \qquad E_s = \begin{bmatrix} 0 & 0 \\ 0 & I \end{bmatrix} \quad \text{for } s = 2, \dots, q,$$

where the identity block in E_s has dimension r_s. Note that, due to the cumulative definitions of W_s and Z_s, every Φ_{s-1}, $s = 2, \dots, q$, is the upper left block of Φ_s.

An important observation to be made at this point is that, although the Jordan decomposition of A has been presented in its full generality, and all results presented below are valid for the general case (6), we may assume with no loss of generality that λ_0 is the only eigenvalue of A, i.e that $\widehat{J}$ is empty. The reason is that, *since we are only interested in first order results*, we may disregard quadratic terms in ε. More precisely, if we write the characteristic polynomial of $A(\varepsilon)$ as $P(\lambda,\varepsilon) = \det(\lambda I - \operatorname{diag}(J,\widehat{J}) - \varepsilon\widetilde{B})$, with

$$\widetilde{B} = \left[\begin{array}{c} Q \\ \hline \widehat{Q} \end{array}\right] B \left[\begin{array}{c|c} P & \widehat{P} \end{array}\right] \tag{11}$$

then one can use Schur's formula to factorize P as

$$P(\lambda,\varepsilon) = \det\left(\begin{bmatrix} \lambda I - J - \varepsilon\widetilde{B}_{11} & -\varepsilon\widetilde{B}_{12} \\ -\varepsilon\widetilde{B}_{21} & \lambda I - \widehat{J} - \varepsilon\widetilde{B}_{22} \end{bmatrix}\right) = \widehat{\pi}(\lambda,\varepsilon)\,\pi(\lambda,\varepsilon),$$

with

$$\begin{aligned} \widehat{\pi}(\lambda,\varepsilon) &= \det(\lambda I - \widehat{J} - \varepsilon\widetilde{B}_{22}) \\ \pi(\lambda,\varepsilon) &= \det(\lambda I - J - \varepsilon\widetilde{B}_{11} - \varepsilon^2\widehat{S}(\lambda,\varepsilon)), \end{aligned}$$

where $\widehat{S}$ is the matrix $\widehat{S}(\lambda,\varepsilon) = \widetilde{B}_{12}(\lambda I - \widehat{J} - \varepsilon\widetilde{B}_{22})^{-1}\widetilde{B}_{21}$. If λ is an eigenvalue of $A(\varepsilon)$ close to λ_0, then it cannot be a root of the polynomial $\widehat{\pi}(\cdot,\varepsilon)$, so it must be a root of the rational function $\pi(\cdot,\varepsilon)$, which depends on $\widehat{J}$ only through terms of the second order in ε. Hence, $\widehat{J}$ has no influence whatsoever on the first order terms and it is sufficient to study $\det(\lambda I - J - \varepsilon\widetilde{B}_{11})$ to characterize the first order behavior. For a rigorous proof, more formal than this plausibility argument, see Lidskii's original paper [14, pp. 83-84] , or [1, § 3.9.1], where spectral projections on the appropriate invariant subspaces are used to completely decouple J from the influence of $\widehat{J}$ up to first order.

A second important simplification is that we may take $\lambda_0 = 0$ at our convenience, since the shift $A \to A - \lambda_0 I$ does not change either the Jordan block structure or the Jordan chains of A. Hence, all results below are invariant under that transformation and λ_0 may be set to zero.

With these simplifying assumptions, one can easily see that the particular form of the leading terms of the eigenvalue expansions will depend mainly on the Jordan structure of A. To see it for the leading exponents, let $J = P^{-1}AP$ be a Jordan form of A, where we assume, as explained above, that A has no eigenvalue other than $\lambda_0 = 0$. In that case, the characteristic polynomial $P(\lambda,\varepsilon) = \det(\lambda I - J - \varepsilon\widetilde{B})$ of $A(\varepsilon)$, with $\widetilde{B} = P^{-1}BP$, is a polynomial of the form (2). If the eigenvalue is semisimple, then J is zero and each $\alpha_k(\varepsilon)$ equals ε^k multiplied by a certain sum of k-dimensional principal minors of $\widetilde{B}$. In this case, the Newton diagram is formed by one single segment of slope $s = 1$. If the eigenvalue is not semisimple, some nontrivial Jordan block appears in J, so, besides the $O(\varepsilon^k)$ terms, each $\alpha_k(\varepsilon)$ contains lower order terms produced by the -1s appearing above the diagonal of J. This clearly shows that the effect of nontrivial Jordan blocks is to introduce in the Newton diagram line segments with slopes less than 1. The smallest possible slope corresponds to a nonderogatory eigenvalue (i.e. one single segment of slope $1/n$) and the largest possible one to the semisimple case. All possible Newton diagrams for the given multiplicity n lie between these two extremal segments.

This said, we are going to determine the *lowest possible Newton diagram compatible with the given Jordan structure* (8). We do it by fixing on the vertical axis an arbitrary height l, ranging between 1 and $f_q = g$. For each height l we look for the rightmost possible point $(k(l), l)$ which may be in the generating set Π of the Newton diagram for some suitable perturbation B. In other words, we are interested in

$$k(l) = \max\{k : \exists B \in \mathbb{C}^{n\times n} \text{ such that } a_k = l\}$$

as a function of $l \in \{1, \ldots, f_q\}$. The following theorem gives us the values of $k(l)$ for the exponents l which are relevant to our argument (recall that f_j is given by formula (10)).

Theorem 1. *For every $l \in \{1, \ldots, f_q\}$ the corresponding $k(l)$ is equal to the sum of the dimensions of the l largest Jordan blocks of J. More precisely, write $f_0 = 0$ and suppose $l = f_{j-1} + \rho$, for some $j = 1, \ldots, q$ and $0 < \rho \leq r_j$. Then,*

$$k(l) = r_1 n_1 + \ldots + r_{j-1} n_{j-1} + \rho n_j$$

and the coefficient of ε^l in $\alpha_{k(l)}$ is equal to $(-1)^l$ multiplied by the sum of all principal minors of Φ_j corresponding to submatrices of dimension l containing the upper left block Φ_{j-1} of Φ_j (if $j = 1$, all *principal minors of dimension l are to be considered). If, in particular, $l = f_j$ for some $j \in \{1, \ldots, q\}$, then the coefficient of ε^{f_j} in $\alpha_{k(f_j)}$ is $(-1)^{f_j} \det \Phi_j$.*

PROOF: As explained above, we may assume that $\lambda_0 = 0$ is the only eigenvalue of the matrix $A \in \mathbb{C}^{n\times n}$. First, recall that given a n by n matrix, the coefficient of λ^{n-k} in its characteristic polynomial is, except for a sign, the sum of all k-dimensional principal minors of the matrix. In our case $P(\lambda, \varepsilon) = \det(\lambda I - A(\varepsilon))$ and each $\alpha_k(\varepsilon)$ in (3) is, up to a sign factor, the sum of all k-dimensional principal minors of $J + \varepsilon\widetilde{B}$. Notice that the only elements of this matrix which are not of order ε are the ones occupying the positions of the 1s above the diagonal of J.

Now, given $l \in \{1, \ldots, f_q\}$, we must determine the largest possible k such that the sum of all k-dimensional principal minors is exactly of order ε^l.

In particular, the minor with the least order must be precisely of that order, implying that the corresponding principal submatrix must contain $k - l$ of the supradiagonal 1s. In other words, maximizing k for the given l is equivalent to including as many 1s as possible in the principal submatrices, while still keeping the order ε^l. Clearly, this is achieved by choosing the rows and columns where the 1s are from the l largest Jordan blocks in J.

If $l = f_j$ for some $j = 1, \ldots, q$, then there is only one way of choosing these blocks. Furthermore, one can easily check our claim on the coefficient of ε^{f_j} in $\alpha_{k(f_j)}$, since the leading term of $\alpha_{k(f_j)}$ is just the determinant of what is left from the k by k principal submatrix once the rows and columns of the chosen $k(f_j) - f_j$ supradiagonal 1s have been removed. The remaining matrix $f_j \times f_j$ matrix is precisely $\varepsilon\Phi_j$.

Finally, if $l = f_j + \rho$ with $\rho < r_{j+1}$, there is more than one way of choosing the l blocks: once the largest f_j Jordan blocks have been exhausted, each one corresponds to a different choice of ρ blocks among the r_{j+1} Jordan blocks of dimension n_{j+1}. As to the coefficient of ε^l, the argument goes much in the same way as above. □

As a consequence of Theorem 1, we conclude that *the lowest possible Newton diagram compatible with the Jordan structure (8) is the concatenation of the segments* S_j, $j = 1, \ldots, q$ *connecting the points* $\mathcal{P}_{j-1}$ *and* $\mathcal{P}_j$, *where* $\mathcal{P}_j = (k(f_j), f_j)$ *for each* $j = 1, \ldots, q$. This diagram, which has been called [16] the *Newton envelope* associated with the Jordan structure (8), is not only the lowest possible, but also the *most likely* diagram, since the actual diagram corresponding to a specific perturbation B will coincide with the Newton envelope unless some of the $\det \Phi_j$ vanishes, i.e. unless B satisfies an algebraic condition which confines it to an algebraic manifold (i.e. to a set of zero Lebesgue measure) in the set of matrices. In other words, the Newton envelope displays the *generic* behavior of the eigenvalues of A under perturbation, in the sense that it coincides with the Newton diagram in all situations except in those nongeneric cases in which the perturbation B causes one of the Φ_j to be singular. The next subsection is devoted to describe in detail this generic behavior.

3.1 The generic case: Lidskii's Theorem

We begin with the main result for generic perturbations, due to Lidskii [14] and, in a more restrictive version, to Vishik and Lyusternik [41].

Theorem 2 (Lidskii [14]). *Let* $j \in \{1, \ldots, q\}$ *be given, and assume that, if* $j > 1$, Φ_{j-1} *is nonsingular. Then there are* $r_j n_j$ *eigenvalues of the perturbed matrix* $A + \varepsilon B$ *admitting a first-order expansion*

$$\lambda_j^{kl}(\varepsilon) = \lambda_0 + (\xi_j^k)^{1/n_j}\, \varepsilon^{1/n_j} + o(\varepsilon^{1/n_j}) \qquad (12)$$

for $k = 1, \ldots, r_j$, $l = 1, \ldots, n_j$, *where*

(i) *the* ξ_j^k, $k = 1, \ldots, r_j$, *are the roots of equation*

$$\det(\Phi_j - \xi\, E_j) = 0 \tag{13}$$

or, equivalently, the eigenvalues of the Schur complement of Φ_{j-1} *in* Φ_j *(if* $j = 1$, *the* ξ_1^k *are just the* r_1 *eigenvalues of* Φ_1*);*

(ii) *the different values* $\lambda_j^{kl}(\varepsilon)$ *for* $l = 1, \ldots, n_j$ *are defined by taking the* n_j *distinct* n_j*-th roots of* ξ_j^k.

If, in addition, the r_j *solutions* ξ_j^k *of* (13) *are all distinct, then the eigenvalues* (12) *can be expanded locally in power series of the form*

$$\lambda_j^{kl}(\varepsilon) = \lambda_0 + (\xi_j^k)^{1/n_j}\, \varepsilon^{1/n_j} + \sum_{s=2}^{\infty} a_{js}^{kl}\, \varepsilon^{s/n_j}, \tag{14}$$

$k = 1, \ldots, r_j$, $l = 1, \ldots, n_j$.

PROOF: Once the Newton diagram technique is at hand, the proof is a consequence of Theorem 1. First, suppose that both Φ_{j-1} and Φ_j are nonsingular. Then, both $\mathcal{P}_{j-1}$ and $\mathcal{P}_j$ are in the set Π generating the Newton diagram, i.e. the segment S_j of slope $1/n_j$ connecting both points is one of the segments in the diagram (recall that no point (k, a_k) can lie below S_j). This gives us the leading exponent of expansion (12). The leading coefficient comes from carefully examining equation (5). One can check that

$$\sum_{(k,a_k)\in S_j} \mu^{n-k}\hat{\alpha}_k$$

$$= \mu^{n-k(f_j)} \left[\mu^{n_j r_j}\hat{\alpha}_{k(f_{j-1})} + \sum_{t\in T} \hat{\alpha}_{k(f_j-t)}\mu^{t\,n_j} + \hat{\alpha}_{k(f_j)} \right] = 0, \tag{15}$$

where

$$T = \{t \in \{1, \ldots, r_j - 1\} : \mathcal{Q}_t = (k(f_j - t), f_j - t) \in \Pi\}, \tag{16}$$

i.e. T is the set of indices corresponding to the intermediate points $\mathcal{Q}_t \in \Pi$ eventually lying on S_j. Notice that bracketed expression in (15) depends on μ only through μ^{n_j}. Now, recall from Theorem 1 that for each $l = f_j - t$ with $t \in T$, the corresponding $\hat{\alpha}_{k(l)}$ is (up to the sign) the sum of all principal minors of Φ_j of dimension l containing Φ_{j-1}. This is precisely the way the coefficients of the powers of ξ are obtained in $\det(\Phi_j - \xi E_j)$. Hence, the nonzero solutions of (15) are solutions of $\det(\Phi_j - \mu^{n_j} E_j) = 0$ as well.

Now, suppose that Φ_j is singular. Then, the corresponding point $\mathcal{P}_j$ no longer belongs to the diagram, implying the loss of some of the expansions (12) or, equivalently, the loss of part of the segment S_j. Let β be either $\beta = r_j$ if no point $\mathcal{Q}_t = (k(f_j - t), f_j - t)$ is in Π or $\beta = \min T$ for T defined as in (16). In either case $\mathcal{Q}_\beta = (k(f_j - \beta), f_j - \beta)$ is the rightmost point of Π on S_j. Then, the part of S_j which remains on the Newton diagram is the segment connecting $\mathcal{P}_{j-1}$ with $\mathcal{Q}_\beta$. This accounts for $(r_j - \beta)n_j$ expansions (12), whose leading coefficients are, reasoning as above, the n_j-th roots of the $r_j - \beta$ nonzero solutions of equation (13). As to the βn_j remaining eigenvalues, they correspond to segments whose slope is strictly larger than $1/n_j$. Hence, expansion (12) is still valid, since they correspond to the β null solutions of equation (13).

As to expansion (14), one can check (see the original reference [14] or [16, § 2]) that, after the change of variables

$$z = \varepsilon^{1/n_j}$$
$$\mu = \frac{\lambda}{z}$$

and an appropriate scaling in z, the polynomial equation $P(\lambda, \varepsilon) = \det(\lambda I - A - \varepsilon B) = P(\mu, z) = 0$ has, for any given $z \neq 0$, the same roots as a new polynomial equation $Q(\mu, z) = 0$ with

$$Q(\mu, 0) = \pm\mu^\alpha \det(\Phi_j - \mu^{n_j} E_j)$$

for some suitable $\alpha \geq 0$. Hence, if all r_j roots of equation (13) are known to be distinct, the implicit function theorem can be applied to $Q(\mu, z) = 0$, implying that the $r_j n_j$ roots

$$\mu_j^{kl}(z) = (\xi_j^k)^{1/n_j} + o(1), \qquad k = 1, \ldots, r_j; \quad l = 1, \ldots, n_j,$$

of $Q(\mu(z), z) = 0$ for small enough z are analytic functions of $z = \varepsilon^{1/n_j}$.

□

Two special cases of Theorem 2 are well known. If λ_0 is semisimple, i.e. $q = n_1 = 1$ with multiplicity r_1, equation (12) reduces to

$$\lambda_1^{k1}(\varepsilon) = \lambda_0 + \xi_1^k\, \varepsilon + o(\varepsilon), \tag{17}$$

where the ξ_1^k are the eigenvalues of the r_1 by r_1 matrix $Y_1 B X_1$ (cf. [9, § II.2.3]). On the other hand, if λ is nonderogatory, i.e. $q = r_1 = 1$ with multiplicity n_1, equation (12) reduces to

$$\lambda_j^{1l}(\varepsilon) = \lambda_0 + (\xi_1^1)^{1/n_1}\, \varepsilon^{1/n_1} + o(\varepsilon^{1/n_1}),$$

where $\xi_1^1 = y_1^1 B x_1^1$. These two cases coincide when λ is simple.

Theorem 2 does not address either the convergence or the ultimate form of the $o(\varepsilon^{1/n_j})$ term in expansion (12), since these issues are beyond the reach of a purely algebraic tool like the Newton diagram. One can show by other means (see [1, § 9.3.1], [9, § II.1.2]) that whenever both Φ_{j-1} and Φ_j are nonsingular, the $r_j n_j$ eigenvalues (12) acn be written as convergent power series in the variable ε^{1/n_j}. This is no longer true if only Φ_{j-1} is nonsingular, unless some additional information, as in the last part of Theorem 2, is available (see, for instance, the perturbation matrix (18) in Example 3 below, for which $\det \Phi_2 = 0$ and two out of the four eigenvalues corresponding to Φ_2 are of order $\varepsilon^{2/3}$)

However, one important special case deserves to be mentioned: if both the unperturbed matrix A and the perturbation matrix B are *normal*, then all eigenvalues of $A + \varepsilon B$ are analytic functions of ε [1, § 7.2], i.e. they have a convergent representation (14) with $n_j = 1$. This property will be crucial in Section 4, when dealing with the perturbation of singular values.

We conclude this subsection by referring the reader interested in eigenvector perturbation results to [14, Theorem 2], the eigenvector perturbation theorem in [14] analogous to the eigenvalue result above, which essentially amounts to replacing (14) in the eigenvalue-eigenvector equation $A(\varepsilon)v(\varepsilon) = \lambda(\varepsilon)v(\varepsilon)$ (the same result appears in [16] as Theorem 2.2).

3.2 Nongeneric perturbations

If the perturbation is nongeneric, i.e. when the matrix B is such that Φ_{j-1} is singular, Theorem 2 does not apply. The question is what can we say about the eigenvalues of $A + \varepsilon B$ in this case. The answer is: not much, at least in such a systematic way as in Theorem 2. Although the Newton diagram is still a powerful instrument which allows us to deal with each particular case, it is not easy to give a clear, global picture of the wide variety of different possible behaviors. To give an idea of the difficulties, consider the following example:

Example 3: Let $A \in \mathbb{R}^{8\times 8}$ be already in Jordan form,

$$A = J = \Gamma_1^1 \oplus \Gamma_2^1 \oplus \Gamma_2^2 \oplus \Gamma_3^1$$

with $n_1 = 3$, $n_2 = 2$, $n_3 = 1$ and $r_1 = r_3 - 1$, $r_2 = 2$, i.e.

$$A = \left[\begin{array}{ccc|cc|cc|c} 0 & 1 & 0 & & & & & \\ 0 & 0 & 1 & & & & & \\ 0 & 0 & 0 & & & & & \\ \hline & & & 0 & 1 & & & \\ & & & 0 & 0 & & & \\ \hline & & & & & 0 & 1 & \\ & & & & & 0 & 0 & \\ \hline & & & & & & & 0 \end{array}\right].$$

According to Theorem 2, a generic perturbation, i.e. any 8 by 8 matrix

$$B = \left[\begin{array}{ccc|cc|cc|c} * & * & * & * & * & * & * & * \\ ** & * & * & * & * & * & * & * \\ *\square & * & * & \clubsuit & * & \clubsuit & * & \heartsuit \\ \hline * & * & * & * & * & * & * & * \\ \clubsuit & * & * & \clubsuit & * & \clubsuit & * & \heartsuit \\ \hline * & * & * & * & * & * & * & * \\ \clubsuit & * & * & \clubsuit & * & \clubsuit & * & \heartsuit \\ \hline \heartsuit & * & * & \heartsuit & * & \heartsuit & * & \heartsuit \end{array}\right]$$

with all three submatrices

$$\Phi_1 = [\ \square\ ], \qquad \Phi_2 = \left[\begin{array}{c|cc} \square & \clubsuit & \clubsuit \\ \hline \clubsuit & \clubsuit & \clubsuit \\ \clubsuit & \clubsuit & \clubsuit \end{array}\right], \qquad \Phi_3 = \left[\begin{array}{c|cc|c} \square & \clubsuit & \clubsuit & \heartsuit \\ \hline \clubsuit & \clubsuit & \clubsuit & \heartsuit \\ \clubsuit & \clubsuit & \clubsuit & \heartsuit \\ \hline \heartsuit & \heartsuit & \heartsuit & \heartsuit \end{array}\right]$$

nonsingular gives rise to a Newton diagram coinciding with the Newton envelope. In that case, Π contains the points $\mathcal{P}_0, \mathcal{P}_1, \mathcal{P}_2$ and $\mathcal{P}_3$, with $\mathcal{P}_0 = (0,0)$, $\mathcal{P}_1 = (3,1)$, $\mathcal{P}_2 = (7,3)$ and $\mathcal{P}_3 = (8,4)$ (see Figure 2a below). Hence, the eigenvalues of A split typically into three eigenvalues of order $\varepsilon^{1/3}$, four eigenvalues of order $\varepsilon^{1/2}$ and one eigenvalue of order ε.

A good number of different possibilities arises whenever some of the matrices Φ_j turn out to be singular. As a first example, consider the perturbation matrix

$$B_1 = \left[\begin{array}{ccc|cc|cc|c} 0 & 0 & 0 & 0 & 0 & 0 & 0 & 0 \\ 0 & 0 & 0 & 0 & 0 & 0 & 0 & 0 \\ 1 & 0 & 0 & 0 & 0 & 1 & 0 & 0 \\ \hline 0 & 0 & 0 & 0 & 0 & 0 & 0 & 0 \\ 0 & 0 & 0 & 0 & 0 & 1 & 0 & 0 \\ \hline 0 & 0 & 0 & 0 & 0 & 0 & 0 & 0 \\ 1 & 0 & 0 & 0 & 0 & 0 & 0 & 1 \\ \hline 0 & 0 & 0 & 1 & 0 & 0 & 0 & 0 \end{array}\right], \tag{18}$$

for which

$$\Phi_3 = \left[\begin{array}{c|cc|c} 1 & 0 & 1 & 0 \\ \hline 0 & 0 & 1 & 0 \\ 1 & 0 & 0 & 1 \\ \hline 0 & 1 & 0 & 0 \end{array}\right].$$

Both Φ_1 and Φ_3 are invertible, but Φ_2 is singular. In this case, the set Π contains the points $\mathcal{P}_0, \mathcal{P}_1$ and $\mathcal{P}_3$, together with the point $\mathcal{Q}_1 = (5,2)$. According to the Newton diagram, shown in Figure 2b, the matrix $A + \varepsilon B_1$

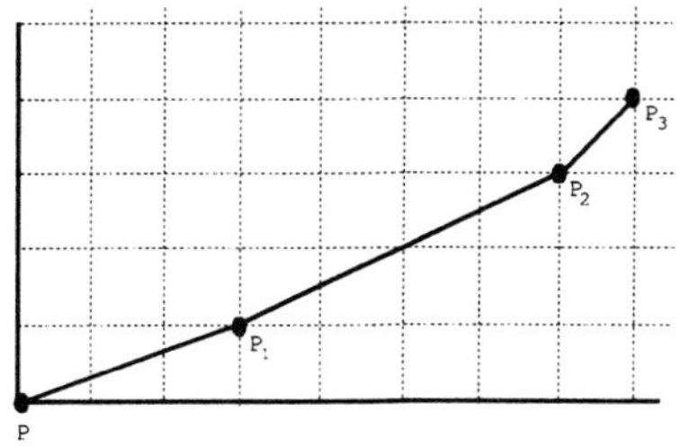

Figure 2a. *Newton diagram associated with the matrix $A+\varepsilon B$ for generic B in Example 3.*

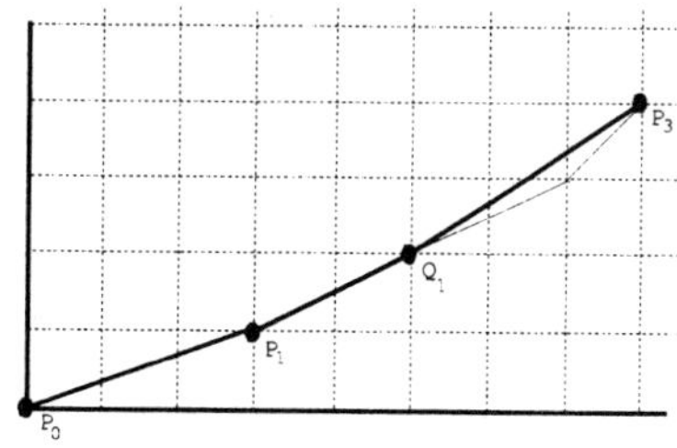

Figure 2b. *Newton diagrams associated with the matrix $A+\varepsilon B_1$ for the matrix B_1 in (18). The Newton envelope is shown as a dashed line, the Newton diagram as a solid one*

has three eigenvalues of order $\varepsilon^{1/3}$, two eigenvalues of order $\varepsilon^{1/2}$ and three eigenvalues of order $\varepsilon^{2/3}$.

Now, take the perturbation

$$B_2 = \left[\begin{array}{ccc|cc|cc|c} 0 & 0 & 0 & 0 & 0 & 0 & 0 & 0 \\ 0 & 0 & 0 & 0 & 0 & 0 & 0 & 0 \\ 1 & 0 & 0 & 1 & 0 & 0 & 0 & 0 \\ \hline 0 & 0 & 0 & 0 & 0 & 0 & 0 & 0 \\ 0 & 0 & 0 & 0 & 0 & 0 & 0 & 1 \\ \hline 0 & 0 & 0 & 0 & 0 & 0 & 0 & 0 \\ 0 & 0 & 0 & 1 & 0 & 0 & 0 & 0 \\ \hline 0 & 0 & 0 & 0 & 0 & 1 & 0 & 0 \end{array}\right], \tag{19}$$

producing

$$\Phi_3 = \left[\begin{array}{c|cc|c} 1 & 1 & 0 & 0 \\ \hline 0 & 0 & 0 & 1 \\ 0 & 1 & 0 & 0 \\ \hline 0 & 0 & 1 & 0 \end{array}\right].$$

Again, both Φ_1 and Φ_3 are invertible, but now neither $\mathcal{P}_2$ nor $\mathcal{Q}_1$ are on the Newton diagram (see Figure 3a). Thus, the matrix $A+\varepsilon B_2$ has three eigenvalues of order $\varepsilon^{1/3}$ and five eigenvalues of order $\varepsilon^{3/5}$.

If we take as our next nongeneric perturbation

$$B_3 = \left[\begin{array}{ccc|cc|cc|c} 0 & 0 & 0 & 0 & 0 & 0 & 0 & 0 \\ 0 & 0 & 0 & 0 & 0 & 0 & 0 & 0 \\ 0 & 0 & 0 & 1 & 0 & 0 & 0 & 0 \\ \hline 0 & 0 & 0 & 0 & 0 & 0 & 0 & 0 \\ 1 & 0 & 0 & 0 & 0 & 0 & 0 & 0 \\ \hline 0 & 0 & 0 & 0 & 0 & 0 & 0 & 0 \\ 0 & 0 & 0 & 0 & 0 & 1 & 0 & 0 \\ \hline 0 & 0 & 0 & 0 & 0 & 0 & 0 & 1 \end{array}\right], \qquad (20)$$

so that

$$\Phi_3 = \left[\begin{array}{c|cc|c} 0 & 1 & 0 & 0 \\ \hline 1 & 0 & 0 & 0 \\ 0 & 0 & 1 & 0 \\ \hline 0 & 0 & 0 & 1 \end{array}\right],$$

then both Φ_2 and Φ_3 are invertible, but Φ_1 is singular and the set Π contains $\{\mathcal{P}_0, \mathcal{Q}_1, \mathcal{P}_2, \mathcal{P}_3\}$. The corresponding Newton diagram is the one in Figure 3b, and $A+\varepsilon B_3$ has five eigenvalues of order $\varepsilon^{2/5}$, two eigenvalues of order $\varepsilon^{1/2}$ and one eigenvalue of order ε.

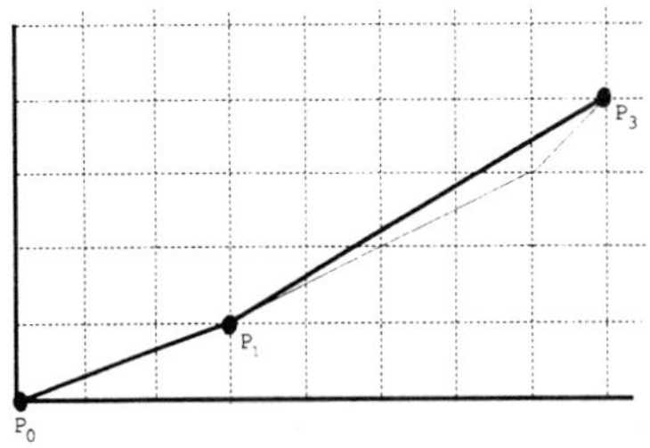

Figure 3a. *Newton diagram associated with $A + \varepsilon B_2$ for the matrix B_2 in (19). The Newton envelope is shown as a thin solid line, the Newton diagram as a thick one*

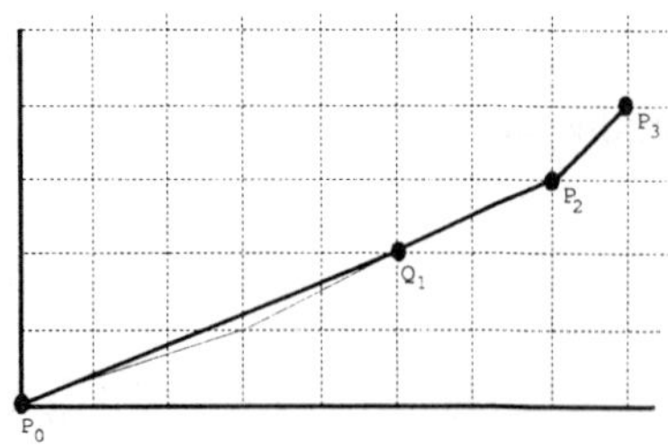

Figure 3b. *Newton diagram associated with $A + \varepsilon B_3$ for the matrix B_3 in (20). The Newton envelope is shown as a thin solid line, the Newton diagram as a thick one*

Now, consider

$$B_4 = \left[\begin{array}{ccc|cc|cc|c} 0 & 0 & 0 & 0 & 0 & 0 & 0 & 0 \\ 0 & 0 & 0 & 0 & 0 & 0 & 0 & 0 \\ 0 & 0 & 0 & 1 & 0 & 1 & 0 & 0 \\ \hline 0 & 0 & 0 & 0 & 0 & 0 & 0 & 0 \\ 1 & 0 & 0 & 0 & 0 & 1 & 0 & 1 \\ \hline 0 & 0 & 0 & 0 & 0 & 0 & 0 & 0 \\ -1 & 0 & 0 & 1 & 0 & 1 & 0 & 0 \\ \hline 2 & 0 & 0 & 0 & 0 & 1 & 0 & 0 \end{array}\right], \tag{21}$$

with

$$\Phi_3 = \left[\begin{array}{c|cc|c} 0 & 1 & 1 & 0 \\ \hline 1 & 0 & 1 & 1 \\ -1 & 1 & 1 & 0 \\ \hline 2 & 0 & 1 & 0 \end{array}\right].$$

In this case, Φ_1 is singular, while Φ_2 and Φ_3 are not. The point $\mathcal{Q}_1$ is not in the set Π and the Newton diagram is the one in Figure 4a, so $A + \varepsilon B_4$ has seven eigenvalues of order $\varepsilon^{3/7}$ and one eigenvalue of order ε.

Finally, consider

$$B_5 = \left[\begin{array}{ccc|cc|cc|c} 0 & 0 & 0 & 0 & 0 & 0 & 0 & 0 \\ 0 & 0 & 0 & 0 & 0 & 0 & 0 & 0 \\ 0 & 0 & 0 & 1 & 0 & 1 & 0 & 0 \\ \hline 0 & 0 & 0 & 0 & 0 & 0 & 0 & 0 \\ 0 & 0 & 0 & 0 & 0 & 0 & 0 & 1 \\ \hline 0 & 0 & 0 & 0 & 0 & 0 & 0 & 0 \\ 0 & 0 & 0 & 1 & 0 & 0 & 0 & 0 \\ \hline 1 & 0 & 0 & 0 & 0 & 1 & 0 & 0 \end{array}\right], \tag{22}$$

with

$$\Phi_3 = \left[\begin{array}{c|cc|c} 0 & 1 & 1 & 0 \\ \hline 0 & 0 & 0 & 1 \\ 0 & 1 & 0 & 0 \\ \hline 1 & 0 & 1 & 0 \end{array}\right].$$

In this case, both Φ_1 and Φ_2 are singular and Π contains $\mathcal{P}_0$ and $\mathcal{P}_3$. Again, the point $\mathcal{Q}_1$ does not belong to Π, so the Newton diagram is a single segment of slope $1/2$ (see Figure 4b) and $A+\varepsilon B_5$ has eight eigenvalues of order $\varepsilon^{1/2}$.

We conclude the example here, although the five instances above do not exhaust all possible behaviors: notice that Φ_3 has been assumed to be invertible in all four cases. Other possibilities arise if Φ_3 becomes singular, but we do not pursue the matter further for the sake of brevity. In any case, the example above

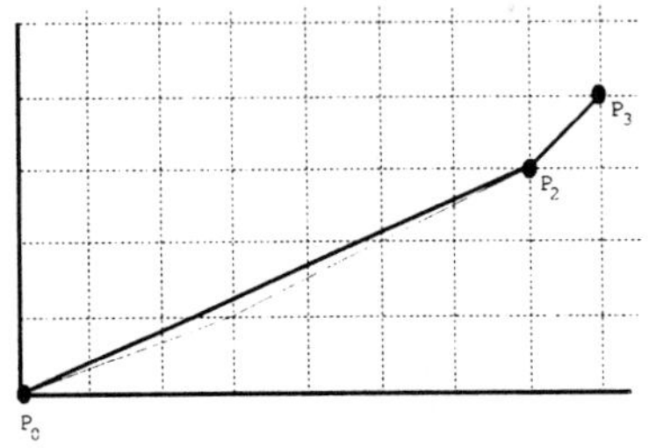

Figure 4a. *Newton diagram associated with $A + \varepsilon\, B_4$ for the matrix B_4 in (21). The Newton envelope is shown as a thin solid line, the Newton diagram as a thick one*

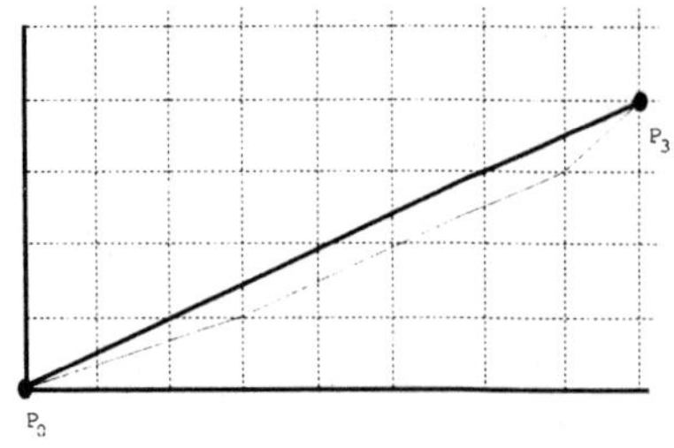

Figure 4b. *Newton diagram associated with $A + \varepsilon\, B_5$ for the matrix B_5 in (22). The Newton envelope is shown as a thin solid line, the Newton diagram as a thick one*

displays the richness of different behaviors to be expected from nongeneric perturbations to matrices with nontrivial Jordan forms.

In fact, the nongeneric case presents a very important additional difficulty which is absent for the generic case: namely that one can no longer make the simplifying assumption of disregarding the part $\widehat{J}$ of the Jordan canonical form corresponding to eigenvalues other than λ_0. The following example illustrates this:

Example 4: Consider the matrix

$$A = \left[\begin{array}{c|c|c} 0 & & \\ \hline & 0 & \\ \hline & & 1 \end{array}\right] \in \mathbb{R}^{3\times 3}, \tag{23}$$

which is already in Jordan form, but having two different eigenvalues, and take the perturbation

$$B = \begin{bmatrix} 0 & 1 & 0 \\ 0 & 0 & 1 \\ 1 & 0 & 0 \end{bmatrix}. \tag{24}$$

Obviously, if we disregard the part corresponding to the nonzero eigenvalue 1, the perturbed matrix would reduce to $\begin{bmatrix} 0 & \varepsilon \\ 0 & 0 \end{bmatrix}$ which still has two zero eigenvalues. However, the characteristic polynomial of the 3 by 3 matrix $A + \varepsilon\, B$ is $P(\lambda, \varepsilon) = \lambda^3 - \lambda^2 - \varepsilon^3$. The corresponding Newton diagram, in Figure 5 below, shows that $A + \varepsilon\, B$ has one eigenvalues of the order of unity, and two eigenvalues of order $\varepsilon^{3/2}$.

Some attempts have been made to analyze the behavior for certain specific classes of nongeneric perturbations. Some preliminary results may be found

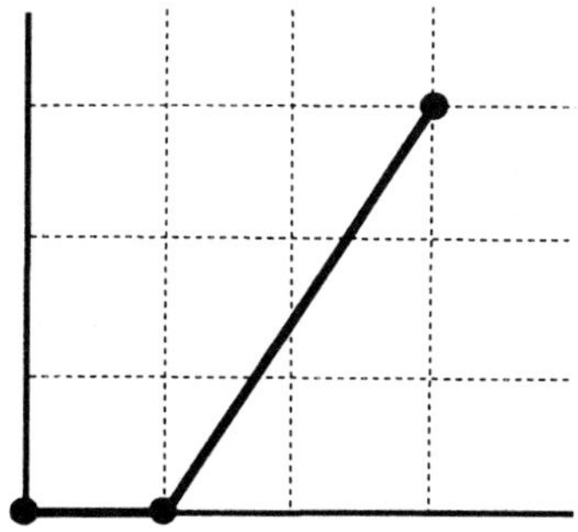

Figure 5. Newton diagram associated with the matrix $A + \varepsilon B$ for the matrices defined in (23) and (24).

in [16, § 3], but the first systematic description of structured perturbations is obtained in Ma and Edelman [15] for upper k-Hessenberg perturbations of Jordan blocks. Following the ideas both in [3] and [16], they show that if a n by n Jordan block is perturbed by an upper k-Hessenberg matrix (i.e. a matrix having the k subdiagonals closest to the main diagonal not zero), then the eigenvalue splits typically into $\lfloor n/k \rfloor$ eigenvalues of order $\varepsilon^{1/k}$ and, if k does not divide n, into r additional eigenvalues of order $\varepsilon^{1/r}$, where $r \equiv n \mod k$. The leading coefficients are also characterized. Again, as in Lidskii's theorem, the typical behavior (in this non-typical situation) is given by certain algebraic conditions which in this case depend only on the elements of the k-th subdiagonal of B.

More recently, Jeannerod [8] has extended Lidskii's results by obtaining explicit formulas for both the leading exponents and leading coefficients of the Puiseux expansions of the eigenvalues of *analytic* perturbations $J + B(\varepsilon)$ of a Jordan matrix J, provided the powers of ε in the perturbation matrix $B(\varepsilon)$ conform in a certain way to the Jordan structure given by J. To be more precise, the least power of ε in either of any two columns of B corresponding to a same Jordan block of J must be the same. Lidskii's Theorem 2 is recovered when that least power is ε for all blocks, but certain nongeneric cases are covered in this more general setting.

Nonetheless, in both cases [15; 8] the particular structure of the perturbations to the Jordan blocks is not preserved by undoing the change of basis leading to the Jordan form. Hence, not much information is provided for nongeneric perturbations of *arbitrary* matrices. Therefore, it is safe to say that, so far, the problem of categorizing all possible behaviors of the eigenvalues of $A + \varepsilon B$ as a function of the perturbation matrix B is a very difficult open problem, which is still quite far from being solved.

However, there is still another specific class of nongeneric perturbations for which quite some information can be obtained

3.2.1 Low rank perturbations. A relevant class of nongeneric perturbations are perturbations of low rank, which frequently arise when a system has to be controlled using less parameters than the number of degrees of freedom defining the system [40]. We focus on how do the eigenvalues and the Jordan structure of a matrix change when it is perturbed by a matrix of low rank, with particular attention to how many Jordan blocks are destroyed for each eigenvalue.

Although some of the used techniques have a similar flavor to the Newton diagram technique, an important difference with the rest of the paper has to be highlighted: *the perturbations studied here are not infinitesimal. Thus we will be able to obtain absolute, instead of first order, results holding for perturbations of any magnitude.* This said, we begin with an example which illustrates how a small rank perturbation typically behaves.

Example 5: Let us denote by $J_s(\lambda_0)$ a Jordan block of dimension s corresponding to the eigenvalue λ_0. Consider the following matrix

$$\begin{aligned} J &= J_3(1) \oplus J_3(1) \oplus J_3(1) \oplus J_1(1) \oplus J_3(-1) \oplus J_2(-1) \oplus J_2(-1) \oplus \\ &\quad J_3(2) \oplus J_3(2). \end{aligned}$$

We have generated 200 random matrices B with rank 2 and computed the Jordan canonical form [2] of $A+B$. The maximum and minimum values obtained for the ratios $\|B\|/\|J\|$ have been 305 and 130, i.e. the matrices $A+B$ are not by any means small perturbations of A. Surprisingly enough, in 171 out of the 200 cases analysed the Jordan canonical form of $A+B$ had one block $J_3(1)$, one block $J_1(1)$ and one block $J_2(-1)$. The rest of the blocks were one-dimensional blocks corresponding to eigenvalues different from $1, -1, 2$. This experiment suggests the following *generic* behavior: *If λ_0 is an eigenvalue of A having g (geometric multiplicity) Jordan blocks and B is a matrix with rank $m < g$, then λ_0 is also an eigenvalue of $A+B$ having exactly $g-m$ Jordan blocks equal to the $g-m$ smallest Jordan blocks of λ_0 in A*, i.e., when perturbing a matrix by a generic rank m perturbation, the m largest Jordan blocks for each eigenvalue dissapear and the rest of the Jordan blocks remain. This "usual" behavior has been observed in many other experiments, but notice that it is easy to build

[2] This was done with the command `'jordan'` of the Symbolic Math Toolbox in MATLAB 5.3. To work properly this command requires the elements of the matrix argument to be integers or ratios of small integers. Therefore, the random matrices B had moderately large integer elements, and are not as random as one might think. This is the reason why about 15% of the generated matrices did not follow the generic behavior described in the example.

m-rank perturbations which do not follow this rule. If $m \geq g$, the eigenvalue λ_0 is typicaly no longer an eigenvalue of $A + B$, as expected.

Now we will try to understand this generic behavior. The next Lemma gives necessary and sufficient algebraic conditions for it.

Lemma 1. *Let A be a matrix with Jordan form (6), i.e. having an eigenvalue λ_0 with Jordan blocks of dimensions $n_1 > n_2 > \ldots > n_q$ repeated $r_1, r_2, \ldots, r_q$ times. Let $m \in \{1, \ldots, g\}$ be an integer number, where g is the geometric multiplicity (9). Suppose $m = f_{j-1} + \rho$ for some $j = 1, \ldots, q$, $0 < \rho \leq r_j$ with $f_0 = 0$. Then λ_0 is an eigenvalue of the matrix $A+B$ with Jordan blocks of dimensions $n_j > n_{j+1} \ldots > n_q$ repeated $r_j - \rho, r_{j+1}, \ldots, r_q$ times (if $\rho = r_j$ there are no blocks of dimension n_j) if and only if*

a) $\operatorname{rank}(A + B - \lambda_0 I)^k = \operatorname{rank}(A - \lambda_0 I)^k + k\, m$, *for all* $k = 1, 2, \ldots, n_j$, *and*

b) $\operatorname{rank}(A + B - \lambda_0 I)^k = \operatorname{rank}(A + B - \lambda_0 I)^{n_j}$ *for all* $k \geq n_j$.

PROOF: Just remember [7, pag. 127] that for an arbitrary matrix C the number $d_k(C, \lambda) = \operatorname{rank}(C - \lambda I)^{k-1} - \operatorname{rank}(C - \lambda I)^k$ is the total number of Jordan blocks of C of dimensions greater than or equal to k associated with the eigenvalue λ. Therefore $d_k(C, \lambda) - d_{k+1}(C, \lambda)$ is the number of Jordan blocks of C associated to λ with dimension equal to k. Thus Conditions a) and b) of the Lemma are equivalent to $d_k(A + B, \lambda_0) = d_k(A, \lambda_0) - m$ for $k = 1, \ldots, n_j$ and $d_k(A + B, \lambda_0) = 0$ for $k > n_j$. The rest is trivial. □

Notice that Lemma 1 does not force B to be a matrix of rank m. In fact, our goal is to justify why most matrices of rank m fulfill the previous equations a) and b) on the ranks. The following plausibility argument shows, under mild restrictions on a matrix B with rank $m = f_{j-1} + \rho$, that something very unlikely has to happen for conditions a) and b) not to be fulfilled. In the first place, notice that for any k the matrix $(A - \lambda_0 I + B)^k$ can be expanded as a sum of $(A - \lambda_0 I)^k$ plus products of two matrices, with the left factor being of the form $(A - \lambda_0 I)^s B$ with s ranging from 0 to $k - 1$. Secondly, elementary linear algebra results show that if $\operatorname{col}(B) \cap \ker(A - \lambda_0 I)^{n_j - 1} = \{0\}$ (here col and ker stand for column and null spaces respectively) then $\operatorname{rank}(A - \lambda_0 I)^s B = \operatorname{rank}(B) = m$ for all $s = 0, 1, \ldots, n_j - 1$. Therefore, the equalities of item a) in Lemma 1 hold unless some "rank cancellation" occurs. It is well known that rank cancellations are very uncommon when the involved matrices have small rank, as in our case. With regard to the assumption $\operatorname{col}(B) \cap \ker(A - \lambda_0 I)^{n_j - 1} = \{0\}$, it is really very mild, since $\dim\, \operatorname{col}(B) = m$ and $\dim\, \ker(A - \lambda_0 I)^{n_j - 1} \leq a - m$. Hence, it follows that the set of matrices B of rank m not satisfying this condition has zero Lebesgue measure.

No need to say that the previous argument does not prove any rigorous result, but it sugests that the problem of understanding why the behavior described in

Example 5 is the most likely one is a problem of determining the Lebesgue measure of the set of matrices B of rank m not fulfilling the conditions of Lemma 1. The most convenient way to deal with this problem is to rewrite conditions a), b) in a simpler but equivalent form. All of these questions are beyond the scope of this survey because different techniques than the ones used here are required. However an argument much in the same spirit of the Newton diagram approach allows us to prove the following theorem which is an important step towards the complete solution of this question. More on this will be said in [17].

Theorem 3. *Let A be a $n \times n$ matrix with Jordan form (6), i.e. having an eigenvalue λ_0 with Jordan blocks of dimensions $n_1 > n_2 > \ldots > n_q$ repeated $r_1, r_2, \ldots, r_q$ times, and algebraic and geometric multiplicities given by (9). Let B be a $n \times n$ matrix with rank $m = f_{j-1} + \rho$ for some $j = 1, \ldots, q$, $0 < \rho \leq r_j$ with $f_0 = 0$. Then the characteristic polynomial of $A + B$ is of the form*

$$p(\lambda) = (\lambda - \lambda_0)^{\tilde{a}} \, t(\lambda - \lambda_0),$$

where

$$\tilde{a} = (r_j - \rho)n_j + r_{j+1}n_{j+1} + \ldots + r_q n_q \tag{25}$$

and $t(\lambda - \lambda_0)$ is a monic polynomial of degree $n - \tilde{a}$. Moreover the constant coefficient of t is

$$t(0) = (-1)^{m+n-a} \, C_0 \, \det(\widehat{Q}A\widehat{P} - \lambda_0 I), \tag{26}$$

where $\widehat{Q}$, $\widehat{P}$ are as in (6) and C_0 is the sum of all principal minors of Φ_j corresponding to submatrices of dimension m containing the upper left block Φ_{j-1} of Φ_j (if $j = 1$, all *principal minors of dimension m are to be considered). If, in particular, $m = f_j$ for some $j \in \{1, \ldots, q\}$, then C_0 is simply* $\det \Phi_j$.

PROOF: This proof resembles that of Theorem 1, but is more involved, since B is not a small perturbation of A. Thus, the simplifying assumption that A has only one eigenvalue cannot be made and the block $\widehat{J}$ of the Jordan canonical form of A containing the eigenvalues different from λ_0 plays also a role. Similarly, all four blocks of the matrix $\widetilde{B}$ defined in (11) have to be taken into account.

We begin by writing the characteristic polynomial of $A + B$ as

$$p(\lambda) = \det((\lambda - \lambda_0)I - \operatorname{diag}(J - \lambda_0 I, \widehat{J} - \lambda_0 I) - \widetilde{B}).$$

For the sake of simplicity we define $\tilde{\lambda} \equiv \lambda - \lambda_0$ and $p_0(\tilde{\lambda}) \equiv p(\lambda)$, so the coefficient of $\tilde{\lambda}^{n-k}$ in $p_0(\tilde{\lambda})$ is $(-1)^k$ times the sum of all k-dimensional principal minors of $\operatorname{diag}(J - \lambda_0 I, \widehat{J} - \lambda_0 I) + \widetilde{B}$. Notice that all principal minors having more than m rows containing *only* elements of $\widetilde{B}$ are zero because

$\mathrm{rank}(\widetilde{B}) = \mathrm{rank}(B) = m$. This simple remark is the key to prove the Theorem. The next step is to find the largest dimension of principal minors having m rows which contain only elements of $\widetilde{B}$. If the dimension of these minors is denoted by $k_{\max}$ then

$$p_0(\tilde{\lambda}) = \tilde{\lambda}^{n-k_{\max}}\, t(\tilde{\lambda}),$$

with t a monic polynomial of degree $k_{\max}$.

Let α be an index set included in $\{1, 2, \ldots, n\}$ and denote by $(\mathrm{diag}(J - \lambda_0 I, \widehat{J} - \lambda_0 I) + \widetilde{B})(\alpha, \alpha)$ the principal matrix of $\mathrm{diag}(J - \lambda_0 I, \widehat{J} - \lambda_0 I) + \widetilde{B}$ that lies in the rows and columns indexed by α. In order to construct the largest principal minors having m rows with only elements of $\widetilde{B}$, the set α has to be of the form

$$\alpha = \{i_1, \ldots, i_s, a+1, a+2, \ldots, n\} \quad \text{with} \quad 1 \le i_1 < i_2 < \ldots < i_s \le a,$$

because the diagonal elements of $\widehat{J} - \lambda_0 I + \widetilde{B}_{22}$, which have indices $a+1, a+2, \ldots, n$, are all of them different from the diagonal elements of $\widetilde{B}_{22}$ and the largest admissible size is desired. Suppose now that $i_1, \ldots, i_s$ are chosen among the indices corresponding to r Jordan blocks of $J - \lambda_0 I$. The row with the largest index chosen from a given block, say the i_b-th row, contributes to the principal minor only with elements of $\widetilde{B}$, either because it is the bottom row of the block or because $i_b + 1$ does not belong to α, and thus the element in the position $(i_b, i_b + 1)$ where $J - \lambda_0 I$ has a superdiagonal 1 is not in the minor. This imposes the restriction $r \le m$ on r. Hence, to obtain the maximum number of elements in α, i.e $k_{\max}$, the indices $i_1 < \ldots < i_s$ have to correspond to a set of m complete largest Jordan blocks of $J - \lambda_0 I$. The number of possible choices is $r_j!/(\rho!(r_j - \rho)!)$, which is simply one when $m = f_j$. In any case

$$k_{\max} = r_1 n_1 + \ldots + r_{j-1} n_{j-1} + \rho n_j + n - a,$$

$\tilde{a} = n - k_{\max}$ which is equation (25).

Now we prove (26). Remember that $t(0)$ is $(-1)^{k_{\max}}$ times the sum of all $k_{\max}$-dimensional principal minors of $\mathrm{diag}(J - \lambda_0 I, \widehat{J} - \lambda_0 I) + \widetilde{B}$. Moreover the only non-zero $k_{\max}$-dimensional principal minors are of the kind described in the previous paragraph. Consider one of these minors and call it M. Let us denote by $1 = j_1 < j_2 < \ldots < j_h$ ($h = k_{\max} - (n-a) - m$) the indices of the rows of this minor where $J - \lambda_0 I$ has superdiagonal 1s. The j_k-th row of this minor is the sum of two rows: one is the $j_k + 1$-th row e_{j_k+1} of the identity matrix, the other a piece of a row of $\widetilde{B}$. Using this fact, we can expand M as a sum of determinants whose j_k-th row is either e_{j_k+1} or a row with only elements of $\widetilde{B}$. With the exception of the determinant with all the vectors $e_{j_1+1}, e_{j_2+1}, \ldots, e_{j_h+1}$, the rest of these determinants are zero because each contains more than m rows with elements of $\widetilde{B}$. A similar argument on the last $n - a$ rows of M allows us to replace every element of $\widetilde{B}$ in these rows by zero without changing the value of M. The cofactor expansion of the remaining determinant along the rows $1 = j_1 < j_2 < \ldots < j_h$ leads to a value for M equal to $(-1)^h \det(\widehat{J} - \lambda_0 I)$ times a minor of Φ_j corresponding to a submatrix of dimension m containing the upper left block Φ_{j-1}. Extending this argument to all non-zero $k_{\max}$-dimensional principal minors of $\mathrm{diag}(J - \lambda_0 I, \widehat{J} - \lambda_0 I) + \widetilde{B}$ leads to (26). □

Notice that Theorem 3 is valid for any matrix B of rank m. If the following two additional restrictions are imposed on B: $t(0) \neq 0$ and $\mathrm{rank}(A - \lambda_0 I + B) = \mathrm{rank}(A - \lambda_0 I) + m$, then $A + B$ has m Jordan blocks less than A for λ_0, and the sum of the dimensions of the remaining blocks is precisely the sum of the dimensions of the $g - m$ smallest Jordan blocks of A for λ_0. It only remains to prove that changes between the dimensions of theses smallest blocks do not happen. This seems intuitively clear because the rank of B has already been used in imposing $t(0) \neq 0$. More on this will be said in [17].

In any case, the question raised in this section is, as far as we know, a new one in the literature. The only two references we are aware of [31, 32] are still unpublished work. In [31] the problem of rank one perturbations is adressed and it is proved that the condition $t(0) \neq 0$ is necessary and sufficient for $A + B$ to have one Jordan block less than A for λ_0, and for the dimensions of the remaining blocks to be precisely the dimensions of the $g - 1$ smallest Jordan blocks of A for λ_0. A similar behavior appears when adding one new row and one new column to a given matrix. This has been studied in [32, Section 2].

4. First order perturbation bounds for singular values

Given an arbitrary matrix $A \in \mathbb{C}^{m \times n}$, one can take advantage of its singular values being eigenvalues of an associated Hermitian matrix in order to obtain first order singular value perturbation results via Theorem 2. The key is using the so-called *Jordan-Wielandt matrix*

$$C = \begin{bmatrix} 0 & A \\ A^* & 0 \end{bmatrix} \in \mathbb{C}^{(m+n)\times(m+n)} \tag{27}$$

associated with A. One can easily check [35, § I.4.1] that if $m \geq n$ and $A = U\Sigma V^*$ is a singular value decomposition with $\Sigma = \mathrm{diag}(\sigma_1, \ldots, \sigma_n) \in \mathbb{R}^{m \times n}$, then the Hermitian matrix C has $2n$ eigenvalues $\pm\sigma_1, \ldots, \pm\sigma_n$ with corresponding normalized eigenvectors

$$\frac{1}{\sqrt{2}} \left[\frac{u_i}{\pm v_i} \right],$$

where u_i is the i-th column of U and v_i the i-th column of V. In addition, C has $m - n$ zero eigenvalues with eigenvectors $\left[\, u_i^T \mid 0 \,\right]^T$, $i = n+1, \ldots, m$.

If the matrix A is perturbed to $A(\varepsilon) = A + \varepsilon B$ as in section 3, the Jordan-Wielandt matrix C is correspondingly perturbed to $C(\varepsilon) = C + \varepsilon D$ with

$$D = \begin{bmatrix} 0 & B \\ B^* & 0 \end{bmatrix}, \tag{28}$$

i.e. to the Jordan-Wielandt matrix of $A(\varepsilon)$ provided the perturbation parameter ε is real. Recall that both C and D are Hermitian and, consequently, normal, so according to our remark right after the proof of Theorem 2, the eigenvalues of $C(\varepsilon)$ are analytic expansions of ε of the form (14) with $n_j = 1$. Of course, this is not necessarily true for the singular values of $A(\varepsilon)$, due to the nonnegativity restriction. However, the loss of analyticity can only be caused by some transversal crossing $\sigma(0) = 0$ of the two eigenvalues $\lambda(\varepsilon) = \pm\sigma(\varepsilon)$ of $C(\varepsilon)$ through the ε-axis at $\varepsilon = 0$. Thus, to recover the singular values all we have to do is to take the *nonnegative* branch. In other words, to obtain the leading term of a singular value expansion of $A(\varepsilon)$ we compute the expansion of the eigenvalue $\lambda(\varepsilon) = \xi\,\varepsilon + O(\varepsilon^2)$ of $C(\varepsilon)$. Then, the corresponding singular value of $A(\varepsilon)$ is just $\sigma(\varepsilon) = |\xi|\varepsilon + O(\varepsilon^2)$.

Some results on singular value perturbation expansions have been obtained by Stewart [34] via the ε-expansions for the eigenvalues $\sigma(\varepsilon)^2$ of $A(\varepsilon)^*A(\varepsilon)$. Sun, on the other hand, deals in [36] with the case of simple nonzero singular values, while the case of zero and multiple singular values is treated in [37]. Both cases are completely described in the following theorem, which, to our knowledge, is new.

Theorem 4. *Let $A \in \mathbb{C}^{m\times n}$, $m \geq n$ be a matrix of rank r, and let σ_0 be a singular value of A.*

If $\sigma_0 > 0$, let $U_0 \in \mathbb{C}^{m\times k}$ and $V_0 \in \mathbb{C}^{n\times k}$ be matrices whose columns span simultaneous bases of the respective left and right singular subspaces of A associated with σ_0. Then, for any $B \in \mathbb{C}^{m\times n}$, the matrix $A(\varepsilon) = A + \varepsilon B$ has k singular values analytic in ε which can be expanded as

$$\sigma_j(\varepsilon) = \sigma_0 + \xi_j\ \varepsilon + O(\varepsilon^2), \tag{29}$$

where the ξ_j, $j = 1, \ldots, k$ are the eigenvalues of the matrix

$$\frac{1}{2}(U_0^* B V_0 + V_0^* B^* U_0).$$

If $\sigma_0 = 0$ let

$$A = \left[\, U_r \;\middle|\; U_0 \;\middle|\; U_z \,\right] \begin{bmatrix} \Sigma_r & 0 \\ 0 & 0 \\ 0 & 0 \end{bmatrix} \left[\frac{V_r^*}{V_0^*} \right]$$

be a singular value decomposition of A with $U_r \in \mathbb{C}^{m\times r}$, $U_0 \in \mathbb{C}^{m\times(n-r)}$, $U_z \in \mathbb{C}^{m\times(m-n)}$, $V_r \in \mathbb{C}^{n\times r}$ and $V_0 \in \mathbb{C}^{n\times(n-r)}$. Then, for any $B \in \mathbb{C}^{m\times n}$, the matrix $A(\varepsilon) = A + \varepsilon B$ has $n-r$ singular values analytic in ε which can be expanded as

$$\sigma_j(\varepsilon) = \xi_j\, \varepsilon + O(\varepsilon^2), \tag{30}$$

where the ξ_j, $j = 1, \ldots, k$ are the singular values of the $(m-r) \times (n-r)$ matrix

$$\left[\frac{U_0^*}{U_z^*}\right] B\, V_0.$$

PROOF: As previously observed, we view the singular values of $A(\varepsilon) = A + \varepsilon B$ as the nonnegative eigenvalues of its Jordan-Wielandt matrix $C(\varepsilon) = C + \varepsilon D$ with C and D given, respectively, by (27) and (28).

In the simplest case when the singular value σ_0 is not zero, the columns of

$$\frac{1}{\sqrt{2}}\left[\frac{U_0}{V_0}\right] \in \mathbb{C}^{(m+n)\times k}$$

form an orthonormal basis of the space of eigenvectors associated with the semisimple eigenvalue σ_0 of the Hermitian matrix C. Hence, formula (17) with $Y_1 = X_1^*$ applied to the perturbed matrix $C + \varepsilon D$ leads to expansion (29).

The situation is slightly more complicated when $\sigma_0 = 0$, since one has to keep track of the $m-n$ additional null eigenvalues of C: the multiplicity of zero as an eigenvalue of C is now $m - n + 2(n-r) = m + n - 2r$, and the columns of

$$\mathcal{Z} = \frac{1}{\sqrt{2}}\begin{bmatrix} U_0 & U_0 & \sqrt{2}U_z \\ V_0 & -V_0 & 0 \end{bmatrix} \in \mathbb{C}^{(m+n)\times(m+n-2r)}$$

form an orthonormal basis of the null space of C. Again, applying Theorem 2 to the perturbation $C + \varepsilon D$ leads to the expansion (30), where the ξ_j are the nonnegative eigenvalues of the matrix

$$\begin{aligned} \mathcal{M} &= \mathcal{Z}^* \begin{bmatrix} 0 & B \\ B^* & 0 \end{bmatrix} \mathcal{Z} \\ &= \frac{1}{2}\begin{bmatrix} M_0 + M_0^* & M_0 - M_0^* & \sqrt{2}M_z \\ M_0^* - M_0 & -(M_0 + M_0^*) & -\sqrt{2}M_z \\ \sqrt{2}M_z^* & -\sqrt{2}M_z^* & 0 \end{bmatrix}, \end{aligned} \tag{31}$$

with $M_0 = V_0^* B^* U_0 \in \mathbb{C}^{(n-r)\times(n-r)}$ and $M_z = V_0^* B^* U_z \in \mathbb{C}^{(n-r)\times(m-n)}$. The proof is completed once we realize that the matrix $\mathcal{M}$ in (31) is unitarily similar to the Jordan-Wielandt matrix of

$$\widetilde{M} = [\ M_0 \mid M_z\ ] = V_0^* B^* [\ U_0 \mid U_z\ ].$$

One can easily check that the unitary matrix

$$\mathcal{Q} = \frac{1}{\sqrt{2}} \begin{bmatrix} I_{n-r} & I_{n-r} & 0 \\ -I_{n-r} & I_{n-r} & 0 \\ 0 & 0 & \sqrt{2} I_{m-n} \end{bmatrix}$$

satisfies that

$$\mathcal{Q}^* \mathcal{M} \mathcal{Q} = \begin{bmatrix} 0 & \widetilde{M} \\ \widetilde{M}^* & 0 \end{bmatrix},$$

i.e. $m-n$ eigenvalues of $\mathcal{M}$ are zero, and the remaining $2(n-r)$ ones are plus/minus the $n-r$ singular values of $\widetilde{M}$. □

5. First order perturbation bounds for generalized eigenvalue problems

In this final section we will just hint the close connections of the results in section 3 with their natural extension for generalized eigenvalue problems obtained by Najman [18, 19], and Langer and Najman [10, 11, 12] in a series of papers devoted to first order eigenvalue perturbation theory for perturbed analytic matrix functions. We focus on these results not only because of their great generality, but also because the main tool in their proofs is the Newton diagram technique, the common umbrella for most of the results in the present survey. This reinforces the resemblance of both the content of the results and the leading ideas in their proofs.

Following the presentation of Langer and Najman, we will analyze the behavior of the eigenvalues $\lambda(\varepsilon)$ of a square $n \times n$ matrix function

$$A(\lambda) + B(\lambda, \varepsilon) \tag{32}$$

for small ε in the neighborhood of an eigenvalue λ_0 of the unperturbed matrix function $A(\lambda)$, i.e. a value such that $\det A(\lambda_0) = 0$. The matrix function is assumed to be analytic around λ_0 with $\det A(\lambda) \not\equiv 0$. The perturbation $B(\lambda, \varepsilon)$ is assumed to be analytic in a neighborhood of $(\lambda_0, 0)$ with $B(\lambda, 0) = 0$ for every λ.

This general framework includes most of the usual spectral problems: taking $A(\lambda) = A - \lambda I$, $B(\lambda, \varepsilon) = \varepsilon B$ leads to the standard eigenvalue problem, the choice $A(\lambda) = \lambda^2 M + \lambda C + K$ with Hermitian positive definite matrices M, K and positive semidefinite C corresponding to perturbed quadratic matrix polynomials appearing in vibrational systems. Many other perturbed generalized eigenvalue problems can be fitted into this framework.

Najman began studying in [18] the case when $A(\lambda)$ is Hermitian and $B(\lambda, \varepsilon) = \varepsilon B(\lambda)$, not necessarily Hermitian. His results were an extension of those in Gohberg, Lancaster and Rodman [5] for the case when both $A(\lambda)$ and

$B(\lambda)$ are Hermitian and $B(\lambda)$ is positive definite. The results in [18] were precised and improved in [10], which already dealt with the general formulation (32). Reference [10] describes the *generic* behavior of the eigenvalues of (32) in the sense of Theorem 2 above, i.e. the behavior which is to be expected unless the perturbation lies on a certain algebraic manifold of zero measure related both with the perturbation and the spectral structure of A. Further work was devoted to exploring nongeneric cases [11, 12], as well as to the study of certain specific quadratic matrix polynomials appearing in damped vibrational systems [13, 19].

We will only state the analogous of Lidskii's Theorem 2 as stated in [11]. As in section 3, this requires some preliminary notations. Also as in section 3 we simplify the presentation *by assuming henceforth that* $\lambda_0 = 0$: let $A(\lambda)$ be an analytic $n \times n$ matrix function with $\det A(0) = 0$ and $\det A(\lambda) \not\equiv 0$. The *geometric multiplicity* of $\lambda_0 = 0$ is

$$g = \dim \ker A(0).$$

To define its partial multiplicities we make use of the Smith local form: one can show (see [20]) that there exist $n \times n$ matrix functions $E(\lambda)$, $F(\lambda)$ analytic and invertible close to $\lambda_0 = 0$ such that

$$A(\lambda) = E(\lambda)D(\lambda)F(\lambda), \tag{33}$$

with $D(\lambda) = \operatorname{diag}(\lambda^{\nu_1}, \ldots, \lambda^{\nu_n})$. Notice that since the ranks of $A(0)$ and $D(0)$ coincide, $n - g$ exponents ν_i must be zero. With no loss of generality we may assume that

$$D(\lambda) = \operatorname{diag}(\lambda^{n_1}, \ldots, \lambda^{n_1}, \lambda^{n_2}, \ldots, \lambda^{n_2}, \ldots, \lambda^{n_q}, \ldots, \lambda^{n_q}, 1, \ldots, 1),$$

where $n_1 < n_2 < \ldots < n_q$, each exponent n_j is repeated r_j times for $j = 1, \ldots, q$, and $r_1 + \ldots + r_q = g$, the geometric multiplicity of $\lambda_0 = 0$. The exponents n_j are called the *partial multiplicities* for $\lambda_0 = 0$ and

$$a = \sum_{j=1}^{q} r_j n_j$$

is its *algebraic multiplicity* as an eigenvalue of $A(\lambda)$. As in section 3, we define the auxiliary quantities

$$f_j = r_1 + \ldots + r_j, \quad j = 1, \ldots, q.$$

Now, consider the perturbed matrix function $A(\lambda) + B(\lambda, \varepsilon)$, with $B(\lambda, \varepsilon)$ analytic around $(0, 0)$ and $B(\lambda, 0) \equiv 0$. Then , using (33), $\lambda(\varepsilon)$ is an eigenvalue of $A(\lambda) + B(\lambda, \varepsilon)$ if and only if it is an eigenvalue of $D(\lambda) + \widehat{B}(\lambda, \varepsilon)$

for $\widehat{B}(\lambda,\varepsilon) = E(\lambda)^{-1}B(\lambda,\varepsilon)F(\lambda)^{-1}$. Now, partition

$$D(\lambda) = \begin{bmatrix} D_1(\lambda) & 0 \\ 0 & I \end{bmatrix}, \qquad \widehat{B}(\lambda,\varepsilon) = \begin{bmatrix} \widehat{B}_{11}(\lambda,\varepsilon) & \widehat{B}_{12}(\lambda,\varepsilon) \\ \widehat{B}_{21}(\lambda,\varepsilon) & \widehat{B}_{22}(\lambda,\varepsilon) \end{bmatrix},$$

where both $D_1(\lambda)$ and $\widehat{B}_{11}(\lambda,\varepsilon)$ are $g \times g$, and denote

$$H = \frac{\partial \widehat{B}_{11}}{\partial \varepsilon}(0,0). \tag{34}$$

For each $j = 1,\ldots,q$, $k = 1,\ldots,r_j - 1$ we define Δ_{jk} as the sum of all the $(g - f_j + k)$-dimensional principal minors of the submatrix $H(\alpha_{j-1},\alpha_{j-1})$ of H which contain the submatrix $H(\alpha_j,\alpha_j)$, where

$$\alpha_{j-1} = \{f_{j-1}+1, f_{j-1}+2,\ldots,g\}, \qquad \alpha_j = \{f_j+1, f_j+2,\ldots,g\}$$

and we have used the same notation as in the proof of Theorem 3 to represent submatrices. We also define

$$\Delta_j = \det H(\alpha_{j-1},\alpha_{j-1})$$

for each $j = 1,\ldots,q$, and for convenience we set $\Delta_{q+1} = 1$.

Theorem 5 (Langer & Najman [11]). *If for some $j \in \{1,\ldots,q\}$ both Δ_j and Δ_{j+1} do not vanish, then there are $r_j n_j$ eigenvalues of the perturbed matrix polynomial $A(\lambda) + B(\lambda,\varepsilon)$ near $\lambda_0 = 0$ admitting a first-order expansion*

$$\lambda_j^{kl}(\varepsilon) = (\xi_j^k)^{1/n_j}\, \varepsilon^{1/n_j} + o(\varepsilon^{1/n_j}) \tag{35}$$

for $k = 1,\ldots,r_j$, $l = 1,\ldots,n_j$, where the ξ_j^k, $k = 1,\ldots,r_j$, are the roots of the equation

$$\Delta_j + \sum_{k=1}^{r_j-1} \Delta_{jk}\xi^k + \Delta_{j+1}\xi^{r_j} = 0. \tag{36}$$

The resemblance of Theorems 2 and 5 is obvious. Disregarding some notational changes, like the partial multiplicities n_j being ordered increasingly, or the shift of one in the subindex j, it is clear that the quantities Δ_j play exactly the same role in this case as $\det \Phi_{j-1}$ in section 3. In particular, the generic behavior corresponds to nonvanishing Δ_j, $j = 1,\ldots,q$. The expansion (35) is exactly the same as (12), and the equation (36) determining the leading coefficients is the analogue of (15) or, equivalently, of (13) in this more general context. The surprising thing is that essentially the same result is obtained in this much more general context without much additional complication, just by replacing the Jordan canonical form with the Smith local form.

References

[1] H. BAUMGÄRTEL, *Analytic Perturbation Theory for Matrices and Operators*, Birkhäuser, Basel, 1985.

[2] E. BRIESKORNA AND H. KNÖRRER, *Plane Algebraic Curves*, Birkhäuser, Basel, 1986.

[3] J. V. BURKE AND M. L. OVERTON, *Stable perturbations of nonsymmetric matrices*, Linear Algebra Appl., 171 (1992), pp. 249–273.

[4] K. O. FRIEDRICHS, *On the perturbation of continuous spectra*, Comm. Pure Appl. Math. 1 (1948), pp. 361–406.

[5] I. GOHBERG, P. LANCASTER AND L. RODMAN, *Perturbations of analytic Hermitian matrix functions*, Applicable Analysis 20 (1985), pp. 23–48.

[6] E. HILLE AND R. S. PHILLIPS, *Functional Analysis and Semigroups*, Am. Math. Soc. Colloq. Publ. vol. 31, Providence, 1957.

[7] R. HORN AND C. R. JOHNSON, *Matrix Analysis*, Cambridge University Press, Cambridge, 1990.

[8] C.-P. JEANNEROD, *On some nongeneric perturbations of an arbitrary Jordan structure*, preprint.

[9] T. KATO, *Perturbation Theory for Linear Operators*, Springer, Berlin, 1980.

[10] H. LANGER AND B. NAJMAN, *Remarks on the perturbation of analytic matrix functions II*, Integr. Equat. Oper. Th., 12 (1989), pp. 392–407.

[11] H. LANGER AND B. NAJMAN, *Remarks on the perturbation of analytic matrix functions III*, Integr. Equat. Oper. Th., 15 (1992), pp. 796–806.

[12] H. LANGER AND B. NAJMAN, *Leading coefficients of the eigenvalues of perturbed analytic matrix functions*, Integr. Equat. Oper. Th., 16 (1993), pp. 600–604.

[13] H. LANGER, B. NAJMAN AND K. VESELIĆ, *Perturbation of the eigenvalues of matrix polynomials*, SIAM J. Matrix Anal. Appl. 13 (1992) pp. 474–489.

[14] V. B. LIDSKII, *Perturbation theory of non-conjugate operators*, U.S.S.R. Comput. Maths. Math. Phys., 1 (1965), pp. 73–85 (*Zh. vychisl. Mat. mat. Fiz., 6 (1965) pp. 52–60*).

[15] Y. MA AND A. EDELMAN, *Nongeneric Eigenvalue Perturbations of Jordan Blocks*, Linear Algebra Appl., 273, (1998) pp. 45-6-3

[16] J. MORO, J. V. BURKE and M. L. OVERTON, *On the Lidskii–Vishik–Lyusternik perturbation theory for eigenvalues of matrices with arbitrary Jordan structure*, SIAM J. Matrix Anal. Appl., 18 (1997), pp. 793–817.

[17] J. MORO AND F. M. DOPICO, *Low rank perturbation of eigenvalues of matrices with arbitrary Jordan canonical form*, in preparation.

[18] B. NAJMAN, *Remarks on the perturbation of analytic matrix functions*, Integr. Equat. Oper. Th., 9 (1986), pp. 592–599.

[19] B. NAJMAN, *The asymptotic behavior of the eigenvalues of a singularly perturbed linear pencil*, SIAM J. Matrix Anal. Appl., 20 (1998), pp. 420–427.

[20] M. NEWMAN, *The Smith normal form*, Linear Algebra Appl. 254 (1997), pp. 367–381.

[21] I. NEWTON, *The correspondence of Isaac Newton vol. 2 (1676-1687)*, Cambridge University Press, 1960.

[22] I. NEWTON, Methodus fluxionum et serierum infinitorum. In *The mathematical works of Isaac Newton*, D. T. Whiteside (ed.), Johnson Reprint Corp., New York, 1964.

[23] R. S. PHILLIPS, *Perturbation theory for semi-groups of linear operators*, Trans. Am. Math. Soc. 74 (1954), pp. 199–221.

[24] V. PUISEUX, *Recherches sur les fonctions algébriques*, J. Math Pures Appl., 15 (1850).

[25] L. RAYLEIGH, *The Theory of Sound, vol. I*, London 1894.

[26] F. RELLICH, *Störungstheorie der Spektralzerlegung, I. Mitteilung. Analytische Störung der isolierten Punkteigenwerte eines beschränkten Operators*, Math. Ann. 113 (1937), pp. 600-619.

[27] F. RELLICH, *Störungstheorie der Spektralzerlegung, II*, Math. Ann. 113 (1937), pp. 677-685.

[28] F. RELLICH, *Störungstheorie der Spektralzerlegung, III*, Math. Ann. 116 (1939), pp. 555-570.

[29] F. RELLICH, *Störungstheorie der Spektralzerlegung, IV*, Math. Ann. 117 (1940), pp. 356-382.

[30] F. RELLICH, *Störungstheorie der Spektralzerlegung, V*, Math. Ann. 118 (1942), pp. 462-484.

[31] S. SAVCHENKO, *The typical change of the spectral properties of a fixed eigenvalue under a rank one perturbation*, preprint (private communication).

[32] S. SAVCHENKO, *The Perron root of a principal submatrix of co-order one as an eigenvalue of the original nonnegative irreducible matrix and the submatrix itself*, preprint.

[33] E. SCHRÖDINGER, *Quantisierung als Eigenwertproblem III. Störungstheorie, mit Anwendung auf den Starkeffekt der Balmer-Linien*, Ann. Phys. 80 (1926), pp. 437–490.

[34] G. W. STEWART, *A note on the perturbation of singular values*, Linear Algebra Appl. 28 (1979), pp. 213-216.

[35] G. W. STEWART AND J.-G. SUN, *Matrix Perturbation Theory*, Academic Press, New York, 1990.

[36] J.-G. SUN, *A note on simple non-zero singular values*, Journal of Computational Mathematics 6 (1988), pp. 258-266.

[37] J.-G. SUN, *Sensitivity analysis of zero singular values and multiple singular values*, Journal of Computational Mathematics 6 (1988), pp. 325-335.

[38] B. v. SZ.-NAGY, *Perturbations des transformations autoadjointes dans l'espace de Hilbert*, Comment. Math. Helv. 19 (1946/47), pp. 347-366.

[39] M. M. VAINBERG AND V. A. TRENOGIN, *Theory of Branching of Solutions of Non-linear Equations*, Noordhoff, Leyden, 1974.

[40] K. VESELIĆ, *On linear vibrational systems with one dimensional damping*, Integr. Equat. Oper. Th., 13 (1990), pp. 883–897.

[41] M. I. VISHIK AND L. A. LYUSTERNIK, *The solution of some perturbation problems for matrices and selfadjoint or non-selfadjoint differential equations I*, Russian Math. Surveys, 15 (1960), pp. 1–74 (*Uspekhi Mat. Nauk, 15 (1960), pp. 3–80*).

II

CONTRIBUTED LECTURES

MICROLOCAL ENERGY DENSITY FOR HYPERBOLIC SYSTEMS*

Nenad Antonić
Department of Mathematics, University of Zagreb
Bijenička cesta 30, 10000 Zagreb, Croatia

Martin Lazar
Department of Mathematics, University of Zagreb
Bijenička cesta 30, 10000 Zagreb, Croatia

Abstract Starting from the method for computing microlocal energy density, which was developed independently by Francfort and Murat, and Gérard for the linear wave equation, we compute that very density for the hyperbolic system

$$\mathbf{A}^0\partial_0\mathsf{v} + \sum_1^d \mathbf{A}^k\partial_k\mathsf{v} + \mathbf{B}\mathsf{v} = \mathsf{f}.$$

We express the energy limit for the sequence of initial problems in terms of the energy of initial conditions. The basic tool we use are H-measures (also known as microlocal defect measures). We associate an H-measure to the sequence of gradients of solutions to our system and it represents the desired microlocal energy density.

We determine the system of equations satisfied by the corresponding H-measure. In the case of constant coefficients it reduces to a hyperbolic system similar to the initial one. Finally, we give a few examples related to the wave equation.

1. Introduction

In this paper we consider a hyperbolic symmetric system

$$\mathbf{A}^0\partial_0\mathsf{v} + \mathbf{A}^k\partial_k\mathsf{v} + \mathbf{B}\mathsf{v} = \mathsf{f}.$$

*This work is supported in part by the Croatian Ministry of Science and Technology through project 037015–Oscillatory solutions of partial differential equations.

Whenever there is no fear of ambiguity, Einstein's convention for repeated indices is assumed, with summation going from 1. The energy associated to the above system is

$$E := \frac{1}{2}\langle \mathbf{A}^0 \mathsf{v} \mid \mathsf{v} \rangle_{\mathrm{L}^2(\mathbf{R}^d)}.$$

The existence and uniqueness for the above hyperbolic system is given by the next theorem (see [2]).

Theorem 1. Existence and uniqueness
Let $\mathsf{f} \in \mathrm{H}^1(\mathbf{R}^{d+1};\mathbf{R}^r), \mathsf{g} \in \mathrm{H}^1(\mathbf{R}^d;\mathbf{R}^r)$ *and* $\mathbf{A}^k \in \mathrm{C}^2(\mathbf{R}^{d+1};\mathbf{M}^{r\times r})$ *symmetric,* $k = 0,\dots,d$ *and* $\mathbf{A}_0$ *be positive definite. Then there exists unique weak solution* $\mathsf{v} \in \mathrm{H}^1(\mathbf{R}^{d+1};\mathbf{R}^r)$ *of the initial problem*

$$\begin{cases} \mathbf{A}^0\partial_0\mathsf{v} + \mathbf{A}^k\partial_k\mathsf{v} + \mathbf{B}\mathsf{v} &= \mathsf{f} \\ \mathsf{v}(0,\cdot) &= \mathsf{g}. \end{cases}$$

∎

We consider a sequence of problems

$$\begin{cases} \mathbf{A}^0\partial_0\mathsf{v}_n + \mathbf{A}^k\partial_k\mathsf{v}_n + \mathbf{B}\mathsf{v}_n &= 0 \\ \mathsf{v}_n(0,\cdot) &= \mathsf{g}_n, \end{cases} \tag{1}$$

where we take $\mathsf{g}_n \rightharpoonup 0 \quad \text{in} \quad \mathrm{L}^2(\mathbf{R}^d;\mathbf{R}^r)$.

The convergence of initial conditions implies the convergence of solutions

$$\mathsf{v}_n \longrightarrow 0 \quad \text{in} \quad \mathrm{L}^2(\mathbf{R}^{d+1};\mathbf{R}^r).$$

We introduce the energy density function

$$d_n = \frac{1}{2}\mathbf{A}_0\mathsf{v}_n \cdot \mathsf{v}_n.$$

Our final goal is to compute the distributional limit of d_n. The basic tool in our calculus are H-measures. Their basic properties are presented in the next section. We express the desired limit of d_n through an H-measure and derive the system of equations satisfied by it. This is our main result given in Theorem 5. Furthermore, we apply this result to the wave equation. Rewriting the wave equation as a hyperbolic system, we calculate the associated H-measure for an oscillating sequence of initial conditions. The result is in accordance with the result obtained by a different method in the literature, and in a particular case to the one obtained by direct calculation of the H-measure from D'Alembert's formula for the solution of wave equation.

2. On H-measures

An H-measure is a Radon measure on the cospherical bundle over a domain Ω in consideration (in general, the base space of the fibre bundle is a manifold Ω, while the fibre is the unit sphere S^{d-1}). For a single parametrisation (suppose $\Omega \subseteq \mathbf{R}^d$ is an open domain) it is a measure on the product $\Omega \times S^{d-1}$. In order to apply the Fourier transform, functions defined on whole of $\mathbf{R}^d$ should be considered and this can be achieved by extending the functions by zero outside the domain. Such an extension preserves the weak convergence in L^2. After such adjustment, the following theorem can be stated (see [8, 4]):

Theorem 2. (existence of H-measures) *If* (u_n) *is a sequence in* $\mathrm{L}^2(\mathbf{R}^d;\mathbf{R}^r)$, *such that* $\mathsf{u}_n \xrightarrow{\mathrm{L}^2} 0$ *(weakly), then there exists a subsequence* $(\mathsf{u}_{n'})$ *and a complex matrix Radon measure* $\boldsymbol{\mu}$ *on* $\mathbf{R}^d \times \mathrm{S}^{d-1}$ *such that for all* $\varphi_1, \varphi_2 \in \mathrm{C}_0(\mathbf{R}^d)$ *and* $\psi \in \mathrm{C}(\mathrm{S}^{d-1})$*:*

$$\lim_{n'} \int_{\mathbf{R}^d} \mathcal{F}\Big(\varphi_1 \mathsf{u}_{n'}\Big) \otimes \mathcal{F}\Big(\varphi_2 \mathsf{u}_{n'}\Big)\psi\left(\frac{\boldsymbol{\xi}}{|\boldsymbol{\xi}|}\right) d\boldsymbol{\xi} = \langle \boldsymbol{\mu}, (\varphi_1\bar{\varphi}_2)\otimes\psi\rangle$$

$$= \int_{\mathbf{R}^d\times \mathrm{S}^{d-1}} \varphi_1(\mathbf{x})\bar{\varphi}_2(\mathbf{x})\psi(\boldsymbol{\xi})\, d\boldsymbol{\mu}(\mathbf{x},\boldsymbol{\xi})\,.$$

∎

In the above, $\mathcal{F}\phi$ denotes the Fourier transform of the function ϕ given by the relation

$$(\mathcal{F}\phi)\,(\boldsymbol{\xi}) = \int e^{-2\pi i\mathbf{x}\cdot\boldsymbol{\xi}}\phi(\mathbf{x})d\mathbf{x}.$$

As an immediate consequence of the last theorem we have that the H-measure associated to a strongly convergent sequence is necessarily zero.

The H-measure could be found explicitly in the following particular case, where T^{d-1} is a torus (the domain of periodic functions).

Theorem 3. *Let* $\mathsf{v} \in \mathrm{C}(\mathbf{R}^d \times T^{d-1})$ *and* $\mathsf{v}_n(\mathbf{x}) = \mathsf{v}(\mathbf{x}, n\mathbf{x})$. *The H-measure associated to the sequence* v_n *is given by*

$$\mu(\mathbf{x},\boldsymbol{\eta}) = \sum_{\mathbf{k}\in\mathbf{Z}^d\setminus\{0\}} \mathsf{v}^{\mathbf{k}}(\mathbf{x})\otimes\mathsf{v}^{\mathbf{k}}(\mathbf{x})\lambda(\mathbf{x})\times\delta_{\frac{\mathbf{k}}{|\mathbf{k}|}}(\boldsymbol{\eta}),$$

where $\mathsf{v}^{\mathbf{k}}$ *represents the Fourier coefficient with respect to the second variable. i.e.*

$$\mathsf{v}^{\mathbf{k}} = \int e^{-2\pi i\mathbf{k}\cdot\mathbf{y}}\mathsf{v}(\mathbf{x},\mathbf{y})d\mathbf{y}.$$

∎

A *classical pseudodifferential operator* is a linear operator

$$A: \mathcal{S}'(\mathbf{R}^d) \longrightarrow \mathcal{S}'(\mathbf{R}^d)$$

such that there is a function a, *an amplitude*, in the space $C^\infty(\mathbf{R}^d \times \mathbf{R}^d)$, with additional boundedness properties on derivatives specified below, such that:

$$Au(\mathbf{x}) = \int_{\mathbf{R}^d} e^{2\pi i \mathbf{x}\cdot\boldsymbol{\xi}} a(\mathbf{x}, \boldsymbol{\xi})\hat{u}(\boldsymbol{\xi})d\boldsymbol{\xi}.$$

The space of pseudodifferential operators on $\mathbf{R}^d$ of order $m \in \mathbf{Z}$, denoted by $\Psi^m(\mathbf{R}^d)$, consists of those operators whose amplitudes (symbols) satisfy the following bounds

$$(\forall\, \boldsymbol{\alpha}, \boldsymbol{\beta} \in \mathbf{N}_0^d) \quad |\partial_{\boldsymbol{\alpha}}\partial^{\boldsymbol{\beta}} a(\mathbf{x}, \boldsymbol{\xi})| \leq C_{\boldsymbol{\alpha\beta}}\Big(\sqrt{1 + 4\pi^2|\boldsymbol{\xi}|^2}\Big)^{m-|\boldsymbol{\beta}|} ,$$

i.e. belong to the Hörmander class $S^m_{0,1}$ (shortly denoted by S^m). In the above $\partial_k = \frac{\partial}{\partial x^k}$, while derivatives with respect to *dual* variable are denoted by $\partial^l = \frac{\partial}{\partial \xi_l}$. The same convention applies for Schwartz' multiindices. The principal symbol is defined as the homogeneous part of function a of degree m. By $\Psi^m_c(\mathbf{R}^d)$ we denote the subspace of $\Psi^m(\mathbf{R}^d)$ containing operators with compactly supported symbols in $\mathbf{x}$, resulting in an operator $A : \mathcal{S}' \to \mathcal{E}'$.

The statement of Theorem 2 can be rephrased as (complex scalar products of vectors and matrices are assumed)

$$(\forall\, \mathbf{P} \in \Psi^0_c(\mathbf{R}^d; \mathbf{M}^{r\times r})) \quad \lim_{n'} \int_{\mathbf{R}^d} \mathbf{P}\mathsf{u}_{n'} \cdot \mathsf{u}_{n'}\, d\mathbf{x} = \langle \boldsymbol{\mu}, \mathbf{p}\rangle ,$$

where $\mathbf{p}$ is the principal symbol of $\mathbf{P}$.

The following two results are classical (see e.g. [6, 7]).

Lemma 1. *Let P and Q be two pseudodifferential operators in $\Psi^m(\mathbf{R}^d)$ and $\Psi^n(\mathbf{R}^d)$ respectively, whose principal symbols are p and q. Then P^* (the adjoint of P), PQ (the composition) and $[P, Q] := PQ - QP$ (the commutator) are pseudodifferential operators which belong to $\Psi^m(\mathbf{R}^d)$, $\Psi^{m+n}(\mathbf{R}^d)$ and $\Psi^{m+n-1}(\mathbf{R}^d)$ respectively. Their principal symbols are given by $\overline{p}, pq$ and $\frac{1}{2\pi i}\{p, q\} = \frac{1}{2\pi i}\Big(\nabla_{\boldsymbol{\xi}} p \cdot \nabla_{\mathbf{x}} q - \nabla_{\mathbf{x}} p \cdot \nabla_{\boldsymbol{\xi}} q\Big)$.* ∎

Lemma 2. *If $P \in \Psi^m_c(\mathbf{R}^d)$ for some $m \in \mathbf{Z}^-$, then P is a compact operator on $L^2(\mathbf{R}^d)$.* ∎

Assume that $\mathsf{u}_n \xrightarrow{L^2} 0$ (weakly) and satisfy the following differential relations in divergence form

$$\partial_k(\mathbf{A}^k \mathsf{u}_n) = \mathsf{f}_n ,$$

where each $\mathbf{A}^k$ is a $(r \times r)$ matrix function, while $\mathsf{f}_n \xrightarrow{H^{-1}_{loc}} 0$ (strongly). Denoting (after passing to a subsequence if needed) the H-measure determined by

(u_n) by $\boldsymbol{\mu}$, we have, with the principal symbol in the differential relation being $\mathbf{p}(\mathbf{x}, \boldsymbol{\xi}) := \xi_k \mathbf{A}^k(\mathbf{x})$, the following *localisation identity* on $\mathbf{R}^d \times S^{d-1}$:

$$\mathbf{p}\boldsymbol{\mu} = \mathbf{0} .$$

(This result implies that the support of the H-measure $\boldsymbol{\mu}$ is contained in the set $\{(\mathbf{x}, \boldsymbol{\xi}) \in \mathbf{R}^d \times S^{d-1} : \det \mathbf{p}(\mathbf{x}, \boldsymbol{\xi}) = 0\}$ of points where $\mathbf{p}(\mathbf{x}, \boldsymbol{\xi})$ is a singular matrix. For the proof see [8] or [1], Theorem 2.)

Theorem 4. (propagation property for symmetric systems) *Let the hermitian matrices $\mathbf{A}^k$ be of class $\mathrm{C}^1_0(\Omega; \mathbf{M}^{r\times r})$. If for every n the pair $(\mathsf{u}_n, \mathsf{f}_n)$ satisfies the system*

$$\mathbf{A}^k \partial_k \mathsf{u} + \mathbf{B}\mathsf{u} = \mathsf{f} \;\; ,$$

and both sequences (u_n) and (f_n) converge to zero weakly in $\mathrm{L}^2(\Omega)$, then for every ψ of class C^1_0 on $\Omega \times S^{d-1}$, the H-measure associated to the sequence $(\mathsf{u}_n, \mathsf{f}_n)$:

$$\boldsymbol{\mu} = \begin{bmatrix} \boldsymbol{\mu}_{11} & \boldsymbol{\mu}_{12} \\ \boldsymbol{\mu}_{21} & \boldsymbol{\mu}_{22} \end{bmatrix} ,$$

satisfies the equation:

$$\langle \boldsymbol{\mu}_{11}, \{\mathbf{p}, \psi\} + \psi \partial_k \mathbf{A}^k - 2\psi \mathbf{S} \rangle + \langle 2 \mathsf{Re}\, \mathsf{tr} \boldsymbol{\mu}_{12}, \psi \rangle = 0 \; ,$$

where $\mathbf{S}$ is the hermitian part of the matrix $\mathbf{B}$, i.e. $\mathbf{S} := \frac{1}{2}(\mathbf{B} + \mathbf{B}^*)$. (The H-measure $\boldsymbol{\mu}$ is associated to the pair of sequences $(\mathsf{u}_n, \mathsf{f}_n)$, with block $\boldsymbol{\mu}_{11}$ corresponding to u_n and $\boldsymbol{\mu}_{22}$ to f_n, while nondiagonal blocks correspond to products of u_n and f_n.) ∎

3. Main result

Let us define function d_n^t representing energy density at moment t, i.e.

$$d_n^t = d_n(t, \cdot) = \frac{1}{2} \mathbf{A}_0 \mathsf{v}_n(t, \cdot) \cdot \mathsf{v}_n(t, \cdot).$$

Using the results of previous section the distributional limit of energy density function d_n^t is given by

$$\lim \langle d_n^t, \phi \rangle = \frac{1}{2} \langle \boldsymbol{\mu}^t, \phi \mathbf{A}_0 \rangle \; ,$$

where $\phi \in \mathrm{C}_c^\infty(\mathbf{R}^d)$ and $\boldsymbol{\mu}^t$ is the H-measure associated to the sequence $(\mathsf{v}_n(t, \cdot))$ of solutions of hyperbolic systems (1). Therefore, the desired energy is fully described through the H-measure $\boldsymbol{\mu}^t$ and this is what we are going to determine using the next theorem. For physical reasons we consider coefficients $\mathbf{A}^k$ to be independent of time, but a similar calculation can easily be derived in general case as well.

In order to clearly distinguish time and space components of variables, we use the following notation: variables in $\mathbf{R} \times \mathbf{R}^d$ are denoted by $\mathbf{x} = (x^0, x^1, \dots, x^d) = (x^0, x) = (t, x^1, \dots, x^d)$, whichever is more convenient. Similarly for the *dual* variable $\boldsymbol{\xi} = (\xi_0, \xi_1, \dots, \xi_d) = (\tau, \xi)$. We conveniently write $\nabla_{\mathbf{x}} = (\partial_t, \nabla_x)$ and use prime or dot for ∂_t.

Theorem 5. *The H-measure $\boldsymbol{\mu}^t$ associated to the sequence $(\mathsf{v}_n(t,\cdot))$ of solutions of hyperbolic systems (1) satisfies the transport equation*

$$\begin{aligned}\mathbf{A}^0\partial_t\boldsymbol{\mu}^t =& \{\mathbf{S}, \boldsymbol{\mu}^t\} - \nabla_x\mathbf{S} \otimes \nabla_\xi\boldsymbol{\mu}^t\xi \cdot \xi \\ &- \left((d-1)\nabla_x\mathbf{S}\cdot\xi + [(\nabla_\xi \otimes \nabla_x)\mathbf{S}]\xi\cdot\xi\right)\boldsymbol{\mu}^t \\ &- \mathbf{B}\boldsymbol{\mu}^t - \boldsymbol{\mu}^t\mathbf{B}^t - (\partial_k\mathbf{A}^k)\boldsymbol{\mu}^t,\end{aligned}$$

where $\mathbf{S}$ *stands for* $\sum \mathbf{A}^k\xi_k$.

Dem. Let P be a pseudodifferential operator of order zero with principal symbol $p(x,\xi)$ depending only on space and dual space variables. Using (1) we have

$$\begin{aligned}\frac{d}{dt}\langle P\mathbf{A}^0\mathsf{v}_n, \mathsf{v}_n\rangle &= \langle P\mathbf{A}^0\partial_0\mathsf{v}_n, \mathsf{v}_n\rangle + \langle P\mathbf{A}^0\mathsf{v}_n, \partial_0\mathsf{v}_n\rangle \\ &= -\langle P(\mathbf{A}^k\partial_k + \mathbf{B})\mathsf{v}_n, \mathsf{v}_n\rangle + \langle [P, \mathbf{A}^0]\mathsf{v}_n, \partial_0\mathsf{v}_n\rangle \\ &\quad + \langle P\mathsf{v}_n, -(\mathbf{A}^k\partial_k + \mathbf{B})\mathsf{v}_n\rangle \\ &= \langle -[P, \mathbf{A}^k\partial_k]\mathsf{v}_n, \mathsf{v}_n\rangle + \langle (\partial_k\mathbf{A}^k)P\mathsf{v}_n, \mathsf{v}_n\rangle \\ &\quad + \langle [P, \mathbf{A}^0]\mathsf{v}_n, \partial_0\mathsf{v}_n\rangle - \langle P\mathbf{B}\mathsf{v}_n, \mathsf{v}_n\rangle - \langle \mathbf{B}^*P\mathsf{v}_n, \mathsf{v}_n\rangle.\end{aligned}$$

In order to pass to the limit in the above equality, we take the advantage of the fact that the sequence of solutions of (1) belongs to the space $\mathrm{H}^1(\mathbf{R}^{d+1}; \mathbf{R}^r)$ and we introduce H-measure $\boldsymbol{\mu}^t$ associated to the sequence $\mathsf{v}_n(t,\cdot)$. As the commutator $[P, \mathbf{A}^0]$ is of negative order it does not contribute to the limit. The limit equation is

$$\begin{aligned}\frac{d}{dt}\int p\mathbf{A}^0 d\boldsymbol{\mu}^t =& -\int \sigma_0[P, \mathbf{A}^k\partial_k]d\boldsymbol{\mu}^t \\ &- \int \left(p\mathbf{B} + \mathbf{B}^*p - (\partial_k\mathbf{A}^k)p\right) d\boldsymbol{\mu}^t.\end{aligned} \tag{2}$$

The symbol of the commutator turns out to be

$$\sigma_0[P, A^k\partial_k] = \nabla_\xi(p)\cdot\nabla_x(\mathbf{S}) - \nabla_x(p)\cdot(\mathbf{A}^1, \dots, \mathbf{A}^d),$$

where we denote $\mathbf{S} = \sum A^k \xi_k$. Integration by parts in (2) yields the transport equation for the measure $\boldsymbol{\mu}^t$

$$\begin{aligned}\mathbf{A}^0 \partial_t \boldsymbol{\mu}^t =& \{\mathbf{S}, \boldsymbol{\mu}^t\} - \nabla_x \mathbf{S} \otimes \nabla_\xi \boldsymbol{\mu}^t \xi \cdot \xi \\ &- \big((d-1)\nabla_x \mathbf{S} \cdot \xi + [(\nabla_\xi \otimes \nabla_x)\mathbf{S}]\xi \cdot \xi\big)\boldsymbol{\mu}^t \\ &- \mathbf{B}\boldsymbol{\mu}^t - \boldsymbol{\mu}^t \mathbf{B}^t - (\partial_k \mathbf{A}^k)\boldsymbol{\mu}^t.\end{aligned} \tag{3}$$

Q.E.D.

For constant coefficient case the above equation simplifies to a system similar to the initial one

$$\mathbf{A}^0 \partial_0 \boldsymbol{\mu}^t + \mathbf{A}^k \partial_k \boldsymbol{\mu}^t + \mathbf{B}\boldsymbol{\mu}^t + \boldsymbol{\mu}^t \mathbf{B}^\tau = \mathbf{0}. \tag{4}$$

The equation (3) is similar to the one obtained in [1]. The main difference is that here we use a pseudodifferential operator independent of time and its dual variable. This enables us to determine the H-measure from the initial conditions for the system (1).

We would like to apply the above result to the wave equation

$$(\rho u')' - \mathsf{div}\,(\mathbf{A}\nabla u) = 0,$$

which can be rewritten as a hyperbolic system

$$\begin{aligned}\partial_0 u - v_0 &= 0 \\ \rho\partial_0 v_0 - a^{ij}\partial_i v_j + b^0 v_0 + b^j v_j &= 0 \\ a^{ij}\partial_0 v_i - a^{ij}\partial_i v_0 &= 0\ ,\end{aligned}$$

where we denote $b^0 := \partial_0 \rho, \quad b^j := -\partial_i a^{ij} = [-\mathsf{div}\,\mathbf{A}^\top]^j, \quad j = 1, \ldots, d.$ The function u appears explicitly only in the first equation. Thus, we can solve the system for v_i first, and later use the solution in order to obtain u. This reduces the system to $d+1$ unknowns $\mathsf{v} = (v_0, \ldots, v_d)$ and $d+1$ equations:

$$\begin{bmatrix} \rho & 0 & \cdots & 0 \\ 0 & & & \\ \vdots & & \mathbf{A} & \\ 0 & & & \end{bmatrix} \partial_0 \mathsf{v} + \sum_{i=1}^{d} \begin{bmatrix} 0 & -a^{i1} & \cdots & -a^{id} \\ -a^{i1} & & & \\ \vdots & & \mathbf{0} & \\ -a^{id} & & & \end{bmatrix} \partial_i \mathsf{v}$$

$$+ \begin{bmatrix} b^0 & b^1 & \cdots & b^d \\ 0 & & & \\ \vdots & & \mathbf{0} & \\ 0 & & & \end{bmatrix} \mathsf{v} = \begin{bmatrix} 0 \\ \vdots \\ 0 \end{bmatrix}.$$

In the particular case where $\rho = 1$ and $\mathbf{A} = \mathbf{I}$, the transport equation (4) for H-measure $\boldsymbol{\mu}^t$ reads

$$\partial_t \boldsymbol{\mu}^t = -(\mathbf{A}^1, \ldots, \mathbf{A}^d) \cdot \nabla_x \boldsymbol{\mu}^t,$$

or in components

$$\begin{aligned} \partial_t \mu^t_{0j} &= \sum_{m=1}^{d} \partial_m \mu^t_{mj}, \qquad j = 0, \dots, d \\ \partial_t \mu^t_{ij} &= \partial_i \mu^t_{0j}, \qquad i = 1, \dots, d, \qquad j = 0, \dots, d. \end{aligned} \tag{5}$$

Taking the time derivatives we get

$$\begin{aligned} \partial^2_t \mu^t_{0j} &= \sum_{m=1}^{d} \partial^2_m \mu^t_{0j} \\ \partial^2_t \mu^t_{ij} &= \partial_i \partial_m \mu^t_{mj}. \end{aligned} \tag{6}$$

The H-measure associated to the derivatives of solutions of wave equation is of special form

$$\begin{aligned} \boldsymbol{\mu}^t &= (\boldsymbol{\xi} \otimes \boldsymbol{\xi}) \nu^t \\ \nu^t &= \mathrm{tr} \mu^t = \mu^t_{00} + \mu^t_{11}, \end{aligned}$$

where ν^t is a scalar Radon measure. Therefore it suffices to write measure ν^t explicitly. More precisely, we want to express ν^t through the measure $\boldsymbol{\mu}^0$ associated to a sequence of oscillating initial conditions

$$\begin{aligned} u^n(0) &= \gamma_n(x) = \frac{1}{n} \alpha(x, nx) \\ \partial_t u^n(0) &= \beta_n(x) = \beta(x, nx), \end{aligned}$$

i.e.

$$(v^n_0, v^n_1)(0) = (\beta_n, \partial_2 \alpha(x, nx)).$$

Such a measure associated to an oscillating sequence can easily be calculated by using Theorem 3, and it is given by

$$\begin{aligned} \mu^0_{00} &= \sum_{k \in \mathbf{Z} \setminus \{0\}} |\beta_k(x, \xi)|^2 \lambda(x) \otimes \delta_{\frac{k}{|k|}}(\xi) \\ \mu^0_{01} &= -2\pi i \sum_{k \in \mathbf{Z} \setminus \{0\}} k \beta_k(x, \xi) \overline{\alpha_k(x, \xi)} \lambda(x) \otimes \delta_{\frac{k}{|k|}}(\xi) \\ \mu^0_{11} &= 4\pi^2 \sum_{k \in \mathbf{Z} \setminus \{0\}} k^2 |\alpha_k(x, \xi)|^2 \lambda(x) \otimes \delta_{\frac{k}{|k|}}(\xi), \end{aligned}$$

where α_k, β_k denote the Fourier transform with respect to the second variable.

Further simplification can be achieved by taking $d = 1$. In such a way (6) becomes

$$\partial^2_t \mu^t_{ij} = \partial^2_1 \mu^t_{ij} \qquad i, j \in \{0, 1\}. \tag{7}$$

We have got a wave equation for each entry of the H-measure μ^t. By (7) and (5) we get the expression for μ^t_{00}

$$\begin{aligned}\mu^t_{00} &= \frac{1}{2}\left[\mu^0_{00}(x+t) + \mu^0_{00}(x-t) + \int_{x-t}^{x+t} (\partial_t \mu^t_{00})_0(y)dy\right] \\ &= \frac{1}{2}\left[\mu^0_{00}(x+t) + \mu^0_{00}(x-t) + \int_{x-t}^{x+t} (\partial_1 \mu^t_{10})_0(y)dy\right] \\ &= \frac{1}{2}\left[\mu^0_{00}(x+t) + \mu^0_{00}(x-t) + \mu^0_{10}(x+t) + \mu^0_{10}(x-t)\right].\end{aligned}$$

Analogously

$$\mu^t_{11} = \frac{1}{2}[\mu^0_{11}(x+t) + \mu^0_{11}(x-t) + \mu^0_{01}(x+t) + \mu^0_{01}(x-t)].$$

Thus

$$\begin{aligned}\nu^t &= \mu^t_{00} + \mu^t_{11} \\ &= \frac{1}{2}\sum_{\pm}[\mu^0_{00}(x\pm t) + \mu^0_{10}(x\pm t) + \mu^0_{01}(x\pm t) + \mu^0_{11}(x\pm t)] \\ &= \frac{1}{2}\sum_{\pm}\sum_{k}\Big(|\beta_k(x\pm t)|^2 \pm 4\pi k \mathrm{Im}\,[\beta_k(x\pm t)\overline{\alpha_k(x\pm t)}] \\ &\quad + 4\pi^2 k^2 |\alpha_k(x\pm t)|^2\Big)\lambda(x)\otimes\delta_{\frac{k}{|k|}}(\xi).\end{aligned} \tag{8}$$

4. Another approach

In this section we are going to calculate the H-measure associated to the sequence of solutions of the wave equation in one space dimension, using D'Alembert's formula for the explicit solution.

More precisely, we consider a sequence of problems

$$\begin{aligned}\partial_t^2 u_n - \partial_x^2 u_n &= 0 \\ u_n(0) = \gamma_n(x) &= \frac{1}{n}\alpha(x, nx) \\ \partial_t u^n(0) = \beta_n(x) &= \beta(x, nx),\end{aligned}$$

where α and β are smooth functions on $\mathbf{R}^d \times T^{d-1}$. The solution of the above problem is

$$u_n(t,x) = \frac{1}{2}\left(\gamma_n(x+t) + \gamma_n(x-t)\right) + \frac{1}{2}\int_{x-t}^{x+t} \beta_n(y)dy. \tag{9}$$

We are interested in the H-measure $\boldsymbol{\mu}$ associated to the sequence of gradients $(\partial_t u_n, \partial_x u_n)$, because the energy density function is

$$d_n = \frac{1}{2}\Big((\partial_t u_n)^2 + (\partial_x u_n)^2\Big).$$

From (9) we have

$$\begin{aligned}\partial_t u_n &= \frac{1}{2}\Big(\partial_y \alpha\big(x+t, n(x+t)\big) - \partial_y \alpha\big(x-t, n(x-t)\big) \\ &\quad + \beta\big(x+t, n(x+t)\big) + \beta\big(x-y, n(x-t)\big)\Big) \\ \partial_x u_n &= \frac{1}{2}\Big(\partial_y \alpha\big(x+t, n(x+t)\big) + \partial_y \alpha\big(x-t, n(x-t)\big) \\ &\quad + \beta\big(x+t, n(x+t)\big) - \beta\big(x-y, n(x-t)\big)\Big).\end{aligned}$$

As H-measure $\boldsymbol{\mu}$ is of the form

$$\boldsymbol{\mu} = (\boldsymbol{\xi} \otimes \boldsymbol{\xi})\nu$$

where $\nu = \mathrm{tr}\boldsymbol{\mu}$ is a scalar Radon measure, it suffices to find entries μ_{00} and μ_{11}.

To this effect we first calculate the H-measures μ_{00}^{+} and μ_{11}^{-} associated to the sequences $\partial_y \alpha\big(x+t, n(x+t)\big) + \beta\big(x+t, n(x+t)\big)$ and $\partial_y \alpha\big(x-t, n(x-t)\big) - \beta\big(x-y, n(x-t)\big)$ respectively. Using Theorem 3 we get

$$\begin{aligned}\mu_{00}^{+}(\mathbf{x}, \boldsymbol{\xi}) = \frac{1}{4}\sum_{k\in\mathbf{Z}\setminus\{0\}}\bigg(4\pi^2|\alpha_k(x+t)|^2 + 2\mathrm{Re}\Big(-2\pi i k \beta_k(x+t)\alpha_k(x+t)\Big) \\ + |\beta_k(x+t)|^2\bigg)\lambda(\mathbf{x}) \otimes \delta_{\{\frac{\mathrm{sign}\,(k)}{\sqrt{2}}, \frac{\mathrm{sign}\,(k)}{\sqrt{2}}\}}(\boldsymbol{\xi}).\end{aligned}$$

Analogously for the another measure we get

$$\begin{aligned}\mu_{00}^{-}(\mathbf{x}, \boldsymbol{\xi}) = \frac{1}{4}\sum_{k\in\mathbf{Z}\setminus\{0\}}\bigg(4\pi^2|\alpha_k(x-t)|^2 + 2\mathrm{Re}\Big(-2\pi i k \beta_k(x-t)\alpha_k(x-t)\Big) \\ + |\beta_k(x-t)|^2\bigg)\lambda(\mathbf{x}) \otimes \delta_{\{-\frac{\mathrm{sign}\,(k)}{\sqrt{2}}, \frac{\mathrm{sign}\,(k)}{\sqrt{2}}\}}(\boldsymbol{\xi}).\end{aligned}$$

It is easy to see that $\nu = \mu_{00} + \mu_{11} = 2(\mu_{00}^{+} + \mu_{11}^{+})$. Therefore

$$\nu(\mathbf{x}, \boldsymbol{\xi}) = \frac{1}{2} \sum_{\pm} \sum_{k \in \mathbf{Z} \setminus \{0\}} \left(4\pi^2 |\alpha_k(x \pm t)|^2 \pm 4\pi k \mathsf{Im} \left(\beta_k(x \pm t) \alpha_k(x \pm t) \right) \right.$$
$$\left. + |\beta_k(x \pm t)|^2 \right) \lambda(\mathbf{x}) \otimes \delta_{\{\pm \frac{\text{sign}(k)}{\sqrt{2}}, \frac{\text{sign}(k)}{\sqrt{2}}\}}(\boldsymbol{\xi}). \tag{10}$$

This is exactly the same result as the expression obtained in [3].

We would like to compare the result to the one derived in previous section, i.e. to formula (8). Here we notice significant difference between measures given in formulas (8) and (10). Namely, the latter depends on dual time variable τ as well, while the first one does not. However, this is not a problem, as we are calculating the distributional limit of energy density function d_n, that is

$$\lim_n \langle d_n, \Phi \rangle \ ,$$

where $\Phi \in C_c^\infty(\mathbf{R}^{d+1})$. We assume Φ to be of special form $\Phi(\mathbf{x}) = \psi(t) \otimes \phi(x)$. From (10) we have

$$\lim_n \int d_n(t,x)\psi(t)\phi(x)dtdx = \langle \psi(t)\phi(x), \nu(t,x,\tau,\xi) \rangle, \tag{11}$$

while (8) gives

$$\begin{aligned} \lim_n \int d_n(t,x)\psi(t)\phi(x)dtdx &= \lim_n \int_{\mathbf{R}} \psi(t) \int_{\mathbf{R}^d} d_n(t,x)\phi(x)dxdt \\ &= \int \psi(t) \int_{\mathbf{R}^d} \langle \psi(t)\phi(x), \nu^t(x,\xi) \rangle dt. \end{aligned} \tag{12}$$

The right hand sides of (12) and (11) are equal, and the same result is obtained by the third method for this example.

References

[1] Nenad Antonić: H-measures applied to symmetric systems, Proceedings of the Royal Society of Edinburgh **126A** (1996) 1133-1155.

[2] Lawrence C. Evans: Partial Differential equations, American Mathematical society, Providence, 1991.

[3] Gilles A. Francfort, François Murat: Oscillations and energy densities in the wave equation, Communications in Partial Differential Equations **17** (1992) 1785–1865.

[4] Patrick Gérard: Microlocal defect measures, Communications in Partial Differential Equations **16** (1991) 1761–1794.

[5] Patrick Gérard: Oscillations and concentration effects in semilinear dispersive wave equations, Journal of Functional Analysis **141** (1996) 60–98.

[6] Lars Hörmander: The analysis of linear partial differential Operators I–IV, Springer-Verlag, Berlin, 1983–1985.

[7] Xavier Saint-Raymond: Elementary introduction to the theory of pseudodifferential operators, CRC Press, 1991.

[8] Luc Tartar: H-measures, a new approach for studying homogenisation, oscillations and concentration effects in partial differential equations, Proceedings of the Royal Society of Edinburgh **115A** (1990) 193–230.

APPROXIMATE SOLUTIONS TO SOME SECOND ORDER LINEAR RECURRENCES

Katalin Balla*
Computer and Automation Research Institute
Hungarian Academy of Science,
1518 Budapest, P.O.Box 63. Hungary
balla@sztaki.hu

Vilmos Horvat†
Department of Mathematics,
University of Zagreb,
Bijenička cesta 30, 10000 Zagreb, Croatia
horvat@math.hr

Abstract The paper deals with the computation of bounded solutions for some linear second order recurrences. The problem is transformed to an equivalent problem and then, its approximation is investigated. For this purpose, an associated Riccati difference equation and its approximate solution are studied. An a'priori error estimate is given. As an example, a boundary value problem is analysed.

Keywords: Linear boundary value problems, Riccati difference equation, asymptotic expansions.

1. Introduction

In paper [4] we characterized the bounded solutions to second order linear recurrences

$$y_{n+1} + b_n y_n + h_n y_{n-1} = e_n \tag{1}$$

with real coefficients such that relations

$$\lim_{n\to\infty} b_n = b_\infty, \quad \lim_{n\to\infty} h_n = h_\infty \quad \text{and} \lim_{n\to\infty} e_n = e_\infty \tag{2}$$

* Supported by Hungarian National Scientific Foundation Grant (OTKA) # T031807.
† Supported by grant 037011 of the Ministry of Science and Technology, Croatia.

hold with some constants b_∞, h_∞, e_∞ and the quadratic equation

$$q^2 + b_\infty q + h_\infty = 0$$

have a pair of (real) roots q_1, q_2,

$$|q_1| > 1 > |q_2| > 0. \tag{3}$$

Under these conditions, we stated that a solution of (1) is bounded *iff*, for any sufficiently large n, it satisfies the first order recurrence

$$y_{n+1} = \omega_n y_n + \psi_n, \tag{4}$$

where the pair ω and ψ is the solution of the problem

$$\omega_{n+1}\omega_n + b_{n+1}\omega_n + h_{n+1} = 0,\ \omega_\infty := \lim_{n\to\infty} \omega_n = q_2, \tag{5}$$

$$\psi_{n+1} + (\omega_{n+1} + b_{n+1})\psi_n - e_{n+1} = 0,\ \psi_\infty := \lim_{n\to\infty} \psi_n = -\frac{e_\infty}{q_1 - 1}. \tag{6}$$

In [4] the solvability of problems (5) and (6) was studied. It was shown that ω_n and ψ_n exist for sufficiently large n and are defined uniquely by (5) and (6). A concluding remark indicated that by the help of functions ω and ψ, the problem (1) with

$$y_n = O(1), \quad n \to \infty, \tag{7}$$

can be transformed to the equivalent problem (1), (4) with a fixed, sufficiently large $n = N$ in (4). Note that (7), as it is, cannot be treated numerically.

Further, it was proven that under the assumption that b, h, e have representations in the form of asymptotic series with respect to non-positive powers of n, so have ω and ψ. Moreover, the coefficients in ω and ψ can be computed by algebraic operations.

The aim of this paper is to show that truncated series for ω and ψ may be used in computations. Namely, we show that given an error bound ε and the number S of terms in the truncated series, one can compute N such that approximation $\tilde{\omega}$ to ω at $n \geq N$ be within the given error bound. Function ψ and its approximation $\tilde{\psi}$ may be treated similarly.

The result may be applied as follows. Let problem (1) and (7) be supplied with a left boundary condition

$$y_0 = \bar{y}_0. \tag{8}$$

The approximate boundary value problem consisting of (1), (8) and

$$y_{N+1} = \tilde{\omega}_N y_N + \tilde{\psi}_N \tag{9}$$

furnishes us with an approximate solution if N is chosen sufficiently large. Now, we are able to fix N so that the error be kept within the a'priori prescribed error bound. Note that Olver's algorithm [2] would give us an approximate solution in a different way.

A similar idea for continuous problems, i.e., segregation of bounded solutions for second order linear ordinary differential equations posed on $[a, \infty)$ was considered in [7]. The approximation of boundary conditions and its error estimate was studied in [6] and [1].

2. Riccati difference equation for the error term

Let us consider the Riccati difference equation

$$\xi_{n+1} = c_n \xi_n + f \xi_n \xi_{n+1} + g_n, \quad n \in \mathbf{N} = \{0, 1, 2 \ldots\}, \tag{10}$$

where

$$c_n := \frac{q_1}{q_2} - \frac{b_{n+1} - b_\infty}{q_2},\ f := -\frac{1}{q_2},\ g_n := -\frac{h_{n+1} - h_\infty}{q_2} - (b_{n+1} - b_\infty),$$

and let assumptions (2) and (3) hold. By Theorem 1 [4, pp. 72] equation (10) has exactly one solution ξ such that $\xi_n \to 0$ when $n \to \infty$. If, additionally, the coefficients of equation (10) have asymptotic representation of the form

$$c_n \sim \sum_{i=0}^{\infty} \frac{C_i}{n^i}; \quad g_n \sim \sum_{i=1}^{\infty} \frac{G_i}{n^i}, \tag{11}$$

then, by the Theorem 2 [4, pp. 74], so has solution ξ:

$$\xi_n \sim \sum_{i=1}^{\infty} \frac{\Omega_i}{n^i}, \quad n \to \infty, \tag{12}$$

with coefficients Ω_i which can be computed by formal substitution into (10), i.e., by algebraic operations using the coefficients in (11). Let an approximation $\tilde{\xi}$ to ξ be defined by

$$\tilde{\xi}_n = \sum_{i=1}^{S} \frac{\Omega_i}{n^i} \tag{13}$$

where S is an arbitrarily fixed value. Denote the error for the approximation by τ:

$$\tau_n := \xi_n - \tilde{\xi}_n.$$

Substitution into equation (10) yields

$$\begin{aligned} \tau_{n+1} \quad &- \quad c_n \tau_n - f \tau_n \tau_{n+1} - f(\tau_n \tilde{\xi}_{n+1} + \tau_{n+1} \tilde{\xi}_n) \\ &= \quad -\tilde{\xi}_{n+1} + c_n \tilde{\xi}_n + f \tilde{\xi}_n \tilde{\xi}_{n+1} + g_n. \end{aligned} \tag{14}$$

The defect δ is on the right hand side of equation (14),

$$\delta_n := -\tilde{\xi}_{n+1} + c_n\tilde{\xi}_n + f\tilde{\xi}_n\tilde{\xi}_{n+1} + g_n. \tag{15}$$

It is noteworth that even for the functions t and d defined by

$$t_n := n^S\tau_n, \; d_n := n^S\delta_n, \tag{16}$$

the limiting relations

$$t_n \to 0 \quad \text{and} \quad d_n \to 0 \quad \text{for } n \to \infty$$

hold. As a consequence, Theorems 1 and 2 [4] are applicable to

$$t_{n+1} = \hat{c}_n t_n + \hat{f}_n t_n t_{n+1} + \hat{d}_n, \quad \lim_{n\to\infty} t_n = 0, \tag{17}$$

where

$$\begin{aligned}
\hat{c}_n &= \left(\frac{n+1}{n}\right)^S \frac{c_n + f\tilde{\xi}_{n+1}}{1 - f\tilde{\xi}_n}, \quad \hat{f}_n = n^{-S}\frac{f}{1 - f\tilde{\xi}_n}, \\
\hat{d}_n &= \left(\frac{n+1}{n}\right)^S \frac{1}{1 - f\tilde{\xi}_n} d_n.
\end{aligned}$$

and $\lim_{n\to\infty} \hat{c}_n = c_\infty$, $\lim_{n\to\infty} \hat{f}_n n^S = f$, $\lim_{n\to\infty} \hat{d}_n = 0$. Thus, t defined by (16) is the unique solution of problem (17).

3. A simple error estimation

Let us define functions T and D by

$$T_{\tilde{n}} := \max_{n\geq\tilde{n}} \mid t_n \mid \quad \text{and} \quad D_{\tilde{n}} := \max_{n\geq\tilde{n}} \mid d_n \mid.$$

Clearly, T and D are monotone, non-increasing w.r. to their argument $\tilde{n}$ and

$$T_{\tilde{n}} \to 0 \quad \text{and} \quad D_{\tilde{n}} \to 0, \quad \text{when} \quad \tilde{n} \to \infty. \tag{18}$$

The constant variation formula may be applied to (17) and, after some computation, we arrive at inequality

$$T_{\tilde{n}} \leq \frac{1}{\mid c_\infty \mid -1}\left[\bar{f}_{\tilde{n}}\left(|f|\tilde{n}^{-S}T_{\tilde{n}}^2 + \left(\frac{\tilde{n}+1}{\tilde{n}}\right)^S D_{\tilde{n}}\right) + \bar{c}_{\tilde{n}}T_{\tilde{n}}\right], \tag{19}$$

where

$$\bar{f}_{\tilde{n}} = \max_{n\geq\tilde{n}}\left|\frac{1}{1 - f\tilde{\xi}_n}\right|, \quad \bar{c}_{\tilde{n}} = \max_{n\geq\tilde{n}}\left|\left(\frac{n+1}{n}\right)^S \frac{c_n + f\tilde{\xi}_{n+1}}{1 - f\tilde{\xi}_n} - c_\infty\right|.$$

Due to assumptions, the discriminant of the quadratic inequality is positive for sufficiently large $\tilde{n}$. This claim becomes transparent if rewritten as

$$D_{\tilde{n}} \leq \frac{(1- \mid c_\infty \mid + \bar{c}_{\tilde{n}})^2}{4\bar{f}_{\tilde{n}}^2 |f|} \left(\frac{\tilde{n}}{\tilde{n}+1} \right)^S \tilde{n}^S. \tag{20}$$

Clearly, inequality

$$T_{\tilde{n}} \leq r_{\tilde{n}} \quad \text{where} \quad r_{\tilde{n}} := \frac{2\left(\frac{\tilde{n}+1}{\tilde{n}}\right)^S \bar{f}_{\tilde{n}} D_{\tilde{n}}}{\mid c_\infty \mid -1 + \bar{c}_{\tilde{n}}}, \tag{21}$$

is sufficient for (19). In most applications, one easily constructs an upper estimate $r'_{\tilde{n}}$ of simple form for $r_{\tilde{n}}$, say, by replacing $\bar{f}_{\tilde{n}}$ and $D_{\tilde{n}}$ for some of their upper estimate and $\bar{c}_{\tilde{n}}$ for some of its nonnegative lower estimate such that (20) remain valid. Then, for the maximal error $T_{\tilde{n}}$, the inequality $T_{\tilde{n}} \leq \varepsilon$ will be ensured if $r'_{\tilde{n}} \leq \varepsilon$ holds.

4. Example

For illustration, let us consider a boundary value problem posed for a second order differential equation on interval $[0, \infty)$:

$$y'' = s(x)y, \quad y(0) = \bar{y}_0, \quad \lim_{x\to\infty} y(x) = 0,$$

assuming that

$$s(x) = s^\infty + \frac{s^1}{x+1}, \qquad s^\infty, s^1 \text{ - real, } s^\infty > 0,\ s^1 > 0.$$

The standard discretization scheme yields an infinite system of linear equations with an infinite tridiagonal matrix:

$$-(2 + s_1 h^2) y_1 + y_2 = -\bar{y}_0, \tag{22}$$

$$y_{n-1} - (2 + s_n h^2) y_n + y_{n+1} = 0, \tag{23}$$

n=2,3,.... Here, for simplicity, we used a fixed stepsize h, $s_n := s(nh)$ and y_n denotes the approximation of the solution value $y(nh)$ by the scheme.

The problem satisfies assumptions (2) and (3) with $b_n = -(2 + s_n h^2)$ and $h_n \equiv 1$. The equivalent problem reduced to interval $[0, X]$, $X = (N+1)h$, consists of equations (22), (23) for $n = 1, 2, ...N$ and

$$-\omega_N y_N + y_{N+1} = 0, \tag{24}$$

where ω is the solution of

$$\omega_{n+1}\omega_n - (2 + s_{n+1}h^2)\omega_n + 1 = 0, \tag{25}$$

$$\lim_{n\to\infty} \omega_n = q_2, \tag{26}$$

$$q_2 = \frac{1}{2}\left(2 + s^\infty h^2 - \sqrt{s^\infty h^2(s^\infty h^2 + 4)}\right) > 0, \quad q_1 = \frac{1}{q_2} > 1.$$

Note, that here ω_n exists for arbitrary n and thus, N may be chosen so that y_n appear in system only up to practical interest. The problem (25)-(26), however, cannot be solved in closed form. By the help of the process described in §3, we could compute, however, an approximation to ω for sufficiently large $n \geq \tilde{n}$ by taking a few first terms, say, two terms from the asymptotic expansion and obtain $|\tilde{\omega}_n - \omega_n| \leq \varepsilon$ for all $n \geq \tilde{n}$:

$$\tilde{\omega}_n = q_2 + \frac{\Omega_1}{n}, \quad \Omega_1 = \frac{s_1 h}{1 - \frac{q_1}{q_2}}.$$

It may happen, however, that $\tilde{n} > N$. In order to avoid superfluous computation of values y_n we rather obtain $\tilde{\omega}_N$.

For this purpose we observe that for any solution of equation (25) such that $|\omega_{\bar{N}}| < 1$, one has $|\omega_n| < 1$ for $0 \leq n < \bar{N}$. This statement can be verified by induction argument beginning from $n = \bar{N} - 1$ in backward direction. At each step, one can check that $\omega_n \geq 1$ would involve $\omega_{n+1} > 1$ while $\omega_n \leq -1$ would yield $\omega_{n+1} > 2$.

Let $\bar{N} = \max\left(\tilde{n}, \left[\frac{-\Omega_1}{1+q_2}\right] + 1\right)$ and ω_n be the exact solution of (25), (26) and $\tilde{\omega}_n$ be the solution of (25) for $n = \bar{N}, \bar{N} - 1, \ldots, N$ computed backward with the initial value $\tilde{\omega}_{\bar{N}}$. Obviously, the above observation applies to both ω_n and $\tilde{\omega}_n$. Then, for the error term $\varepsilon_n := \omega_n - \tilde{\omega}_n$, we can derive equation $\varepsilon_n = \varepsilon_{n+1}\omega_n\tilde{\omega}_n$, i.e., $|\varepsilon_n| \leq |\varepsilon_{n+1}| \leq \cdots \leq |\varepsilon_{\bar{N}}| = \varepsilon$. Thus, during the backward computation of $\tilde{\omega}$, the initial error in $\tilde{\omega}_N$ is damped.

This observation confirms also that the linear system obtained for the exact boundary value problem on $[0, N + 1]$ is well conditioned and thus, small perturbation of the entry ω_n causes small perturbation in the solution. Moreover, the factorization methods work well for the system [8, 9]. It means that the error in the solution is governed by the error in the boundary condition and it is of the same order.

If $\tilde{n} < N$, then, it is enough to solve the reduced system on $[0, \tilde{n} + 1]$. As noted above, it is well conditioned. For $n \in (\tilde{n} + 1, N + 1]$, one may apply recursion $y_{n+1} = \omega_n y_n$.

References

[1] A. A. Abramov, K. Balla. *O priblizhennyh resheniyah, osnovannyh na teoremah sravneniya, skalyarnyh i matrichnih uravnenij Riccati na beskonechnom intervale.* Zhurnal Vychislitel'noj Matematiki i Matematicheskoj Fiziki, 33 (1): 35-51, 1993; English transl. in Comput. Math. Math. Phys. 33, 1993.

[2] R. P. Agarwal. *Difference Equations and Inequalities. Theory, Methods and Applications. Second Edition, Revised and Expanded.* New York- Basel: Marcel Dekker Inc., 1999.

[3] K. Balla. *Asymptotic behaviour of certain Riccati difference equations.* Computers Math. Applic. 36 (10-12): 243-250, 1998.

[4] K. Balla. *On asymptotic behaviour of solutions to some difference equations.* In: Advances in Difference Equations, Proceedings of the 2nd Int. Conf. on Difference Equations, Veszprém, Hungary, 1995 (eds.: S. Elaydi, G. Ladas and I. Győri) Amsterdam: Gordon and Beach Sci. Publ., pp.67-80, 1997.

[5] N. S. Bahvalov. *Chislennye metody.* Moscow: Nauka, 1973.

[6] E. S. Birger. *Ob ocenke pogreshnosti zameny usloviya ogranichennosti resheniya linejnogo differencial'nogo uravneniya na beskonechnom intervale.* Zhurnal Vychislitel'noj Matematiki i Matematicheskoj Fiziki, 8 (3): 674-678, 1968; English transl. in USSR J. Comput. Math. Math. Phys. 8, 1968.

[7] E. S. Birger, N. B. Lyalikova. *O nahozhdenii dlya nekotoryh sistem obyknovennyh differencial'nyh uravnenij reshenij s zadannym usloviem na beskonechnosti. I.* Zhurnal Vychislitel'noj Matematiki i Matematicheskoj Fiziki, 5 (6): 979-990, 1965; English transl. in USSR J. Comput. Math. Math. Phys. 5, 1965.

[8] S. K. Godunov, V. S. Ryaben'kij. *Raznostnye shemy.* Moscow: Nauka, 1977.

[9] A. A. Samarskij. *Vvedenie v teoriyu raznostnyh shem.* Moscow: Nauka, 1971.

ASYMPTOTIC BEHAVIOUR OF TENSION SPLINE COLLOCATION MATRIX*

Ivo Beroš
Department of Mathematics, University of Zagreb
Bijenička cesta 30, 10000 Zagreb, Croatia
ivo.beros@math.hr

Miljenko Marušić
Department of Mathematics, University of Zagreb
Bijenička cesta 30, 10000 Zagreb, Croatia
miljenko.marusic@math.hr

Abstract Tension spline of order k is a function that, for given partition $x_0 < x_1 < \ldots < x_n$, on each interval $[x_i, x_{i+1}]$ satisfies differential equation $(D^k - p_i^2/h_i^2 D^{k-2})u = 0$, where p_i's are prescribed nonnegative real numbers.

Many articles deal with tension splines of order four applied to collocation method for solving singularly perturbed boundary value problem $(Lu)(x) = \varepsilon u''(x) + b(x)u'(x) + c(x)u(x) = f(x)$, $(0 \le x \le 1)$ with Dirichlet boundary conditions $u(0) = u_0$, $u(1) = u_1$. Accuracy of considered approximations is $\mathcal{O}(h)$ or $\mathcal{O}(h^2)$ for small perturbation parameter ε, depending on the choice of collocation points.

Here we present an algorithm for a collocation method with higher order tension splines ($k = 6, 8$) and, for the case $k = 6$, examine behaviour of the collocation matrix when perturbation parameter ε tends to 0. Our results show that the accuracy is $\mathcal{O}(h^3)$ for small perturbation parameter.

Keywords: tension spline, collocation method

*This work was supported by grant 037011 by Croatian Ministry of Science and Technology.

1. Preliminaries

We consider numerical solution of the singularly perturbed two-point boundary value problem for the ODE

$$(Lu)(x) \quad := \quad \varepsilon u''(x) + b(x)u'(x) + c(x)u(x) = f(x) \tag{1}$$

$$u(0) = u_0, \; u(1) = u_1. \tag{2}$$

We assume that perturbation parameter ε satisfies $0 < \varepsilon \ll 1$ and coefficient c satisfies $-c(x) \geq c_{min} > 0$ for $x \in [0, 1]$. These conditions guarantee existence and uniqueness of a solution to the problem (1)–(2).

Tension spline of order k is a function that, for a given partition $x_0 < x_1 < \ldots < x_n$, on each interval $[x_i, x_{i+1}]$ satisfies differential equation

$$(D^k - \frac{p_i^2}{h_i^2} D^{k-2})u = 0,$$

where p_i's are prescribed nonnegative real numbers ("tension parameters") and $h_i = x_{i+1} - x_i$.

Almost all papers dealing with tension splines consider tension splines of order four. Applications of such splines are mostly shape preserving approximation [15, 16] and a numerical solution of singularly perturbed two-point boundary value problem for ODE [3, 10]. Higher order tension splines are described in several papers [4, 9], and in [2] an algorithm is described for a collocation method using tension splines of order six or eight.

Here we describe collocation method using tension splines of order six and behavior of the collocation matrix when perturbation parameter $\varepsilon \ll h_i$.

2. B-splines

As for the case of polynomial splines, we use B-spline representation for tension splines [4, 6, 8, 11, 12].

Assume that partition Π of interval $[a, b]$ is given by

$$\Pi \ldots \quad a = x_0 < x_1 < \ldots < x_n = b,$$

and the non-negative real numbers p_i are prescribed for each interval (x_{i-1}, x_i), $i = 1, \ldots, n$. Further, we assume that multiplicity vector $\mathbf{l} = (l_1, \ldots, l_{n-1})$, $l_i \in \mathbf{N}$, is given. Using knots of partition Π and multiplicity vector $\mathbf{l}$, we can define extended partition $\mathbf{t} = (t_{1-k}, \ldots, t_{N+k})$ of $[a, b]$, where $N = l_1 + l_2 + \cdots + l_{n-1}$ and

$$t_i = \begin{cases} x_0, & i = -k-1, \ldots, 0, \\ x_r, & \sum_{j=1}^{r-1} l_j + 1 \leq i \leq \sum_{j=1}^{r} l_j, \\ x_n, & i = N+1, \ldots, N+k. \end{cases}$$

For each non-empty interval (t_{i-1}, t_i), we define nonnegative number ρ_i, where

$$\rho_i := \begin{cases} p_1/h_1, & i = -k+1, \ldots, 1, \\ p_r/h_r, & \text{if } t_i = x_r, \\ p_n/h_n, & i = N, \ldots, N+k-1, \end{cases} \qquad h_r = x_r - x_{r-1}.$$

Tension B-splines of order k on the $\mathbf{t}$ with tension parameters ρ_i are defined recursively, see [4, 13]. In [4], it is proved that B-splines $B_{i,k}$ have the following properties:

(i) supp $B_{i,k} = [t_i, t_{i+k}]$ (local support property)

(ii) $B_{i,k}(x) > 0$ for $t_i < x < t_{i+k}$ (positivity)

(iii) $\sum_i B_{i,k} \equiv 1$, $k \geq 3$ (partition of unity)

(iv) $B_{i,k}$ has $k - 1 - l$ continues derivatives at a knot of multiplicity l.

Moreover, B-splines $B_{j,k}$, $j = -k+1, \ldots n$, are linearly independent and they form the basis for the space $\mathcal{T}_k = \text{span}\{B_{-k+1,k}, \ldots, B_{n,k}\}$. $\mathcal{T}_k$ is a space of tension spline of order k with smoothness at knots t_i defined by the multiplicity of the knots.

We concentrate our attention to the case where knots $x_1, \ldots, x_{n-1}$ of partition Π have multiplicity one, i.e. $\mathbf{1} = (1, 1, \ldots, 1)$. Then, the extended partition is

$$\mathbf{t} = (\underbrace{x_0, \ldots, x_0}_{k}, x_1, \ldots, x_{n-1}, \underbrace{x_n, \ldots, x_n}_{k}) \tag{3}$$

and tension parameters are

$$\rho_i := \begin{cases} p_1/h_1, & i = -k+1, \ldots, 1, \\ p_i/h_i, & \text{if } 1 \leq i \leq n-1, \\ p_n/h_n, & i = n, \ldots, n+k-1, \end{cases} \qquad h_r = x_r - x_{r-1}.$$

With $\mathcal{T}_k^0 = \mathcal{T}_k^0(\mathbf{t}, \rho)$ we denote the space of tension splines of order k with knot sequence $\mathbf{t}$ and tension parameters $\rho = (\rho_1, \ldots, \rho_n)$. Dimension of $\mathcal{T}_k^0$ is $n + k - 1$, and $\mathcal{T}_k^0 \subset C^k(x_0, x_n)$.

Observe B-splines $B_{j,k}^l \in \mathcal{T}_k^0(\mathbf{t}, \rho(s))$, which are defined on the partition $\mathbf{t}$ with tension parameters $\rho(s) = (\rho_1(s), \ldots, \rho_n(s))$. Let

$$\rho_{min}(s) = \min\{\rho_1(s), \ldots, \rho_n(s)\} \text{ and } \rho_{max}(s) = \max\{\rho_1(s), \ldots, \rho_n(s)\}.$$

If

$$\lim_{s \to \infty} \rho_{min}(s) = \infty \tag{4}$$

and

$$\rho_{max}(s)/\rho_{min}(s) \leq \alpha \in \mathbf{R}, \text{ for all } s \tag{5}$$

then we have next theorem, proved in [1]:

Theorem 1. *Let* $\mathbf{t}$ *be partition of interval* $[a,b]$ *as in (3) and let* $\rho(s)$ *be sequence of tension parameters which satisfy (4) i (5). Let* $P_{j,k}$ *denotes polynomial B-spline of order* k *and* $B^s_{j,k}$ *denotes tension B-spline of order* k*,* $B^s_{j,k} \in \mathcal{T}^0_k(\mathbf{t},\rho(s))$*. Then, for* $k \geq 4$*,*

$$\begin{aligned}
\lim_{s\to\infty} B^s_{j,k}(x) &= P_{j+1,k-2}(x),\ j=-k+3,\ldots,n-3,\\
\lim_{s\to\infty} B^s_{-k+2,k}(x) &= P_{-k+3,k-2}(x),\ x\neq a,\\
\lim_{s\to\infty} B^s_{n-2,k}(x) &= P_{n-1,k-2}(x),\ x\neq b,\\
\lim_{s\to\infty} B^s_{-k+1,k}(x) &= \begin{cases} 1, & x=a,\\ 0, & \textit{otherwise,}\end{cases}\\
\lim_{s\to\infty} B^s_{n-1,k}(x) &= \begin{cases} 1, & x=b,\\ 0, & \textit{otherwise.}\end{cases}
\end{aligned}$$

3. Collocation method

Here we briefly describe collocation methods by tension splines $s \in \mathcal{T}^0_k$. Assume that equidistant partition Π of interval $[0,1]$ is given by

$$\Pi\ldots \quad 0 = x_0 < x_1 < \ldots < x_n = 1,\ x_i = i/n,$$

and the nonnegative real numbers ρ_i are prescribed. Further, suppose that x_i, $i=0,\ldots,n$, are also collocation points.

Spline s is the collocation spline for the problem (1)–(2) if it satisfies collocation equations

$$(Ls)(x_i) = f(x_i),\ i=0,\ldots n, \tag{6}$$

and boundary conditions

$$s(0) = u_0 \text{ and } s(1) = u_1. \tag{7}$$

These $n+3$ conditions are sufficient for unique determination of the tension spline of order four, ($s \in \mathcal{T}^0_4$), but to determine $s \in \mathcal{T}^0_6$ (or $s \in \mathcal{T}^0_8$), we need two (four) additional conditions. For the case $k=6$, we add two collocation equations at points $x_{1/2} = (x_0+x_1)/2$ and $x_{n-1/2} = (x_{n-1}+x_n)/2$, and for the case $k=8$ we add four collocation equations at points $x_{1/3} = (2x_0+x_1)/3$, $x_{2/3} = (x_0+2x_1)/3$, $x_{n-2/3} = (2x_{n-1}+x_n)/3$ and $x_{n-1/3} = (x_{n-1}+2x_n)/3$.

If the collocation spline s is written in the terms of the B-spline basis,

$$s(x) = \sum_{i=1-k}^{n-1} c_i B_{i,k}(x),$$

collocation equations yield the matrix equation $Ac = f$. Matrix A has at most k non-zero elements in each row which follows from the local support property of B_{ik}.

In the previously mentioned collocation methods, we use the choice of tension parameters ρ_i given in [7]:

$$\rho_i := \max\{\tilde{\rho_i}^-, \tilde{\rho_i}^+\}, \tag{8}$$

where

$$\tilde{\rho_i}^- = \frac{1}{2\varepsilon}\left(|b(x_i)| + \sqrt{|b(x_i)|^2 + 4\varepsilon|c(x_i)|}\right) \quad \text{and} \tag{9}$$

$$\tilde{\rho_i}^+ = \frac{1}{2\varepsilon}\left(|b(x_{i+1})| + \sqrt{|b(x_{i+1})|^2 + 4\varepsilon|c(x_{i+1})|}\right). \tag{10}$$

4. Asymptotic behavior of the collocation matrix

In [1] it is proved, using the same technique as in [14], that the collocation error is $\mathcal{O}(h^{k-2})$ for large perturbation parameter ε. In this section we consider the case when $\varepsilon \ll h_i$ and $k = 6$. Also, we assume that $b(x) > 0$ and $-c(x) \geq c_{min} > 0$.

Let the partition Π of interval $[0,1]$, tension parameters ρ_i and collocation points be defined as in §3. Let u be the exact solution of (1)–(2), $\tilde{u} = \sum_{i=1}^{n+5} c_i B_{i-6,6}$ the interpolation tension spline of order 6 for the u with knots x_i as a interpolation points, and $\bar{u} = \sum_{i=1}^{n+5} s_i B_{i-6,6}$ the collocation tension spline of order 6 for (1)–(2). By triangle inequality, we get

$$\| u - \bar{u} \|_\infty \leq \| u - \tilde{u} \|_\infty + \| \tilde{u} - \bar{u} \|_\infty. \tag{11}$$

Error for interpolation by tension spline $s \in \mathcal{T}_6^0$ is given in [1], and for large tension parameters (i.e. ε is small) we obtain

$$\| u - \tilde{u} \|_\infty = \mathcal{O}(h^4). \tag{12}$$

Since $\tilde{u} - \bar{u} = \sum_{i=1}^{n+5}(c_i - s_i)B_{i-6,6}$, using properties (ii) and (iii) for the tension B-splines, we get

$$\| \tilde{u} - \bar{u} \|_\infty \leq \| c - s \|_\infty,$$

where $c = (c_1, \ldots, c_{n+5})$ and $s = (s_1, \ldots, s_{n+5})$. Vector $e := c - s$ we obtain by solving the linear system $Ae = r$, where A is the collocation matrix for the problem (1)–(2), and r is the residual vector,

$$r = (0, r_2, \ldots, r_{n+4}, 0), \; r_{i+1} = L\tilde{u}(\eta_i) - L\bar{u}(\eta_i), \tag{13}$$

where η_i are collocation points, i.e. $\eta_1 = x_0$, $\eta_2 = x_{1/2}$, $\eta_i = x_{i-2}$, for $i = 3, \ldots, n+1$, $\eta_{n+2} = x_{n-1/2}$ and $\eta_{n+3} = x_n$. The collocation matrix A

depends on the perturbation parameter ε, so we shall consider the limit matrix

$$\bar{A} = \lim_{\varepsilon \to 0} A.$$

Elements of collocation matrix $A = (a_{i,j}) \in M_{n+5}$ are defined by

$$a_{i,j} = \begin{cases} 1, & i = j = 1 \text{ or } i = j = n+5, \\ 0, & i = 1, j \neq 1 \text{ or } i = n+5, j \neq n+5 \\ LB_{i-6,6}(\eta_{j-1}), & \text{otherwise.} \end{cases}$$

From the choice of tension parameters (8) we see that the tension parameters satisfy conditions of the Theorem 1, and get

$$\lim_{\varepsilon \to 0} LB_{i,6}(\eta) = LP_{i+1,4}(\eta),$$

where $\eta \in \{x_{1/2}, x_1, \ldots, x_{n-1}, x_{n-1/2}\}$.

In [1] it is shown that for large ρ are

$$\begin{array}{ll} B''_{-5,6}(0) \approx \rho^2, & B'_{-5,6}(0) \approx \rho, \\ B''_{-4,6}(0) \approx -\rho^2 - 3\rho/h, & B'_{-4,6}(0) \approx -\rho, \\ B''_{-3,6}(0) \approx 3\rho/h. & \end{array} \tag{14}$$

From (8) it follows that tension parameters depend on ε and tend to infinity when ε tends to 0, so using (14), one can easily obtain that

$$\begin{aligned} \lim_{\varepsilon \to 0} LB_{-5,6}(0) &= \lim_{\varepsilon \to 0} \varepsilon \left(B''_{-5,6}(0) + b(0)B'_{-5,6}(0) + c(0)B_{-5,6}(0) \right) \\ &= \lim_{\varepsilon \to 0} \left(\varepsilon \rho_0^2 + b(0)\rho_0 + c(0) \right)). \end{aligned} \tag{15}$$

Since $b(x) > 0$, our choice of tension parameters gives

$$\rho_0 = \frac{1}{2\varepsilon}(b_0^2 + \sqrt{b_0^2 - 4\varepsilon c_0})$$

where $b_0 = b(0)$ and $c_0 = c(0)$, and (15) becomes

$$\begin{aligned} \lim_{\varepsilon \to 0} LB_{-5,6}(0) &= \lim_{\varepsilon \to 0} \left(\frac{1}{4\varepsilon}(b_0^2 + \sqrt{b_0^2 - 4\varepsilon c_0})^2 - \frac{b_0}{2\varepsilon}(b_0^2 + \sqrt{b_0^2 - 4\varepsilon c_0}) + c_0 \right) \\ &= 0. \end{aligned}$$

In the similar manner, we conclude that

$$\begin{aligned} \lim_{\varepsilon \to 0} LB_{-4,6}(0) &= \lim_{\varepsilon \to 0} \left(-\frac{1}{4\varepsilon}(b_0^2 + \sqrt{b_0^2 - 4\varepsilon c_0})^2 - \frac{b_0}{2h\varepsilon}(b_0^2 + \sqrt{b_0^2 - 4\varepsilon c_0}) \right) \\ &= c_0 - 3\frac{b_0}{h}, \end{aligned}$$

$$\lim_{\varepsilon \to 0} LB_{-3,6}(0) = 3\frac{b_0}{h},$$

and we have analogous results for the values $LB_{i,6}(1)$, $i = -5, \dots, n-1$.

Using definition of polynomial B splines on equidistant partition, we obtain for x_i, $i = 2, \dots, n-2$

$$\begin{aligned} LP_{i-3,4}(x_i) &= c_i/6 - b_i/2h + \varepsilon/h^2, \\ LP_{i-2,4}(x_i) &= 2c_i/3 - 2\varepsilon/h^2, \\ LP_{i-1,4}(x_i) &= c_i/6 + b_i/2h + \varepsilon/h^2 \text{ and} \\ LP_{j,4}(x_i) &= 0, \quad j \notin \{i-3, i-2, i-1\}. \end{aligned}$$

At points x_1 and $x_{1/2}$ we obtain

$$\begin{aligned} LP_{-3,4}(x_1) &= 0, \\ LP_{-2,4}(x_1) &= \frac{c_1}{4} - \frac{3\,b_1}{4\,h} + \frac{3\,\epsilon}{2\,h^2}, \\ LP_{-1,4}(x_1) &= \frac{7\,c_1}{12} + \frac{b_1}{4\,h} - \frac{5\,\epsilon}{2\,h^2}, \\ LP_{0,4}(x_1) &= \frac{c_1}{6} + \frac{b_1}{2\,h} + \frac{\epsilon}{h^2}, \end{aligned}$$

and

$$\begin{aligned} LP_{-3,4}(x_{1/2}) &= \frac{c_{1/2}}{8} - \frac{3\,b_{1/2}}{4\,h} + \frac{3\,\epsilon}{h^2}, \\ LP_{-2,4}(x_{1/2}) &= \frac{19\,c_{1/2}}{32} - \frac{3\,b_{1/2}}{16\,h} - \frac{15\,\epsilon}{4\,h^2}, \\ LP_{-1,4}(x_{1/2}) &= \frac{25\,c_{1/2}}{96} + \frac{13\,b_{1/2}}{16\,h} + \frac{\epsilon}{4\,h^2}, \\ LP_{0,4}(x_{1/2}) &= \frac{c_{1/2}}{48} + \frac{b_{1/2}}{8\,h} + \frac{\epsilon}{2\,h^2}. \end{aligned}$$

In the same way we obtain values for the $LP_{i,j}$ in the collocation points x_{n-1} i $x_{n-1/2}$, so we prove that for each i, j $\lim_{\varepsilon \to 0} a_{i,j}$ exists, and matrix $\bar{A} = (\bar{a}_{i,j})$ given by

$$\bar{a}_{i,j} := \lim_{\varepsilon \to 0} a_{i,j}$$

is well-defined.

Now, we will consider the solution of the system $\bar{A}\bar{e} = r$, where r is defined by (13). Matrix $\bar{A} \in M_{n+5}$ is of the form

$$\bar{A} = \begin{pmatrix} 1 & 0 & 0 \\ B_1 & \hat{A} & B_2 \\ 0 & 0 & 1 \end{pmatrix},$$

where $B_1, B_2 \in M_{n+3,1}$ and $\hat{A} \in M_{n+3}$. Since $r_1 = r_{n+5} = 0$, we conclude that the first and the last components of $\bar{e}$ are 0. Matrix $\hat{A}$ is given by

$$\hat{A} = \begin{pmatrix} \hat{A}_1 \\ \hat{A}_2 \\ \hat{A}_3 \end{pmatrix},$$

where matrices $\hat{A}_1 \in M_{3,n+3}$, $\hat{A}_2 \in M_{n-3,n+3}$ and $\hat{A}_3 \in M_{3,n+3}$ are defined by

$$\hat{A}_1 = \begin{pmatrix} c_0 - \frac{3b_0}{h} & \frac{3b_0}{h} & 0 & \cdots & 0 \\ \frac{19\,c_{1/2}}{32} - \frac{3\,b_{1/2}}{16\,h} & \frac{25\,c_{1/2}}{96} + \frac{13\,b_{1/2}}{16\,h} & \frac{c_{1/2}}{48} + \frac{b_{1/2}}{8\,h} & \cdots & 0 \\ \frac{c_1}{4} - \frac{3\,b_1}{4\,h} & \frac{7\,c_1}{12} + \frac{b_1}{4\,h} & \frac{c_1}{6} + \frac{b_1}{2\,h} & \cdots & 0 \end{pmatrix}$$

$$\hat{A}_2 = \begin{pmatrix} 0 & \frac{c_2}{6} - \frac{b_2}{2h} & \frac{2c_2}{3} & \frac{c_2}{6} + \frac{b_2}{2h} & \cdots & & 0 \\ & \ddots & \ddots & \ddots & & & \\ 0 & 0 & \cdots & \frac{c_{n-2}}{6} - \frac{b_{n-2}}{2h} & \frac{2c_{n-2}}{3} & & \frac{c_{n-2}}{6} + \frac{b_{n-2}}{2h} \end{pmatrix}$$

$$\hat{A}_3 = \begin{pmatrix} 0 & \cdots & \frac{c_{n-1}}{6} + \frac{b_{n-1}}{2\,h} & \frac{7\,c_{n-1}}{12} + \frac{b_{n-1}}{4\,h} & \frac{c_{n-1}}{4} - \frac{3\,b_{n-1}}{4\,h} \\ 0 & \cdots & \frac{c_{n-1/2}}{48} + \frac{b_{n-1/2}}{8\,h} & \frac{25\,c_{n-1/2}}{96} + \frac{13\,b_{n-1/2}}{16\,h} & \frac{19\,c_{n-1/2}}{32} - \frac{3\,b_{n-1/2}}{16\,h} \\ 0 & \cdots & 0 & \frac{3b_n}{h} & c_n - \frac{3b_n}{h} \end{pmatrix},$$

where $b_i = b(x_i)$, $c_i = c(x_i)$.

Our goal is to prove that coefficients of the vector $\bar{e} = (\bar{e}_1, \ldots, \bar{e}_{n+5})$ satisfy

$$\bar{e}_i = \sum_{j=2}^{i-1} hK_{ij}r_j + \mathcal{O}(h^2), i = 5, \ldots, n+1, \tag{16}$$

where K_{ij} are constants that depend only on b_k and c_k.

First, we solve the system

$$\begin{pmatrix} c_0 - \frac{3b_0}{h} & \frac{3b_0}{h} & 0 \\ \frac{19\,c_{1/2}}{32} - \frac{3\,b_{1/2}}{16\,h} & \frac{25\,c_{1/2}}{96} + \frac{13\,b_{1/2}}{16\,h} & \frac{c_{1/2}}{48} + \frac{b_{1/2}}{8\,h} \\ \frac{c_1}{4} - \frac{3\,b_1}{4\,h} & \frac{7\,c_1}{12} + \frac{b_1}{4\,h} & \frac{c_1}{6} + \frac{b_1}{2\,h} \end{pmatrix} \begin{pmatrix} \bar{e}_2 \\ \bar{e}_3 \\ \bar{e}_4 \end{pmatrix} = \begin{pmatrix} r_2 \\ r_3 \\ r_4 \end{pmatrix},$$

and obtain

$$\begin{aligned} \bar{e}_2 &= \frac{h}{3}\left(-\frac{r_2}{b_0} + \frac{4r_3}{b_{1/2}} - \frac{r_4}{b_1}\right) + \mathcal{O}(h^2), \\ \bar{e}_3 &= \frac{h}{3}\left(\frac{4r_3}{b_{1/2}} - \frac{r_4}{b_1}\right) + \mathcal{O}(h^2), \\ \bar{e}_4 &= h\left(-\frac{r_2}{2b_0} + \frac{4r_3}{3b_{1/2}} + \frac{5r_4}{3b_1}\right) + \mathcal{O}(h^2). \end{aligned}$$

Since

$$\bar{e}_i = \frac{1}{\frac{c_{i-3}}{6} + \frac{b_{i-3}}{2h}} \left(r_i - \left(\frac{c_{i-3}}{6} - \frac{b_{i-3}}{2h} \right) e_{i-2} - \frac{2c_{i-3}}{3} e_{i-1} \right),$$

for $i = 5, \ldots, n+1$, we have

$$\bar{e}_5 = h \left(\frac{4r_3}{3b_{1/2}} - \frac{r_4}{3b_1} + \frac{2r_5}{b_2} \right) + \mathcal{O}(h^2),$$

and relation the (16), is satisfied for $i = 5$. In the same way, we prove that (16) is satisfied for other $\bar{e}_i$.

Now, we have to bound the residuals r_i. By using $L\bar{u}(\eta_j) = Lu(\eta_j)$, we derive expressions for the residuals:

$$\begin{aligned} r_i &= L\tilde{u}(\eta_i) - L\bar{u}(\eta_i) = L\tilde{u}(\eta_i) - Lu(\eta_i) \\ &= \varepsilon(\tilde{u}''(\eta_i) - u''(\eta_i)) + b(\eta_i)(\tilde{u}'(\eta_i) - u'(\eta_i)) + c(\eta_i)(\tilde{u}(\eta_i) - u(\eta_i)). \end{aligned}$$

We again use bounds for the interpolation error from [1], and obtain

$$|r_i| \leq C_i h^3.$$

Applying this result on (16), we obtain

$$|\bar{e}_i| \leq h \sum_{j=2}^{i-1} |K_{ij}| \cdot |r_j| \leq Ch^3,$$

and $\| \bar{e} \|_\infty \leq Ch^3$. Since $Ae = r = \bar{A}\bar{e}$, we have

$$\| e - \bar{e} \|_\infty \leq \| A^{-1}\bar{A} - I \|_\infty \| \bar{e} \|_\infty \leq Dh^3.$$

Inequality (11) implies that $\| u - \bar{u} \|_\infty = \mathcal{O}(h^3)$.

References

[1] I. Beroš *Collocation Method For Solving Two-Points Singularly-Perturbed Boundary Value Problem for ODE with High Order Tension Splines, Ms. Thesis*, in Croatian, Zagreb, 2001.

[2] I. Beroš and M. Marušić, *Collocation with High Order Tension Splines*, Proceedings of the 1. Conference on Applied Mathematics and Computation Dubrovnik, Croatia, 1999., M. Rogina, V. Hari, N.Limić, and Z. Tutek, eds., Zagreb, 2001.

[3] J. E. Flaherty and W. Mathon, *Collocation with polynomial and tension splines for singularly-perturbed boundary value problems*, SIAM J. Sci. Stat. Comput., 1 (1980), pp. 260–289.

[4] P. E. Koch and T. Lyche, *Construction of exponential tension B-spline of arbitrary order*, in P. J. Laurent, A. Le Mehaute and L. L Schumaker, eds., Curves and Surfaces, N.Y., 1991, Academic Press, pp. 255–258.

[5] J. KOZAK, *Shape Preserving Approximation*, Computers in Industry, 7 (1986), pp. 435–440.

[6] B. I. KVASOV, P. SATTAYATHAM,*GB-splines of arbitrary order*, J. Comp. Anal. Appl. Math., 104(1999), pp. 63–68.

[7] M. MARUŠIĆ, *A fourth/second order accurate collocation method for singularly perturbed two-point boundary value problems using tension splines*, Numer. Math., 88(2001), pp. 135-158.

[8] M. MARUŠIĆ, *Stable calculation by splines in tension*, Grazer Math. Ber., 328 (1996), pp. 65–76.

[9] M. MARUŠIĆ, *On Interpolation by Hermite Tension Spline of Arbitrary Order*, in K. Chui and L. L Schumaker, eds., Aproximation Theory IX, Nashville, 1998, Vanderbilt University Press, pp. 1–8.

[10] M. MARUŠIĆ AND M. ROGINA, *A collocation method for singularly perturbed two-point boundary value problems with splines in tension*, Adv. Comput. Math., 6 (1996), pp. 65–76.

[11] M. MARUŠIĆ AND M. ROGINA, *B-splines in tension*, in R. Scitovski, ed., Proceedings of VII Conference on Applied Mathematics, Osijek, 1989, pp. 129–134.

[12] B. J. MCCARTIN, *Theory of exponential splines*, J. Approx. Theory, 66 (1991), pp. 1–23.

[13] M. ROGINA, *Basis of splines associated with some singular differential operators*, BIT, 32 (1992), pp. 496–505.

[14] R. D. RUSSEL AND L. F. SHAMPINE, *A Collocation Method for Boundary Value Problems*, Numer. Math., 19 (1972), pp. 1–28.

[15] D. G. SCHWEIKERT, *An interpolating curve using a spline in tension*, J. Math. Physics, 45 (1966), pp. 312–317.

[16] H. SPÄTH, *Exponential spline interpolation*, Computing, 4 (1969), pp. 225–233.

NUMERICAL STABILITY OF KRYLOV SUBSPACE ITERATIVE METHODS FOR SOLVING LINEAR SYSTEMS

Nela Bosner

Department of Mathematics, University of Zagreb
Bijenička cesta 30, 10000 Zagreb, Croatia
nela@math.hr

Abstract Very often used methods for solving linear systems are Krylov subspace iterative methods. Usually, the iterations stop at the moment when some norm of the residual reaches tolerable value. Since all computations are done in finite precision arithmetic, we can check only some approximation of the residual norm. The main goal of our research is, to give estimation for the real residual norm, calculated from this approximation, in order to make stopping criterion more reliable.

1. Introduction

Modern methods for solving linear system

$$Ax = b, \qquad A \in \mathbb{R}^{n\times n},\ b \in \mathbb{R}^n, \tag{1}$$

where A is nonsingular matrix, are iterative methods on Krylov subspaces. In each iteration of such method, a solution approximation is searched in a Krylov subspace, in order to satisfy certain conditions. Because we are dealing with iterative methods, we have to determine when to stop the iterations. The best measure for departure of a solution approximation y from the exact solution $x = A^{-1}b$ is some norm of the error vector $e = x - y$. But, for knowing the error we have to now the exact solution, so this is not feasible approach. The other approach is to control a norm of the residual vector $r = b - Ay$. When this norm is reduced below some prescribed value we stop. We consider the solution approximation obtained in the last iteration to be accurate enough. In finite precision arithmetic, even the computed residual norm won't be correctly computed, so we have to now how to estimate the true residual norm. This article is dealing with this problem. In all iterative methods for solving linear systems, the residual or the residual norm is never calculated directly. Methods based

Edited by Drmač *et al.*, Kluwer Academic/Plenum Publishers, 2002

on Conjugate Gradient (CG) method calculate the residual vector recursively, and methods like Generalized Minimal Residual (GMRES) method compute residual norm as absolute value of one component of an indicative vector.

2. Iterative methods that recursively compute residual

We will first consider iterative methods, which use short recurrence for computing solution approximations and residuals, such as

$$x_k = x_{k-1} + \alpha_{k-1} p_{k-1}, \qquad r_k = r_{k-1} - \alpha_{k-1} A p_{k-1}, \tag{2}$$

for $k = 1, 2, \ldots$. Vector p_{k-1} is some direction vector, and α_{k-1} coefficient, both depending on the method we use. We note that the residual r_k is computed from the recurrence, instead of direct calculation $b - Ax_k$. We will consider now formulae (2) in finite precision arithmetic, with unit roundoff error ϵ. Greenbaum in [2] gives complete numerical analysis for methods which use recurrences (2). Applying results for operations performed in finite precision arithmetic on recurrences (2), we obtain relations

$$x_k = x_{k-1} + \alpha_{k-1} p_{k-1} + \zeta_k, \tag{3}$$

$$r_k = r_{k-1} - \alpha_{k-1} A p_{k-1} + \eta_k. \tag{4}$$

All parameters in relations above are values computed in finite precision arithmetic, like x_k, r_k, p_{k-1} and α_{k-1}. ζ_k and η_k are floating point errors and x_0 is the initial approximation. We are interested in relationship between the true residual in k-th iteration, $b - Ax_k$, and the computed residual r_k. From (3) and (4) we have

$$b - Ax_k - r_k = (b - Ax_0 - r_0) - \sum_{j=1}^{k} (A\zeta_j + \eta_j). \tag{5}$$

Taking the norm in (5) we obtain

$$\frac{\|b - Ax_k - r_k\|_2}{\|A\|_2 \|x\|_2} \le \frac{\|b - Ax_0 - r_0\|_2}{\|A\|_2 \|x\|_2} + \sum_{j=1}^{k} \left(\frac{\|\zeta_j\|_2}{\|x\|_2} + \frac{\|\eta_j\|_2}{\|A\|_2 \|x\|_2} \right), \tag{6}$$

where x is the exact solution, and from [3, chap. 2,3] it follows

$$\|b - Ax_0 - r_0\|_2 \le \epsilon[(1 + c)\|A\|_2 \|x_0\|_2 + \|b\|_2] + \mathcal{O}(\epsilon^2) c \|A\|_2 \|x_0\|_2, \tag{7}$$

where $c = c(n)$ is constant, such that $\|Av - \mathrm{fl}(Av)\|_2 \le c(\epsilon + \mathcal{O}(\epsilon^2))\|A\|_2 \|v\|_2$, and $c \le n^{3/2}$. The final bound for (6) is given in Theorem 2.2 in [2, pp. 541]. In this paper, we present similar results for two methods that are not analyzed in Greenbaum's papers.

2.1 BICGSTAB method

BICGSTAB method also computes residual recursively, but in each iteration the solution approximation is updated by adding a linear combination of two vectors. The solution approximation x_k and the residual r_k for $k = 1, 2, \ldots$ are computed through recurrences

$$x_k = x_{k-1} + \alpha_{k-1} p_{k-1} + \omega_k q_{k-1} + \zeta_k, \tag{8}$$

$$r_k = r_{k-1} - \alpha_{k-1} A p_{k-1} - \omega_k A q_{k-1} + \eta_k, \tag{9}$$

implemented in finite precision arithmetic. All values, again, are computed floating point values. This time, we will monitor the value

$$\|b - Ax_k\|_2 / \|A\|_2 \|x_k\|_2,$$

because this is the measure of backward error for perturbed linear system, where only the matrix A is perturbed, and x_k is its solution approximation. See Wilkinson's Lemma 1.1 [3, pp. 14]. By direct application of relations from [3, chap. 2,3] we get

$$\|\zeta_k\|_2 \le (2\epsilon+\epsilon^2)\|x_{k-1}\|_2 + (3\epsilon+3\epsilon^2+\epsilon^3)\|\alpha_{k-1}p_{k-1}\|_2 + (2\epsilon+\epsilon^2)\|\omega_k q_{k-1}\|_2, \tag{10}$$

and

$$\begin{aligned}\|\eta_k\|_2 \le\ & (2\epsilon + \epsilon^2)\|r_{k-1}\|_2 + (3\epsilon + 3\epsilon^2 + \epsilon^3)\|\alpha_{k-1} A p_{k-1}\|_2 + \\ & (1+\epsilon)^3 \|\alpha_{k-1}(\mathrm{fl}(Ap_{k-1}) - Ap_{k-1})\|_2 + (2\epsilon + \epsilon^2)\|\omega_k A q_{k-1}\|_2 + \\ & +(1+\epsilon)^2 \|\omega_k(\mathrm{fl}(Aq_{k-1}) - Aq_{k-1})\|_2. \end{aligned} \tag{11}$$

For a relation between computed and true residual we can also apply relation (5) and bound (7), but we need bounds on the norm of the remaining terms in (5).

Lemma 2.1. *Define*

$$\Theta_k = \max_{j \le k} \frac{\|x_j\|_2}{\|x_k\|_2}, \tag{12}$$

and

$$\Phi_{k-1} = \max_{j \le k-1} \frac{\|\alpha_j p_j\|_2 + \|\omega_{j+1} q_j\|_2}{\|x_k\|_2}. \tag{13}$$

Then the terms defined in (8) and (9) for BICGSTAB algorithm satisfy

$$\sum_{j=1}^{k} \frac{\|\zeta_j\|_2}{\|x_k\|_2} \le (\epsilon + O(\epsilon^2)) k (2\Theta_{k-1} + 3\Phi_{k-1}), \tag{14}$$

$$\sum_{j=1}^{k} \frac{\|\eta_j\|_2}{\|A\|_2\|x_k\|_2} \leq (\epsilon + \mathcal{O}(\epsilon^2))k\left[2\frac{\|b\|_2}{\|A\|_2\|x_k\|_2} + 2\Theta_{k-1} + (3+c)\Phi_{k-1}\right]. \tag{15}$$

Proof. The proof of this Lemma is similar to the proof of Lemma 2.1 in [2, pp. 539]. From 10 it follows

$$\frac{\|\zeta_j\|_2}{\|x_k\|_2} \leq (\epsilon + \mathcal{O}(\epsilon^2))(2\Theta_{j-1} + 3\Phi_{j-1}), \tag{16}$$

so by summing all such expressions for $j = 1, \ldots, k$, and taking the maximum for Θ_{j-1} and Φ_{j-1} we will obtain the final bound (14). According to [3, chap. 3] and the relation (11), the norm of η_j will have the following form

$$\begin{aligned} \|\eta_j\|_2 \leq\ & (2\epsilon + \epsilon^2)\|r_{j-1}\|_2 \\ & + (\epsilon + \mathcal{O}(\epsilon^2))(3+c)\|A\|_2(\|\alpha_{j-1}p_{j-1}\|_2 + \|\omega_j q_{j-1}\|_2). \end{aligned} \tag{17}$$

Using mathematical induction we can prove that each term $\|\eta_i\|_2$, $i = 1, \ldots,$ $j-1$ is bounded with

$$\mathcal{O}(\epsilon)[\|b\|_2 + \|A\|_2(\max_{l \leq i-1} \|x_l\|_2 + \max_{l \leq i-1}(\|\alpha_l p_l\|_2 + \|\omega_{l+1} q_l\|_2))].$$

Applying the assumption of induction together with (7) and (5), we obtain

$$\begin{aligned} \|r_{j-1}\|_2 \leq\ & \|b\|_2 + \|A\|_2\|x_{j-1}\|_2 + \mathcal{O}(\epsilon)[\|b\|_2 + \\ & + \|A\|_2(\max_{l \leq j-2} \|x_l\|_2 + \max_{l \leq j-2}(\|\alpha_l p_l\|_2 + \|\omega_{l+1} q_l\|_2))]. \end{aligned} \tag{18}$$

Finally, when we substitute (18) in (17) we have

$$\begin{aligned} \|\eta_j\|_2 \leq\ & (\epsilon + \mathcal{O}(\epsilon^2))[2\|b\|_2 + \|A\|_2(2\max_{l \leq j-1} \|x_l\|_2 + \\ & + (3+c)\max_{l \leq j-1}(\|\alpha_l p_l\|_2 + \|\omega_{l+1} q_l\|_2))], \end{aligned} \tag{19}$$

which proves the statement of induction for j. From this, it follows

$$\frac{\|\eta_j\|_2}{\|A\|_2\|x_k\|_2} \leq (\epsilon + \mathcal{O}(\epsilon^2))\left[2\frac{\|b\|_2}{\|A\|_2\|x_k\|_2} + 2\Theta_{j-1} + (3+c)\Phi_{j-1}\right],$$

so after summing and maximizing we will obtain the bound in (15). □

The proof of the following theorem follows from the relations (5), (7), (14) and (15).

Theorem 2.2. *The difference between the true residual $b - Ax_k$ and the computed vector r_k for BICGSTAB algorithm satisfies*

$$\frac{\|b - Ax_k - r_k\|_2}{\|A\|_2\|x_k\|2} \leq (\epsilon + \mathcal{O}(\epsilon^2)) \cdot$$

$$\cdot \left[(2k+1)\frac{\|b\|_2}{\|A\|_2\|x_k\|_2} + (4k+1+c)\Theta_{k-1} + (6+c)k\Phi_{k-1}\right], \quad (20)$$

where Θ_k is defined with (12) and Φ_{k-1} is defined with (13).

It can be shown that, as the method approaches convergence, vector norm $\|x_k\|_2$ starts to stagnate, and in that case $\|\alpha_{k-1}p_{k-1}\|_2$ and $\|\omega_k q_{k-1}\|_2$ approach zero. So, we don't expect Φ_{k-1} to be a large constant. On the other hand, if we control the value of Θ_k in every step of the algorithm, we can restart the recursive computation of the residual when this value becomes to large.

3. Methods that solve least squares problem

The second group of methods of our interest, are the Arnoldi and Lanczos algorithms, which minimize the residual norm or quasi-residual norm by solving least squares problem of lower dimension. The residual norm approximation is obtained through taking absolute value of one component of a vector which is solution of the least squares problem. Typical examples for such methods are GMRES and QMR methods. All values that we will use in our analysis will be computed values in finite precision arithmetic. The dimension of the system is n. The numerical analysis for GMRES method is given in [1].

3.1 QMR method

The basic steps of QMR method in exact arithmetic can be described as follows. The main step of QMR algorithm is two-sided Lanczos algorithm, which produces matrix $V_k = [v_1, \ldots, v_k]$ and $(k+1) \times k$ tridiagonal matrix $T_{k+1,k}$, such that $AV_k = V_{k+1}T_{k+1,k}$. Then, QR factorization of $T_{k+1,k}$ is computed, and triangular system is solved. Solution approximation is computed recursively. In exact arithmetic we also have

$$b - Ax_k = V_{k+1}(\beta e_1 - T_{k+1,k}R_k^{-1}g^{(k)}(1:k))$$

and

$$\begin{aligned}\|b - Ax_k\|_2 &\leq \|V_{k+1}\|_2\|\beta e_1 - T_{k+1,k}R_k^{-1}g^{(k)}(1:k)\|_2 \\ &= \|V_{k+1}\|_2|g^{(k)}(k+1)|,\end{aligned}$$

where $z_k = \beta e_1 - T_{k+1,k}R_k^{-1}g^{(k)}(1:k)$ is quasi-residual, and $g^{(k)}(1:k)$ denotes k-dimensional vector whose components are equal to the first k components of the vector $g^{(k)}$. With e_i we denote the i-th unit vector, whose components are all equal to zero except i-th, which is equal to 1. Now we turn to the finite precision arithmetic, and all further values will be computed values. As we can see, the first step in QMR algorithm is two-sided Lanczos algorithm. For $i > 1$, from [3, chap. 2,3] it follows

$$Av_i = \beta_{i-1}v_{i-1} + \alpha_i v_i + \gamma_i v_{i+1} + f_i, \tag{21}$$

where β_{i-1}, α_i and γ_i are elements in the positions $(i-1,i)$, (i,i) and $(i+1,i)$ of matrix $T_{k+1,k}$, and

$$\|f_i\|_2 \leq \epsilon[(3+c)\|A\|_2 + 4|\alpha_i| + 3|\beta_{i-1}|] + \mathcal{O}(\epsilon^2)(\|A\|_2 + \|T_{k+1,k}\|_1). \tag{22}$$

Constant c was defined in Section 2. If we define matrix F_k as

$$F_k = [f_1\ f_2\ \ldots\ f_k],$$

then two-sided Lanczos algorithm, performed in finite precision arithmetic, can be written as

$$AV_k = V_{k+1}T_{k+1,k} + F_k, \tag{23}$$

where $T_{k+1,k}$ is tridiagonal matrix and for the modest constant $c_1 = c_1(n) = 3 + c$ the bound for the norm of F_k is given with

$$\|F_k\|_2 \leq \epsilon\sqrt{k}[c_1\|A\|_2 + 4\|T_{k+1,k}\|_1] + \mathcal{O}(\epsilon^2)(\|A\|_2 + \|T_{k+1,k}\|_1). \tag{24}$$

Now we are finished with the first step of QMR algorithm. After two-sided Lanczos algorithm Givens QR factorization of the matrix $T_{k+1,k}$ is performed. For the next step in our analysis we will need Theorem 18.9 from [3, pp. 375], which claims that there exists an orthogonal $(k+1)\times(k+1)$ matrix $\widehat{F}^{(k)}$ such that

$$\widehat{F}^{(k)}T_{k+1,k} = \begin{bmatrix} R_k \\ 0 \end{bmatrix} + \Delta\widehat{R}_k, \quad \|\Delta\widehat{R}_k\|_F \leq 6\sqrt{2}k(\epsilon + \mathcal{O}(\epsilon^2))\|T_{k+1,k}\|_F. \tag{25}$$

The next step is computing the elements of matrix $P_k = \mathrm{fl}(V_k R_k^{-1})$, which in fact are computed by backward substitution, as the solution of a triangular linear system. According to Theorem 8.5 from [3, pp. 154], we have

$$P_k R_k = V_k + \Delta V_k, \quad \text{with} \quad |\Delta V_k| \leq (3\epsilon + \mathcal{O}(\epsilon^2))|P_k||R_k|.$$

Finally, we consider the error which appears in computing the matrix P_k alone,

$$P_k = V_k R_k^{-1} + \Delta V_k R_k^{-1} = V_k R_k^{-1} + \Delta P_k, \tag{26}$$

where

$$\|\Delta P_k\|_2 \leq 3\sqrt{k}(\epsilon + \mathcal{O}(\epsilon^2))\|P_k\|_F \kappa(R_k). \tag{27}$$

At the end of QMR iteration, the algorithm checks the norm of the quasi-residual, what is, in the exact arithmetic, equal to examination of the absolute value of $(k+1)$-th component of the vector $F^{(k)}(\beta e_1)$. The matrix $F^{(k)^T}$ is the computed orthonormal factor in the QR factorization, and $\beta = \text{fl}(\|b - Ax_0\|_2)$. In the finite precision arithmetic, at the end of iteration algorithm checks $(k+1)$-th component of the vector $g^{(k)} = \text{fl}(F^{(k)}(\beta e_1))$. Now we need Lemma 18.7 from [3, pp. 373], which gives backward error for application of one Givens rotation on a vector. Since $F^{(k)}$ is a product of k Givens rotations, this lemma is applied k times to get a bound for $g^{(k)}$. We have

$$\|g^{(k)}\|_2 \leq \beta[1 + 6\sqrt{2}k(\epsilon + \mathcal{O}(\epsilon^2))]. \tag{28}$$

Now we will need one more thing, and that is upper bound for β. Let $\hat{r}_0 = \text{fl}(b - Ax_0)$. From [3, chap. 2,3], for modest constants $c_i = c_i(n)$, $i = 2, \ldots, 5$ we have

$$\beta \leq (1 + c_2\epsilon)\,\|b\|_2 + (1 + c_3\epsilon)\,\|A\|_2\|x_0\|_2 + \mathcal{O}(\epsilon^2)(\|A\|_2\|x_0\|_2 + \|b\|_2). \tag{29}$$

The bound on β was necessary for finishing the bound on $\|g^{(k)}\|_2$, hence we have

$$\|g^{(k)}\|_2 \leq (1 + c_4\epsilon)\|b\|_2 + (1 + c_5\epsilon)\,\|A\|_2\|x_0\|_2 + \mathcal{O}(\epsilon^2)(\|A\|_2\|x_0\|_2 + \|b\|_2) \tag{30}$$

Before beginning of the next iteration QMR algorithm calculates solution approximation for the current iteration. The recurrence for solution approximation in finite precision arithmetic has form

$$x_k = x_{k-1} + g^{(k)}(k)p_{k-1} + \Delta x_k, \tag{31}$$

which is equivalent to the relation (3). In the same manner as in the proof of Lemma 2.1 in [2, pp. 540], under the assumption $1 - 2\epsilon - \epsilon^2 > 0$, we have

$$\|\Delta x_k\|_2 \leq (5\epsilon + \mathcal{O}(\epsilon^2))\chi_k, \quad \text{where } \chi_k = \max_{i \leq k} \|x_i\|_2. \tag{32}$$

After k steps of recurrence (31), we will obtain the final form for computed solution approximation

$$x_k = x_0 + P_k g^{(k)}(1:k) + \sum_{i=1}^{k} \Delta x_i, \quad \sum_{i=1}^{k} \|\Delta x_i\|_2 \leq (5\epsilon + \mathcal{O}(\epsilon^2))k\chi_k. \tag{33}$$

where $g^{(k)}(1:k)$ is k-dimensional vector formed from first k components of the vector $g^{(k)}$. We have

$$b - Ax_k - V_{k+1}(\beta e_1 - T_{k+1,k}R_k^{-1}g^{(k)}(1:k)) = \tag{34}$$

$$= (b - Ax_0 - \beta v_1) - A\Delta P_k g^{(k)}(1:k) - F_k R_k^{-1} g^{(k)}(1:k) - \sum_{i=1}^{k} A\Delta x_i.$$

The only bound left to be found is the bound on $\|b - Ax_0 - \beta v_1\|_2$ where $v_1 = \text{fl}(\frac{\hat{r}_0}{\beta})$. We have

$$\|b - Ax_0 - \beta v_1\|_2 \leq (c+2)\epsilon\|A\|_2\|x_0\|_2 + 2\epsilon\|b\|_2 + \mathcal{O}(\epsilon^2)(c\|A\|_2\|x_0\|_2 + \|b\|_2), \tag{35}$$

which follows from [3, chap. 2,3] The only problem is that we do not know the value of true quasi-residual, and instead of that we have the scalar $|g^{(k)}(k+1)|$, which doesn't have to be equal to the norm $\|\beta e_1 - T_{k+1,k}R_k^{-1}g^{(k)}(1:k)\|_2$, as in exact arithmetic. Next task will be finding the relationship between these two values. According to Lemma 18.7 from [3, pp. 373], and from relations (25), (30), (29) and (28) we have

$$\begin{aligned} \|\beta e_1 - T_{k+1,k}R_k^{-1}g^{(k)}(1:k)\|_2 &\leq |g^{(k)}(k+1)| + \\ &+ \mathcal{O}(\epsilon) f(\|T_{k+1,k}\|_F, \|R_k^{-1}\|_2, \|A\|_2, \|b\|_2, \|x_0\|_2), \end{aligned} \tag{36}$$

where f is a polynomial of order 4. From (34) and all given results we obtain the final result.

Theorem 3.1. *Assume that* $1 - 2\epsilon - \epsilon^2 > 0$. *Then the norm of true QMR residual is bounded by*

$$\|b - Ax_k\|_2 \leq \sqrt{k+1}\left(1 + \frac{n+4}{2}\epsilon\right)|g^{(k)}(k+1)| +$$

$$+\mathcal{O}(\epsilon)g(\|A\|_2, \|b\|_2, \|x_0\|_2, \|T_{k+1,k}\|_F, \|R_k^{-1}\|_2, \kappa(R_k), \|P_k\|_F, \chi_k),$$

where g *is a polynomial of order 5, and* χ_k *is defined with (32).*

4. Conclusion

Our intention was to obtain bounds on the true residual norm for all standard iterative methods for solving linear systems. Greenbaum gives such bounds only for some specific methods, and in this paper numerical analysis was performed for methods BICGSTAB and QMR. Analysis for BICGSTAB method was very similar to the analysis shown in [2], but QMR method is, from numerical point of view, hybrid between GMRES method and the methods that recursively compute residual. Because, in QMR method the stopping criterion is the size of quasi-residual norm, we shown what is the relationship between the true residual norm and the component of vector $g^{(k)}$ which in exact arithmetic is equivalent to the quasi-residual norm. From the bound in Theorem 3.1 we can conclude when the true residual norm can be large, for example when condition number of R_k is large. Generally, bounds shown in this paper can be used as an illustration of what happens when iterative methods are performed in finite precision arithmetic. They show how the methods can be improved and in what cases we can expect big difference between computed residual norm and true residual norm.

References

[1] J. Drkošová, A. Greenbaum, M. Rozložník, and Z. Strakoš. **Numerical Stability of GMRES**. *BIT*, 35: pp. 309–330, 1995.

[2] A. Greenbaum. **Estimating the attainable accuracy of recursively computed residual methods**. *SIAM J. Matrix Anal. Appl.*, 18(3): pp. 535–551, July 1997.

[3] N. J. Higham. **Accuracy and Stability of Numerical Algorithms**. SIAM, Philadelphia, 1996.

NONLINEAR PROBLEMS IN DYNAMICS BY THE FINITE ELEMENT IN TIME METHOD

Nenad Kranjčević
Faculty of Mech. Eng. and Nav. Arch., University of Zagreb, Croatia
nenad.kranjcevic@fsb.hr

Milenko Stegić
Faculty of Mech. Eng. and Nav. Arch., University of Zagreb, Croatia
milenko.stegic@fsb.hr

Nikola Vranković
Faculty of Mech. Eng. and Nav. Arch., University of Zagreb, Croatia
nikola.vrankovic@fsb.hr

Abstract The finite element method in time domain is a fast and reliable numerical method for predicting the steady state response of the nonlinear dynamic system under periodic excitation. The stability of solutions can be investigated by the application of Floquet–Lyapunov theorem on the transition matrix. Close agreement is found between obtained results and the published findings of a harmonic balance method.

Keywords: Nonlinear Vibrations, Finite Element in Time, Clearance, Stability

1. Introduction

Nonlinear problems in dynamics are governed by a system of nonlinear differential equations. Exact solutions of nonlinear differential equations are very rare and almost all of the methods for their solving are only approximate. A general solution method of systems with complex and strong nonlinearity under periodic excitations is the numerical time integration. But this procedure can be very time consuming since a great number of excitation periods must be simulated to obtain a steady state response.

A very efficient method for solving nonlinear differential equations in the frequency domain is the harmonic balance method [2, 3]. When the assumption of dominance of primary resonance in the response is satisfied, the harmonic balance method is very accurate and numerically reliable method for obtaining periodic response of nonlinear dynamic systems with harmonic excitation.

Commonly used methods for constructing periodic solutions in the time domain are finite difference schemes, shooting technicques and Poincare map methods [4]. Another approach for solving nonlinear problems is the finite element in time method, which is based on a weak form of Hamilton principle [1, 6]. Similar to the standard finite element technique, the time interval is divided into a finite number of time elements. The solution for all the spatial degrees of freedom at all time steps within a given interval is sought through a set of algebraic equations. The conventional shape functions can be used in the time element because the time element does not impose the periodic boundary conditions.

Furthermore, the straightforward determination of the stability of solution is the second advantageous feature of this method. The stability of periodic solution is investigated considering small perturbations of the solution. According to Floquet-Lyapunov theorem the stability of solutions can be determined by the eigenvalues of the transition matrix that relates the initial and final perturbations.

There are many examples of mechanical systems that possess nonlinear characteristics. One typical example is a nonlinearity due to the presence of gaps and clearances. Vibration of such systems can result in relative motion across the clearance space and impacting between the components.

The objective of this paper is to apply the finite element in time method to a three degrees-of-freedom (3-DOF) semi-definite system with clearances under harmonic excitations. The stability of obtained frequency responses is analyzed using Floquet-Lyapunov theorem. The results are compared with published findings of the harmonic balance method [5].

2. Problem formulation

The model of a 3-DOF semi-definite system, employed in [5], is used to study the feasibility of the finite element in time method for solving nonlinear differential equations. The model consists of three mass elements m_1, m_2, m_3, two linear viscous dampers c_1, c_2, and two nonlinear stiffness elements $k_1 h_1(x_1 - x_2)$, $k_2 h_2(x_2 - x_3)$, as shown in Figure 1. With and $\boldsymbol{q}^T = [q_1, q_2] = \frac{1}{b}[x_1 - x_2, x_2 - x_3]$ being the nondimensional relative displacements, the equation of motion can be expressed as:

$$\boldsymbol{q}'' + \boldsymbol{Z}\boldsymbol{q}' + \boldsymbol{\Omega}\boldsymbol{h}(\boldsymbol{q}) = \boldsymbol{f}_m + \boldsymbol{f}_a \cos(\eta\tau), \tag{1}$$

where:

$$\frac{\mathrm{d}}{\mathrm{d}\tau}(\,) = (\,)', \quad \tau = \omega_k t, \quad \omega_k = \sqrt{k_1\left(\frac{1}{m_1} + \frac{1}{m_2}\right)}, \tag{2}$$

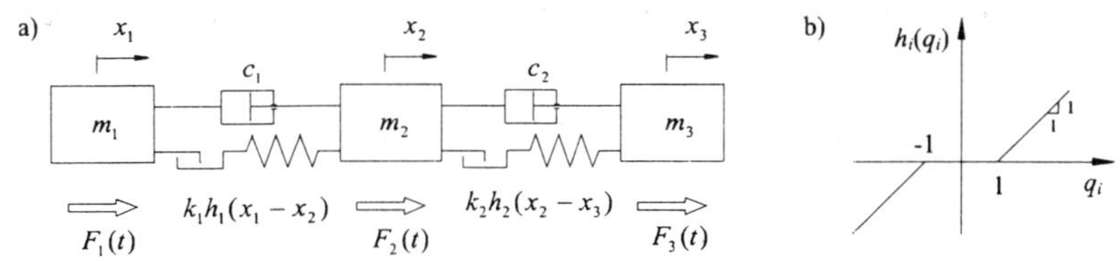

Figure 1. 3-DOF semi-definite nonlinear model: a) mechanical model, b) nonlinear unit displacement function.

$$\boldsymbol{Z} = 2\begin{bmatrix} \zeta_{11} & -\zeta_{12}\omega_{12} \\ -\zeta_{21}\omega_{21} & \zeta_{22}\omega_{22} \end{bmatrix}, \quad \boldsymbol{\Omega} = \begin{bmatrix} 1 & -\omega_{12}^2 \\ -\omega_{21}^2 & \omega_{22}^2 \end{bmatrix}, \tag{3}$$

$$\boldsymbol{h}^T(\boldsymbol{q}) = [h_1(q_1), h_2(q_2)]. \tag{4}$$

Furthermore, b is the characteristic length, η denotes a nondimensional excitation frequency while $\boldsymbol{f}_m$ and $\boldsymbol{f}_a$ are the amplitude vectors of mean and alternating load, respectively. The elements of matrices $\boldsymbol{Z}$ and $\boldsymbol{\Omega}$ are found to be:

$$\zeta_{11} = \frac{c_1\left(\frac{1}{m_1}+\frac{1}{m_2}\right)}{2\omega_k}, \quad \zeta_{12} = \frac{c_2}{2m_2\omega_k\omega_{12}}, \quad \zeta_{21} = \frac{c_1}{2m_2\omega_k\omega_{21}},$$

$$\zeta_{22} = \frac{c_2\left(\frac{1}{m_2}+\frac{1}{m_3}\right)}{2\omega_k\omega_{22}}, \ \omega_{12}^2 = \frac{k_2}{m_2\omega_k^2}, \ \omega_{21}^2 = \frac{k_1}{m_2\omega_k^2}, \ \omega_{22}^2 = \frac{k_2\left(\frac{1}{m_2}+\frac{1}{m_3}\right)}{\omega_k^2}. \tag{5}$$

The nonlinear displacement functions $h_i(q_i)$ for $i = 1, 2$ that describe the clearance of value $2b$ are defined as (Figure 1b):

$$h_i(q_i) = \begin{cases} q_i + 1, & q_i < -1 \\ 0, & -1 \le q_i \le 1 \\ q_i - 1, & q_i > 1 \end{cases}. \tag{6}$$

3. Finite element in time method

The finite element in time method for solving nonlinear dynamic problems is based on Hamilton weak principle which describes the motion of the system between the two known times τ_1 and τ_2:

$$\int_{\tau_1}^{\tau_2} (\delta L + \delta W)\, \mathrm{d}\tau = \delta \boldsymbol{q}^T \boldsymbol{p} \Big|_{\tau_1}^{\tau_2}, \tag{7}$$

where L and W denote the Lagrangian and the work done by the generalized force, respectively, while $\boldsymbol{p}$ denotes the vector of the generalized momenta.

For the equation of motion (1), the Hamilton weak principle (7) takes the form:

$$\int_{\tau_1}^{\tau_2} \left\{ \delta \boldsymbol{q}'^T \boldsymbol{q}' + \delta \boldsymbol{q}^T [-\boldsymbol{Z}\boldsymbol{q}' - \boldsymbol{\Omega} \boldsymbol{h}(\boldsymbol{q}) + \boldsymbol{f}_m + \boldsymbol{f}_a \cos(\eta\tau)] \right\} \mathrm{d}\tau = \delta \boldsymbol{q}^T \boldsymbol{p} \Big|_{\tau_1}^{\tau_2} . \tag{8}$$

3.1 Finite element discretization

The basis of the finite element in time discretization is the division of the time interval into a finite number n_e of time elements. The dependent variable being approximately represented in a piecewise fashion across the domain, time element by time element, in terms of the independent variable τ and the nodal values of the dependent variable $\boldsymbol{q}$. Within a time element i, the $\boldsymbol{q}_i(\tau)$ and $\boldsymbol{q}'_i(\tau)$ are approximated by using the standard finite element technique:

$$\begin{aligned} \boldsymbol{q}_i(\tau) &= \boldsymbol{N}_i(\tau)\bar{\boldsymbol{q}}_i, \quad i = 1(1)n_e \\ \boldsymbol{q}'_i(\tau) &= \boldsymbol{N}'_i(\tau)\bar{\boldsymbol{q}}_i, \quad i = 1(1)n_e \end{aligned} \tag{9}$$

where $\boldsymbol{N}_i$ is the matrix of shape functions. The Lagrange family of interpolation functions, for example, can be used as shape functions. Inserting the above assumed solution into the Hamilton weak principle (8), the variational form for time element i, in terms of nodal values, becomes:

$$\int_{\tau_i}^{\tau_{i+1}} \delta \bar{\boldsymbol{q}}_i^T \Big[\boldsymbol{N}'^T_i \boldsymbol{N}'_i \bar{\boldsymbol{q}}_i - \boldsymbol{N}'^T_i \boldsymbol{Z} \boldsymbol{N}_i \bar{\boldsymbol{q}}_i - \boldsymbol{N}^T_i \boldsymbol{\Omega} \boldsymbol{h}(\boldsymbol{q}) + \boldsymbol{N}^T_i \boldsymbol{f}_m + \boldsymbol{N}^T_i \boldsymbol{f}_a \cos(\eta\tau) \Big] \, \mathrm{d}\tau = \delta \bar{\boldsymbol{q}}_i^T \boldsymbol{b}_i, \tag{10}$$

where $\boldsymbol{b}_i^T = [\boldsymbol{p}(\tau_i), 0, ..., 0, \boldsymbol{p}(\tau_{i+1})]$ and is the vector of linear momenta. Since $\delta \bar{\boldsymbol{q}}_i$ is completely arbitrary, the finite element approximation at the time element level yields:

$$\boldsymbol{A}_i \bar{\boldsymbol{q}}_i + \boldsymbol{g}_i(\bar{\boldsymbol{q}}_i) + \boldsymbol{f}_i = \boldsymbol{b}_i, \tag{11}$$

where:

$$\boldsymbol{A}_i = \int_{\tau_i}^{\tau_{i+1}} \Big[\boldsymbol{N}'^T_i \boldsymbol{N}'_i - \boldsymbol{N}'^T_i \boldsymbol{Z} \boldsymbol{N}_i \Big] \, \mathrm{d}\tau, \tag{12}$$

$$\boldsymbol{g}_i(\bar{\boldsymbol{q}}_i) = - \int_{\tau_i}^{\tau_{i+1}} \left[\boldsymbol{N}^T_i \boldsymbol{\Omega} \boldsymbol{h}(\boldsymbol{q}) \right] \mathrm{d}\tau, \tag{13}$$

$$\boldsymbol{f}_i = \int_{\tau_i}^{\tau_{i+1}} \left[\boldsymbol{N}^T_i \boldsymbol{f}_m + \boldsymbol{N}^T_i \boldsymbol{f}_a \cos(\eta\tau) \right] \mathrm{d}\tau. \tag{14}$$

The Hamilton weak principle (8) can now be represented at the global level as a sum of integrations over the time intervals of all time elements:

$$\boldsymbol{A}\bar{\boldsymbol{q}} + \boldsymbol{g}(\bar{\boldsymbol{q}}) + \boldsymbol{f} = \boldsymbol{0}, \tag{15}$$

where, with a summation notation representing assembly operation:

$$\bar{\boldsymbol{q}} = \sum_{i=1}^{n_e} \bar{\boldsymbol{q}}_i, \quad \boldsymbol{A} = \sum_{i=1}^{n_e} \boldsymbol{A}_i, \quad \boldsymbol{g}(\bar{\boldsymbol{q}}) = \sum_{i=1}^{n_e} \boldsymbol{g}_i(\bar{\boldsymbol{q}}_i), \quad \boldsymbol{f} = \sum_{i=1}^{n_e} \boldsymbol{f}_i, \tag{16}$$

Furthermore, for a periodic steady state response, the momentum vector vanishes as the periodic boundary condition requires both displacements and momenta to be identical at τ_1 and τ_2: $\sum_{i=1}^{n_e} \boldsymbol{b}_i = \boldsymbol{0}$.

3.2 Solution method

The global finite element approximation (15) results in a set of nonlinear algebraic equation in unknown nodal displacement vector $\bar{\boldsymbol{q}}$. Using the Newton-Raphson method for obtaining the solution, the unknown nodal displacement $\bar{\boldsymbol{q}}$ may be expressed in an iteration procedure as:

$$\bar{\boldsymbol{q}}^{(n+1)} = \bar{\boldsymbol{q}}^{(n)} + \Delta\bar{\boldsymbol{q}}^{(n)}. \tag{17}$$

For a small increment $\Delta\bar{\boldsymbol{q}}^{(n)}$, the nonlinear equation (15), can be expanded into Taylor series retaining only the linear terms:

$$\boldsymbol{d}^{(n)} + \boldsymbol{K}^{(n)}\Delta\bar{\boldsymbol{q}}^{(n)} = \boldsymbol{0}, \tag{18}$$

where:

$$\boldsymbol{d}^{(n)} = \boldsymbol{A}^{(n)}\bar{\boldsymbol{q}}^{(n)} + \bar{\boldsymbol{g}}^{(n)}(\bar{\boldsymbol{q}}) + \boldsymbol{f}, \quad \boldsymbol{K}^{(n)} = \boldsymbol{A}^{(n)} + \frac{\partial\bar{\boldsymbol{g}}^{(n)}(\bar{\boldsymbol{q}})}{\partial\bar{\boldsymbol{q}}^{(n)}}. \tag{19}$$

The iteration procedure requires evaluation of the integrals of each time element in matrix $\boldsymbol{A}$ and vectors $\boldsymbol{g}$ and $\boldsymbol{f}$. The integral components of matrix $\boldsymbol{A}$ and vector $\boldsymbol{f}$ are constants while the integral components of vector $\boldsymbol{g}$ depend on nonlinear displacement functions. The evaluation of the vector $\boldsymbol{g}$ and matrix $\partial\bar{\boldsymbol{g}}/\partial\bar{\boldsymbol{q}}$ at time element level:

$$\boldsymbol{g}_i(\bar{\boldsymbol{q}}) = -\int_{\tau_i}^{\tau_{i+1}} \boldsymbol{N}_i^T \boldsymbol{\Omega} \boldsymbol{h}(\boldsymbol{q})\,\mathrm{d}\tau, \quad \frac{\partial\boldsymbol{g}_i}{\partial\bar{\boldsymbol{q}}_i} = -\int_{\tau_i}^{\tau_{i+1}} \boldsymbol{N}_i^T \boldsymbol{\Omega} \frac{\partial\boldsymbol{h}}{\partial\boldsymbol{q}} \boldsymbol{N}_i\,\mathrm{d}\tau, \tag{20}$$

requires a knowing of the zeroes of the function $q_i(\tau) - 1$ within the interval $[\tau_i, \tau_{i+1}]$. Since $q_i(\tau)$ is approximated by polynomials, these zeroes can be easily determined. Applying the Newton-Raphson iteration procedure to the equations (17) and (18) the nodal displacement vector $\bar{\boldsymbol{q}}$ can be determined. For a converged solution the increment of nodal displacement vector $\Delta\bar{\boldsymbol{q}}$ converges towards zero.

3.3 Numerical results

The frequency responses of a 3-DOF semi-definite system with clearances shown in Figure 1 are calculated here using the finite element in time method. The system parameters $\zeta_{11} = \zeta_{12} = \zeta_{21} = \zeta_{22} = 0.05$, $\omega_{12} = \omega_{21} = 0.6$ and $\omega_{22} = 1.1$ are adopted from [5] where the same problem was solved using the harmonic balance method. In reference [5] is assumed that only $h_1(q_1)$ is nonlinear as defined by equation (10) while $h_2(q_2) = q_2$.

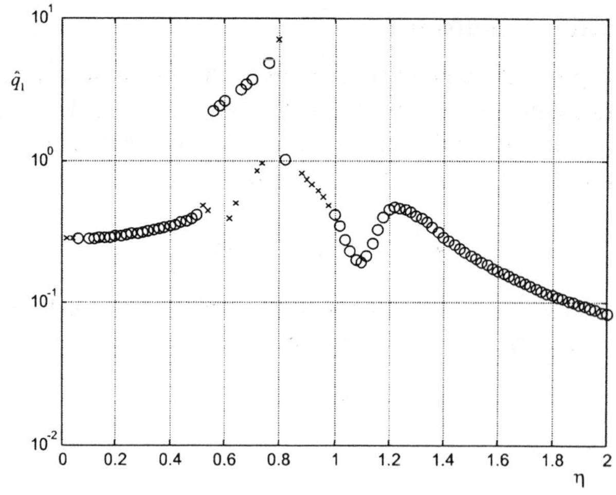

Figure 2. Frequency response of the 3-DOF semi-definite system with clearance for $\boldsymbol{f}_m = [0.25, 0.25]^T$ and $\boldsymbol{f}_a = [0.25, 0]^T$: $\circ$ – converged solution, $\times$ – achieved maximum of 100 iterations.

The finite element in time method cannot solve problems with ideal clearance because the zero stiffness implies singularity of the tangent matrix $\boldsymbol{K}^{(n)}$. In the finite element analysis, the nonlinear displacement function (6) is approximated by the bilinear system at which the slope of the first stage is 1% of the slope of the second stage. Furthermore, the calculation was performed with ten time elements and with the response of the linear system as the starting vector $\bar{\boldsymbol{q}}^{(0)}$. The third order interpolation polynomials are used as the shape functions.

As the results of finite element calculations, the frequency response, in terms of the nondimensional excitation frequency η and the steady state amplitude $\hat{q}_1$, is shown in Figure 2. Obtained solution agrees very well with results in reference [5]. The converged solutions are obtained in maximum 100 iterations (usually only few steps), only at the frequency range where the first harmonic is dominant in the response.

4. Stability analysis

The converged and un-converged solutions in 100 iterations, obtained by the finite element in time method can be either stable or unstable. Because of that, it is of a great importance to determine the stability of solutions obtained by the finite element in time method.

4.1 Stability of steady state solutions

The set of nonlinear incremental equations (18) can be portioned at the time element level as:

$$\begin{bmatrix} \boldsymbol{d}_B \\ \boldsymbol{d}_I \end{bmatrix}_i + \begin{bmatrix} \boldsymbol{K}_{BB} & \boldsymbol{K}_{BI} \\ \boldsymbol{K}_{IB} & \boldsymbol{K}_{II} \end{bmatrix}_i \begin{bmatrix} \Delta\bar{\boldsymbol{q}}_B \\ \Delta\bar{\boldsymbol{q}}_I \end{bmatrix}_i = \begin{bmatrix} \boldsymbol{b}_B \\ \boldsymbol{0} \end{bmatrix}_i, \tag{21}$$

where B and I correspond the temporal nodes at the ends and in the interior of time element i, respectively, while $\boldsymbol{b}_B^T = [-\boldsymbol{p}(\tau_i), \boldsymbol{p}(\tau_{i+1})]$ denotes the vector of momenta at the boundary nodes. Assuming $\boldsymbol{K}_{II}$ invertible, the equation (21) gives the solutions for $\Delta\bar{\boldsymbol{q}}_I$ and $\Delta\bar{\boldsymbol{q}}_B$ as follow:

$$\Delta\bar{\boldsymbol{q}}_I = -\boldsymbol{K}_{II}^{-1}\boldsymbol{d}_I - \boldsymbol{K}_{II}^{-1}\boldsymbol{K}_{IB}\Delta\bar{\boldsymbol{q}}_B, \tag{22}$$

$$\bar{\boldsymbol{d}}_B + \bar{\boldsymbol{K}}_{BB}\Delta\bar{\boldsymbol{q}}_B = \boldsymbol{b}_B, \tag{23}$$

where:

$$\bar{\boldsymbol{d}}_B = \boldsymbol{d}_B - \boldsymbol{K}_{BI}\boldsymbol{K}_{II}^{-1}\boldsymbol{d}_I, \qquad \bar{\boldsymbol{K}}_{BB} = \boldsymbol{K}_{BB} - \boldsymbol{K}_{BI}\boldsymbol{K}_{II}^{-1}\boldsymbol{K}_{IB}. \tag{24}$$

By imposing the compatibility conditions at the common node of two time elements, the equation (23) can be assembled into the global equation:

$$\begin{bmatrix} \bar{\boldsymbol{d}}_1 \\ \bar{\boldsymbol{d}}_2 \end{bmatrix} + \begin{bmatrix} \bar{\boldsymbol{K}}_{11} & \bar{\boldsymbol{K}}_{12} \\ \bar{\boldsymbol{K}}_{21} & \bar{\boldsymbol{K}}_{22} \end{bmatrix} \begin{bmatrix} \Delta\bar{\boldsymbol{q}}_1 \\ \Delta\bar{\boldsymbol{q}}_2 \end{bmatrix} = \begin{bmatrix} -\boldsymbol{p}(\tau_1) \\ \boldsymbol{p}(\tau_2) \end{bmatrix}, \tag{25}$$

where 1 and 2 denote the time nodes at the ends of the stationary period.

4.2 Floquet-Lyapunov theorem

The stability of a periodic stationary solution can be examined by imposing small perturbations of the equation (25) for the initial and final time nodes:

$$\begin{bmatrix} \bar{\boldsymbol{K}}_{11} & \bar{\boldsymbol{K}}_{12} \\ \bar{\boldsymbol{K}}_{21} & \bar{\boldsymbol{K}}_{22} \end{bmatrix} \begin{bmatrix} \delta\bar{\boldsymbol{q}}_1 \\ \delta\bar{\boldsymbol{q}}_2 \end{bmatrix} = \begin{bmatrix} -\delta\boldsymbol{p}(\tau_1) \\ \delta\boldsymbol{p}(\tau_2) \end{bmatrix}. \tag{26}$$

The transition matrix $\boldsymbol{T}$ which relates the final and initial perturbations is readily derived from the equation (26) as:

$$\begin{bmatrix} \delta\bar{\boldsymbol{q}}_2 \\ \delta\boldsymbol{p}(\tau_2) \end{bmatrix} = \boldsymbol{T} \begin{bmatrix} \delta\bar{\boldsymbol{q}}_1 \\ \delta\boldsymbol{p}(\tau_1) \end{bmatrix}, \tag{27}$$

where:

$$T = \begin{bmatrix} -\bar{K}_{12}^{-1}\bar{K}_{11} & -\bar{K}_{12}^{-1} \\ \bar{K}_{21} - \bar{K}_{22}\bar{K}_{12}^{-1}\bar{K}_{11} & -\bar{K}_{22}\bar{K}_{12}^{-1} \end{bmatrix}. \tag{28}$$

According to Floquet-Lyapunov theorem the stability of solution is determined by the eigenvalues λ_i of the matrix $\boldsymbol{T}$. The solution is asymptotically stable if all the moduli of the eigenvalues are less then unity; otherwise the solution is unstable.

4.3 Numerical implementation

In order to investigate the stability of steady state solutions, Floquet-Lyapunov theorem is considered on the 3-DOF semi-definite system with two nonlinear stiffness elements (Figure 1). The stiffness and damping parameters are taken from reference [5]. Applying the finite element incremental procedure and searching the eigenvalues of transition matrix, un-converged or unstable solutions are found in the range of $0.7 < \eta < 0.96$ as shown in Figure 3.

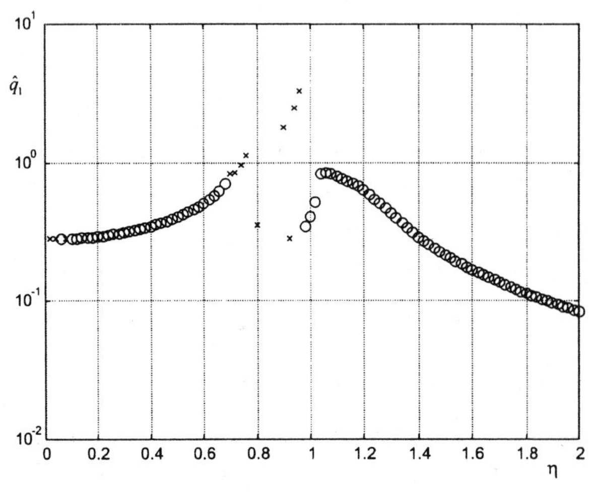

Figure 3. Frequency response of the 3-DOF semi-definite system with 2 clearances for $\boldsymbol{f}_m = [0.25, 0.25]^T$ and $\boldsymbol{f}_a = [0.25, 0]^T$: o – stable solution, × – achieved maximum of 100 iterations or $|\lambda_i| > 1$ (starting vector is linear system response).

Using the previous stable solution as the starting vector, the un-converged solutions in 100 iterations are not found any more (Figure 4).

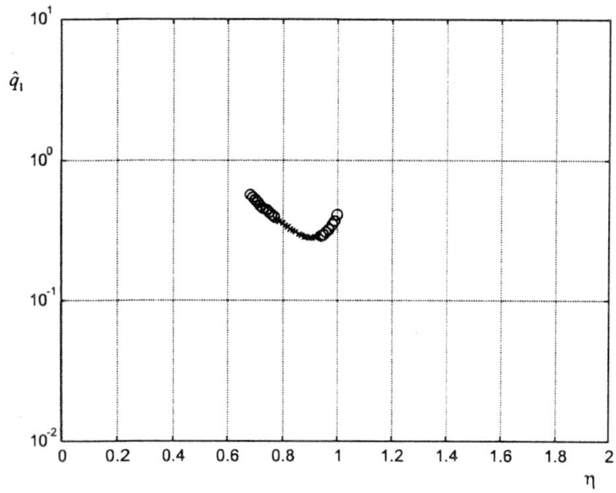

Figure 4. Frequency response of the 3-DOF semi-definite system with 2 clearances for $\boldsymbol{f}_m = [0.25, 0.25]^T$ and $\boldsymbol{f}_a = [0.25, 0]^T$: ∘ – stable solution, × – unstable solution (starting vector is previous stable solution).

5. Concluding remarks

The finite element in time method is an implicit method which can be used for obtaining the steady state response of the nonlinear mechanical system under periodic excitation. The method gives accurate numerical results with only a few time elements and without significant computational effort. The solutions often tend to the first harmonic responses while the range of convergence for sub-periodic responses is very small. The convergence of frequency responses with dominance of subharmonics can be improved by using the previous stable solution as the starting vector. The stability of steady state solution is directly determined using Floquet-Lyapunov theorem on the transition matrix which is a by-product of the finite element procedure.

References

[1] M. Borri, C. Bottasso, and P. Mantegazza. Basic features of the time finite element approach for dynamics. *Meccanica*, 27:119–130, 1992.

[2] S. Chatterjee, A. K. Mallik, and A. Ghosh. Periodic response of piecewise non-linear oscillators under harmonic excitation. *Journal of Sound and Vibration*, 191:129–144, 1996.

[3] R. J. Comparin and R. Singh. Frequency response characteristics of a multi-degree-of-freedom systems with clearances. *Journal of Sound and Vibration*, 142:101–124, 1990.

[4] A. H. Nayfeh and B. Balachandran. *Applied Nonlinear Dynamics*. John Wiley & Sons, New York, first edition, 1995.

[5] C. Padmanabhan and R. Singh. Spectral coupling issues in a two-degree-of-freedom system with clearance non-linearities. *Journal of Sound and Vibration*, 155:209–230, 1992.

[6] Y. Wang. Dynamics of unsymmetric piecewise-linear/non-linear systems using finite elements in time. *Journal of Sound and Vibration*, 185:155–170, 1995.

ON DIRECTIONAL BIAS OF THE L_P-NORM

Tomislav Marošević
Department of Mathematics, University of Osijek
Trg Ljudevita Gaja 6, HR-31000 Osijek, Croatia
Tomislav.Marosevic@etfos.hr

Abstract We look at the so-called directional bias of the L_p-norm ($1 \leq p \leq \infty$) (see [1]). The L_p-norm is a popular distance measure in the problems of estimating the distances among objects. Distance $\|x - y\|_p$ between any two points x and y depends on a particular set of orthogonal reference axes, except for the L_2-norm (i.e. euclidean norm). The axes and the distance function should be chosen in accordance with the directional bias.

Particularly, we consider the directional bias of norms in 3-dimensional space $\mathbb{R}^3$ and in n-dimensional space $\mathbb{R}^n$. (In [1] the concept of the directional bias function of the norm in 2-dimensional space $\mathbb{R}^2$ is analysed.)

Keywords: directional bias, L_p-norm

1. Introduction

Among many applications and other usages (see [2, 3, 4, 5]), the L_p-norm ($1 \leq p \leq \infty$) is also a popular distance measure used in continuous location models [1]. It is defined by the well-known expression:

$$x = (x_1, \ldots, x_n) \in \mathbb{R}^n, \qquad \|x\|_p = \sqrt[p]{\sum_{i=1}^{n} |x_i|^p}, \qquad p \geq 1,$$

$$\|x\|_\infty = \max\{|x_1|, \ldots, |x_n|\}.$$

Three mostly used norms are: L_1-norm, L_2-norm (i.e. euclidean norm) and L_∞-norm (i.e. Chebyshev norm). (When $0 < p < 1$, the triangle inequality is not satisfied; then one has the so-called hyper-rectilinear distances.)

Estimating the distances among objects occurs in many areas of application (e.g. road distances, distances in Geographical Information System models). As a distance predicting function one can use, for instance, the weighted L_2-norm model, or the weighted L_p-norm model. In those cases there appears the problem of fitting the weighted L_p-norm (e.g. unknown parameters of a distance function with respect to the data set on travel distances in a transportation network).

Distance $\|x - y\|_p$ between any two points x and y depends on a particular set of orthogonal reference axes, except for the L_2-norm (i.e. euclidean norm).

The directional bias function of the L_p-norm is defined as a ratio

$$\frac{\|x\|_p}{\|x\|_2}, \qquad x \neq 0 .$$

In a certain sense, the directional bias function could represent a measure of the relative difficulty of travel in any direction [1]. The reference axes and the distance function should be chosen in accordance with the directional bias (e.g. rotating the axes in the space $\mathbb{R}^2$ results in a different directional bias).

1.1 Directional bias of the L_p-norm on $\mathbb{R}^2$

Most location problems appear in the plane. Therefore, in two-dimensional space $\mathbb{R}^2$ the directional bias of the L_p-norm, denoted by r_p, has the following form [1]:

$$r_p(\phi) = \frac{\|x\|_p}{\|x\|_2} = \frac{\sqrt[p]{|x_1|^p + |x_2|^p}}{\|x\|_2} = \sqrt[p]{|\cos\phi|^p + |\sin\phi|^p}\,,$$

$x = (x_1, x_2)^T \neq 0$, where the polar coordinates $\|x\|_2 = \rho$, $\phi \in [0, 2\pi\rangle$ are taken into account: $x_1 = \|x\|_2 \cdot \cos\phi$, $x_2 = \|x\|_2 \cdot \sin\phi$. It is obvious that r_p is a function of ϕ alone.

In order to cover the same L_2/euclidean distance between two points, one must travel a longer distance along a line at angle ϕ_2 with the x_1 axis than along a line at angle ϕ_1 with the x_1 axis provided that $r_p(\phi_2) > r_p(\phi_1)$.

The graph of r_p is illustrated in Figure 1 for various values of p and for $\phi \in [0, \frac{\pi}{2}]$. Let us note some useful properties of the directional bias of the L_p-norm [1].

- If $p \leq 2$, then $r_p \geq 1$;
  if $p \geq 2$, then $r_p \leq 1$.
- $r_p(\phi)$ is periodic with period $T_\phi = \frac{\pi}{2}$ $(= 90^\circ)$.
- For any real ω, $r_p(\frac{\pi}{4} + \omega) = r_p(\frac{\pi}{4} - \omega)$.

From the two properties mentioned above, it follows that r_p is the mirror image of itself about the line $\phi = \frac{\pi}{4}$.

- If $0 < p < 2$, then in the interval $0 \leq \phi \leq \frac{\pi}{4}$ r_p is a strictly increasing function of ϕ, while it is strictly decreasing in the same interval if $p > 2$.

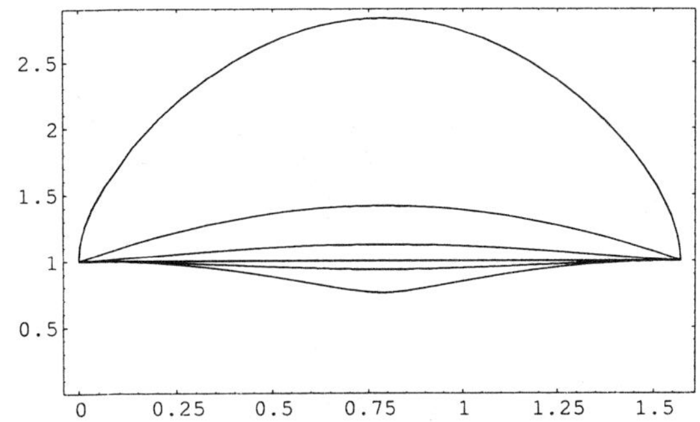

Figure 1: Directional bias function $r_p(\phi)$ for $p = 0.5, 1, 1.5, 2, 2.5, 10$, from the top, respectively

In Figure 1 one can also see that: r_p has its maximum value at $\phi = \frac{\pi}{4}$ and its minimum value at $\phi = 0$, $\phi = \frac{\pi}{2}$, if $0 < p < 2$. (It holds conversely if $p > 2$).

Defining the direction of the greatest difficulty as the value of ϕ which maximizes $r_p(\phi)$, and the direction of the least difficulty as the value of ϕ which minimizes $r_p(\phi)$, one can see that for $0 < p < 2$ the direction of the greatest difficulty is at 45° to the axes ($\phi = \frac{\pi}{4}, \frac{3\pi}{4}, \frac{5\pi}{4}, \frac{7\pi}{4}$), and the direction of the least difficulty is parallel to the axes ($\phi = 0, \frac{\pi}{2}, \pi, \frac{3\pi}{2}$).

In other words, the distance $\|x - y\|_p$ between any two points x and y separated by a straight line segment of the fixed length $\|x - y\|_2 = d$ is maximized if this line segment of the fixed length is at 45° to the axes, and minimized if it is parallel to an axis, provided that $0 < p < 2$. The converse holds when $p > 2$.

2. Directional bias of the L_p-norm on $\mathbb{R}^3$

With respect to the directional bias, we could look at the L_p-norm on three-dimensional space $\mathbb{R}^3$. Then we have

$$r_{p,3}(\phi, \theta_3) = \frac{\|x\|_p}{\|x\|_2} = \sqrt[p]{(|\cos\phi|^p + |\sin\phi|^p) \cdot |\sin\theta_3|^p + |\cos\theta_3|^p} \qquad (1)$$

where $x = (x_1, x_2, x_3)^T \neq (0, 0, 0)$ and sphere coordinates $\|x\|_2$, $\phi \in [0, 2\pi\rangle$, $\theta_3 \in [0, \pi]$ are used:

$$\begin{aligned} x_1 &= \|x\|_2 \cdot \cos\phi \cdot \sin\theta_3\,, \\ x_2 &= \|x\|_2 \cdot \sin\phi \cdot \sin\theta_3\,, \\ x_3 &= \|x\|_2 \cdot \cos\theta_3\,. \end{aligned}$$

The graphs (surfaces) of $r_{p,3}$ for some values p are shown in Figure 2.

We can note some properties of the directional bias $r_{p,3}$.

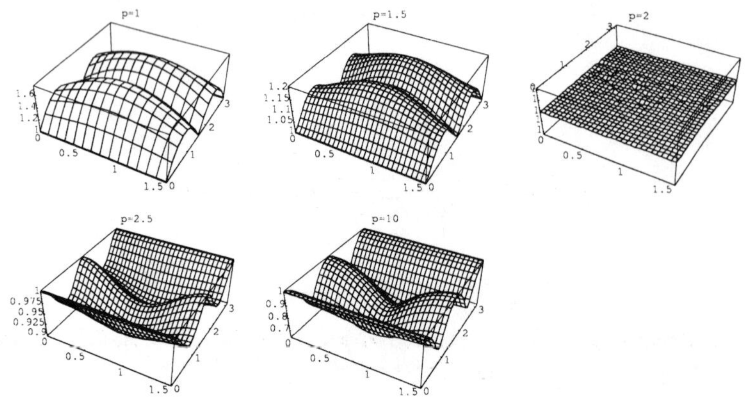

Figure 2: Directional bias function $r_{p,3}(\phi, \theta)$ for $p = 1, 1.5, 2, 2.5, 10$

- $r_{p,3}(\phi, \theta_3)$ is a function of ϕ and θ_3.
- If $p \leq 2$, then $r_{p,3} \geq 1$;
  if $p \geq 2$, then $r_{p,3} \leq 1$.
- $r_{p,3}(\phi, \theta_3)$ is periodic in its first variable ϕ with period $T_\phi = \frac{\pi}{2}$, that is $r_{p,3}(\phi + \frac{\pi}{2}, \theta_3) = r_{p,3}(\phi, \theta_3)$.
- For any real ω there holds

  $$r_{p,3}(\frac{\pi}{4}+\omega, \theta_3) = r_{p,3}(\frac{\pi}{4}-\omega, \theta_3) \text{ and } r_{p,3}(\phi, \frac{\pi}{2}+\omega) = r_{p,3}(\phi, \frac{\pi}{2}-\omega)\,;$$

  that is, with respect to particular/partial variables, $r_{p,3}(\phi, \theta_3)$ is the mirror image of itself about $\phi = \frac{\pi}{4}$, and about $\theta_3 = \frac{\pi}{2}$. (Because of this symmetry, we could look at points $(\phi, \theta_3) \in [0, \frac{\pi}{4}] \times [0, \frac{\pi}{2}]$.)
- For $0 < p < 2$, $r_{p,3}$ has its maximum value at $\phi = \frac{\pi}{4}, \theta_3 = \text{arctg}(\sqrt{2}) \approx 54.7356°$ (and also at $\phi \in \{\frac{3\pi}{4}, \frac{5\pi}{4}, \frac{7\pi}{4}\}, \theta_3 = \pi - \text{arctg}\sqrt{2}$);
  - the direction of the greatest difficulty for $0 < p < 2$ is, for example, at line $\phi = \frac{\pi}{4}, \theta_3 = \text{arctg}\sqrt{2}$;
  - the direction of the least difficulty for $0 < p < 2$ is at (ϕ, θ_3) for which $r_{p,3}(\phi, \theta_3) = 1$; e.g. at line $\theta_3 = 0, \pi$ (parallel to x_3-axis) and also at $(\phi, \theta_3) \in \{(0, \frac{\pi}{2}), (\frac{\pi}{2}, \frac{\pi}{2})\}$ (parallel to x_1-axis, x_2-axis, respectively). It holds conversely for $p > 2$.
- A few other interesting properties:

  $$r_{p,3}(\phi, 0) = 1\,,\ \forall\phi, \qquad r_{p,3}(0, \frac{\pi}{2}) = 1\,.$$

3. Directional bias of the L_p-norm on $\mathbb{R}^n$

Let us generalize this concept of the directional bias of the L_p-norm on n-dimensional space $\mathbb{R}^n$.

For this purpose we could take into account 'generalized' sphere coordinates in $\mathbb{R}^n$ ($n \in \mathbb{N}$, $n > 3$):

$$\begin{aligned}
x &= (x_1, x_2, \ldots, x_n)^T \in \mathbb{R}^n, \qquad x \neq 0, \\
x_1 &= \|x\|_2 \cdot \cos\phi \cdot \sin\theta_3 \cdot \sin\theta_4 \cdot \cdots \cdot \sin\theta_n, \\
x_2 &= \|x\|_2 \cdot \sin\phi \cdot \sin\theta_3 \cdot \sin\theta_4 \cdot \cdots \cdot \sin\theta_n, \\
x_3 &= \|x\|_2 \cdot \cos\theta_3 \cdot \sin\theta_4 \cdot \cdots \cdot \sin\theta_n, \\
x_4 &= \|x\|_2 \cdot \cos\theta_4 \cdot \sin\theta_5 \cdot \cdots \cdot \sin\theta_n, \\
&\vdots \\
x_{n-1} &= \|x\|_2 \cdot \cos\theta_{n-1} \cdot \sin\theta_n, \\
x_n &= \|x\|_2 \cdot \cos\theta_n,
\end{aligned}$$

where $\phi \in [0, 2\pi]$ is an angle with the x_1 axis, $\theta_3 \in [0, \pi]$ is an angle with the x_3 axis, and generally $\theta_k \in [0, \pi]$ is an angle with the x_k axis for $k = 3, \ldots, n$.

It is not difficult to see by means of induction that the directional bias of the L_p-norm has the form

$$r_{p,n}(\phi, \theta_3, \ldots, \theta_n) = \sqrt[p]{[r_{p,n-1}(\phi, \theta_3, \ldots, \theta_{n-1})]^p \cdot |\sin\theta_n|^p + |\cos\theta_n|^p}$$

and $r_{p,3}$ is given by the expression (1).

By means of induction and generalization, we could also draw a conclusion on some properties of function $r_{p,n}(\phi, \theta_3, \ldots, \theta_n)$:

- If $p \leq 2$, then $r_{p,n} \geq 1$;
  if $p \geq 2$, then $r_{p,n} \leq 1$.

- $r_{p,n}(\phi, \theta_3, \ldots, \theta_n)$ is periodic in its first variable ϕ with period $T_\phi = \frac{\pi}{2}$, that is
$$r_{p,n}(\phi + \frac{\pi}{2}, \theta_3, \ldots, \theta_n) = r_{p,n}(\phi, \theta_3, \ldots, \theta_n).$$

- For any real ω there holds
$$r_{p,n}(\frac{\pi}{4} + \omega, \theta_3, \ldots, \theta_n) = r_{p,n}(\frac{\pi}{4} - \omega, \theta_3, , \ldots, \theta_n)$$
  and
$$r_{p,n}(\phi, \frac{\pi}{2} + \omega, \ldots, \theta_n) = r_{p,n}(\phi, \frac{\pi}{2} - \omega, \ldots, \theta_n);$$
  that is, with respect to particular/partial variables, $r_{p,n}(\phi, \theta_3, \ldots, \theta_n)$ is the mirror image of itself about $\phi = \frac{\pi}{4}$, and about $\theta_k = \frac{\pi}{2}$, $k = 3, \ldots, n$.

(Because of this symmetry, we could look at points $(\phi, \theta_3, \ldots, \theta_n) \in [0, \frac{\pi}{4}] \times [0, \frac{\pi}{2}] \times \ldots \times [0, \frac{\pi}{2}]$.)

- For $0 < p < 2$ there holds:

dimension	*maximum value*	*at the points/variable*
$n = 3$	$r_{p,3}(\frac{\pi}{4}, \mathrm{arctg}\sqrt{2})$	$\theta_3^* = \mathrm{arctg}\sqrt{2}$;
$n = 4$	$r_{p,4}(\frac{\pi}{4}, \mathrm{arctg}\sqrt{2}, \theta_4^*)$	$\theta_4^* = \mathrm{arctg}\left(\frac{1}{r_{p,3}(\frac{\pi}{4}, \theta_3^*)^p}\right)^{\frac{1}{p-2}}$;
$\vdots$	$\vdots$	$\vdots$
$n \in \mathbb{N}$	$r_{p,n}(\frac{\pi}{4}, \theta_3^*, \ldots, \theta_n^*)$	$\theta_n^* = \mathrm{arctg}\left(\frac{1}{r_{p,n-1}(\frac{\pi}{4}, \theta_3^*, \ldots, \theta_{n-1}^*)^p}\right)^{\frac{1}{p-2}}$.

Therefore:

  - the direction of the greatest difficulty for $0 < p < 2$ is, for example, at 'line' (in $\mathbb{R}^n$) for which $\phi^* = \frac{\pi}{4}$, $\theta_3^* = \mathrm{arctg}\sqrt{2}$, and θ_k^*, $k = 3, \ldots, n$, defined by the relations mentioned above;
  - the direction of the least difficulty for $0 < p < 2$ is at $(\phi, \theta_3, \ldots, \theta_n)$ for which $r_{p,n}(\phi, \theta_3, \ldots, \theta_n) = 1$; for example, at 'line' $\theta_n = 0, \pi$ (parallel to x_n-axis) and also at 'lines' parallel to x_k-axis $\forall k = 1, \ldots, n$. It holds conversely for $p > 2$.

- A few other interesting properties:

$$r_{p,n}(\phi, \theta_3, \ldots, \theta_{n-1}, 0) = 1\,, \quad \forall \phi, \theta_3, \ldots, \theta_{n-1},$$

$$r_{p,n}(\phi, \theta_3, \ldots, \theta_{n-1}, \frac{\pi}{2}) = r_{p,n-1}(\phi, \theta_3, \ldots, \theta_{n-1})\,.$$

Remark 1. The directional bias of the L_p-norm implicitly belongs to a particular set of orthogonal reference axes, that is, it depends on the reference axes, because an angle is measured relative to the given axes. The choice of axes and the distance function determines a certain form of the directional bias. Therefore, there also appears the problem of determining the reference axes in practical applications.

References

[1] Brimberg, J. and R. F. Love. (1995). Estimating Distances (Chapt. 1.) in: *Facility Location, Survey of Application and Methods*, Z. Drezner (Ed.): Springer Series in Operations Research, Springer-Verlag New York.

[2] Gonin, R. and A. H. Money. (1989). *Nonlinear L_p-Norm Estimation*, Marcel Dekker, New York.

[3] Marošević, T. (1996). A choice of norm in discrete approximation, *Mathematical Communications* **1**, 147–152. Univ. Osijek.

[4] Powell, M. J. D. (1981). *Approximation Theory and Methods*, Cambridge Univ. Press, Cambridge.

[5] Rice, J. R. (1964). *The Approximation of Functions*, Vol. 1, Addison Wesley, Reading.

A NOTE ON SLIP CONDITION ON CORRUGATED BOUNDARY

Eduard Marušić-Paloka
Department of Mathematics, University of Zagreb
Bijenička 30, 10000 Zagreb, Croatia
emarusic@math.hr

Abstract We study the flow of Newtonian fluid in a domain with periodically wrinkled boundary. On the corrugated boundary the slip (Navier's) boundary condition is imposed. Using the method of homogenization we replace the slip condition, posed on the rough boundary, by effective boundary condition posed on the middle surface of oscillating boundary. The effective condition is no-slip in direction transversal to the wrinkles and slip in direction of the wrinkles.

Keywords: Homogenization, rapidly oscillating boundary, Navier's law.

Introduction

Several types of boundary conditions can be physically and mathematically reasonable for the Navier-Stokes system. As the fluid is viscous the most usual boundary condition imposed on the solid wall is the no-slip condition. However, in some situations, a stagnant layer of fluid in vicinity of the wall appears allowing a fluid to slip. In that case Navier's law (see [15]) or the slip condition is applied. It says that the slip velocity, tangential to the surface, is proportional to the shear stress while the normal velocity remains zero (since the wall is impermeable). Such boundary conditions can be caused by the roughness of the boundary, as described in [1, 2, 19, 12], or by porous boundary (it is than called Beavers and Joseph's law, see [12]). Such boundary behaviour can also result from an exterior electric field, as in [7, 8].

Our intention is to consider the stationary flow of Newtonian fluid in domains with one part of the boundary being a rough surface. The rough boundary is supposed to be corrugated (i.e. rough in only one direction, see Fig.1.) and periodic. On that rough part of the boundary we impose the slip boundary condition. The goal of this paper is to study the behaviour of the solution in the vicinity of the boundary in order to smooth out the ruffles, i.e. to replace the slip condition posed on the wrinkled surface by some other boundary condition posed on its middle surface that leads to a good approximation of the original problem. Since the new problem is posed in the domain with smooth boundary without ruffles it is much more convenient for numerical resolution. To describe the result we denote by $(\mathbf{t}_1, \mathbf{t}_2)$, the natural basis of the tangential plane on the boundary. The vector $\mathbf{t}_1$ is transversal on the wrinkles and $\mathbf{t}_2$ has the same direction as the wrinkles (see Fig.1.). The effective condition is unusual but it has a clear physical meaning. It claims that the effective boundary condition is no slip, except in the direction of the vector $\mathbf{t}_2$. In the direction of $\mathbf{t}_2$ we have slip condition with the friction coefficient depending on the geometry of wrinkles.

Our technique is based on the *method of homogenization.* Similar problems

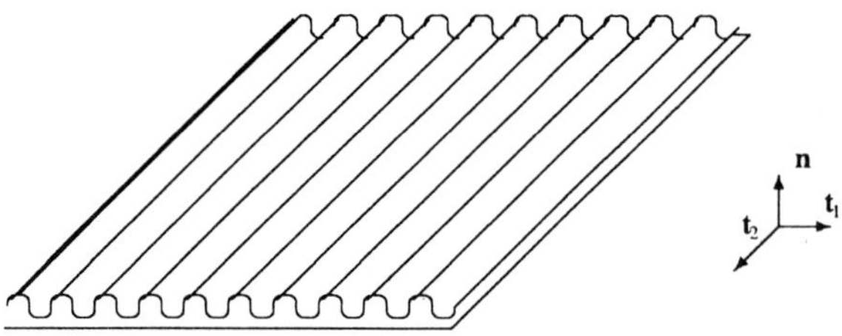

Figure 1. Local geometry of Γ_2^ε

were addressed previously in the existing literature. The problem of rapidly oscillating boundary which oscillates in both variables z_1 and z_2 was already studied in [13, 14] but the particular case of corrugated boundary was left out. The problem of wrinkled boundary has been successfully solved by Sanchez-Palencia [17], Belyaev [4] and Checkin, Friedman and Piatnitski [6] for a heat conduction equation with Neumann and Robin boundary condition.

Some results about the eigenvalue problem for an elliptic operator in a domain with oscillating boundary were stated in the short paper [4].

The influence of boundary roughness on hydrodynamic drag was studied in [3].

1. Description of the geometry

We describe the rough boundary similarly as in [13]. We denote by $\Omega \subset \mathbf{R}^3$ a bounded C^2 domain placed only on one side of its boundary Γ. We suppose that $\Sigma \subset \Gamma$ can be parametrised by a single parametrisation. This is a technical assumption that simplifies the presentation. Let $U \subset \mathbf{R}^2$ and $\mathbf{x} : U \to \mathbf{R}^3$ be the parametrisation of Σ. We suppose that $\mathbf{x} : U \to \mathcal{O} = \mathbf{x}(U)$ is bijective, that $\mathbf{x}$ and $\mathbf{x}^{-1}$ are of class C^2 and that $\nabla\mathbf{x}(z)$ has rank 2 at every point $z \in U$. Function $h \in C^2_{per}(]0, 1[, \mathbf{R}_+)$, is the shape function for cogs. Let $0 < \varepsilon \ll 1$

be a small parameter (we can think of it as of the period and the amplitude of ruffles). Let $\mathbf{y} : U \times \mathbf{R} \to \mathbf{R}^3$ be the mapping defined by

$$\mathbf{y}(z,t) = \mathbf{x}(z) + t\,\mathbf{n}(z) \ ,$$

where $\mathbf{n}$ is the exterior unit normal on Σ. For sufficiently small ε_0 there exists $\mathcal{H} \subseteq U$, such that the mapping $\mathbf{y} : \mathcal{H}\times\,]-\varepsilon_0,\varepsilon_0\,[\to \mathbf{y}(\mathcal{H}\times\,]-\varepsilon_0,\varepsilon_0\,[)$ is a diffeomorphism (see e.g. [16, 9]). For the sake of simplicity, we suppose that $\mathcal{H} = U$ (that is not a loss of generality since we could have replaced, from the beginning, U by, possibly, smaller set $\mathcal{H}$). In order to have the smooth junction between the rough boundary and the rest of the boundary we define the cut-off function $\eta \in C_0^\infty(\mathcal{H})$ such that $\eta = 1$ on $\mathcal{G} \subset \mathcal{H}$. Now we define

$$h_\varepsilon(z) = \varepsilon\,\eta(z)\,h(\frac{z_1}{\varepsilon})$$

By construction, such h_ε is obviously in $C_0^2(\mathcal{H})$. The parametrisation of the wavy boundary Γ_2^ε, $\mathbf{y}_\varepsilon : \mathcal{G} \to \mathbf{R}^3$, for $\varepsilon < \varepsilon_0$, is now given by

$$\mathbf{y}_\varepsilon(z) = \mathbf{y}(z, h_\varepsilon(z)) \ .$$

We now define

$$\Omega_\varepsilon = \Omega \cup \{\mathbf{y}(z,t) \in \mathbf{R}^3 \ ; \ z \in \mathcal{H}\ ,\ 0 \le t < h_\varepsilon(z)\,\}$$

and we pose

$$\begin{aligned} \Gamma_2^\varepsilon &= \{\mathbf{y}_\varepsilon(z) \in \mathbf{R}^3 \ ; \ z \in \mathcal{G}\}, \quad \Gamma_1^\varepsilon = \partial\Omega_\varepsilon \backslash \Gamma_2^\varepsilon \\ \Gamma_2 &= \{\mathbf{y}(z,0) \in \mathbf{R}^3 \ ; \ z \in \mathcal{G}\}, \quad \Gamma_1 = \partial\Omega \backslash \Gamma_2 \ . \end{aligned}$$

Remark 1. *We note that the definition of Γ_2^ε depends on the choice of parametrisation of Γ_2, i.e. on $\mathbf{x}$. If Γ_2 is plain then Γ_2^ε is a periodic union of wrinkles. In general Γ_2^ε is not actually periodic due to the distortion coming from the choice of parametrisation.*

2. The equations of motion

A stationary flow of an incompressible, viscous, Newtonian fluid is governed by the Navier-Stokes system. As described before, we impose a no-slip condition on Γ_1^ε and a non-homogeneous slip condition on the rough part of the boundary Γ_2^ε. Our system can be written as:

$$(u^\varepsilon\nabla)u^\varepsilon - \mu\Delta u^\varepsilon + \nabla p^\varepsilon = f^\varepsilon, \quad \operatorname{div} u^\varepsilon = 0 \ \text{ in } \ \Omega_\varepsilon \tag{1}$$

$$u^\varepsilon = 0 \ \text{ on } \Gamma_1^\varepsilon, \quad u^\varepsilon \cdot \mathbf{n}^\varepsilon = 0 \ \text{ on } \Gamma_2^\varepsilon \tag{2}$$

$$(\mathbf{I} - \mathbf{n}\otimes\mathbf{n})\,\mu\,\frac{\partial u^\varepsilon}{\partial \mathbf{n}} + \kappa u^\varepsilon = g^\varepsilon \ \text{ on } \Gamma_2^\varepsilon \tag{3}$$

To place our problem in an appropriate functional framework we define the following functional spaces:

$$V_\varepsilon = \{ v \in H^1(\Omega_\varepsilon)^3 \ , \ \operatorname{div} v = 0 \ , \ v = 0 \text{ on } \Gamma_1^\varepsilon \ , \ v \cdot \mathbf{n}^\varepsilon = 0 \text{ on } \Gamma_2^\varepsilon \ \}$$

equipped by the norm

$$\|v\|_{V_\varepsilon} = \|\nabla v\|_{L^2(\Omega_\varepsilon)^9} \quad .$$

In order to have a well-posed problem we add the following assumptions on f^ε, g^ε and κ:

(i) $f^\varepsilon \in V_\varepsilon'$.

(ii) $g^\varepsilon \in H^{1/2}(\Gamma_2^\varepsilon)^3$.

(iii) $\mathcal{K} \in C(\overline{\mathcal{G}}), \quad 0 < \kappa_0 \le \mathcal{K} \le \kappa_1, \quad \kappa = \mathcal{K} \circ \mathbf{y}_\varepsilon$.

The variational formulation of (1)-(3) now reads:
Find $u^\varepsilon \in V_\varepsilon$ such that

$$\mu \int_{\Omega_\varepsilon} \nabla u^\varepsilon \nabla \phi + \int_{\Omega_\varepsilon} (u^\varepsilon \nabla) u^\varepsilon \, \phi + \int_{\Gamma_2^\varepsilon} \kappa \, u^\varepsilon \, \phi = {}_{V_\varepsilon'}\langle f^\varepsilon | \phi \rangle_{V_\varepsilon} + \int_{\Gamma_2^\varepsilon} g^\varepsilon \, \phi \ , \quad (4)$$

for any $\phi \in V_\varepsilon$.

3. Some technical results

We denote by $\mathbf{t}_i(z)$ the basis of the tangent plane on Γ_2 at point $z \in \mathcal{G}$ such that

$$(\mathbf{t}_1(z), \mathbf{t}_2(z)) = \nabla \mathbf{x}(z) \quad (5)$$

and by

$$\mathbf{n} = \frac{\mathbf{t}_1 \times \mathbf{t}_2}{|\mathbf{t}_1 \times \mathbf{t}_2|} \quad . \quad (6)$$

Analogously we denote the basis of the tangential space on Γ_2^ε by $\mathbf{t}_i^\varepsilon(z)$, $i = 1, 2$, where these vectors are chosen such that $(\mathbf{t}_1^\varepsilon(z), \mathbf{t}_2^\varepsilon(z)) = \nabla \mathbf{y}_\varepsilon(z)$ and can be computed as

$$\mathbf{t}_1^\varepsilon(z) = \mathbf{t}_1(z) + h'(\frac{z_1}{\varepsilon})\,\mathbf{n} + \varepsilon\, h(\frac{z_1}{\varepsilon}) \frac{\partial \mathbf{n}}{\partial z_1}(z) = \mathbf{t}_1(z) + h'(\frac{z_1}{\varepsilon})\,\mathbf{n} + O(\varepsilon)$$

$$\mathbf{t}_2^\varepsilon(z) = \mathbf{t}_2 + \varepsilon\, h(\frac{z_1}{\varepsilon}) \frac{\partial \mathbf{n}}{\partial z_2}(z) = \mathbf{t}_2(z) + O(\varepsilon) \ .$$

We denote by

$$
\begin{aligned}
\mathbf{n}^\varepsilon = \;& \mathbf{t}_1^\varepsilon \times \mathbf{t}_2^\varepsilon = \mathbf{t}_1 \times \mathbf{t}_2 + h'(\frac{z_1}{\varepsilon})\,(\mathbf{n} \times \mathbf{t}_2) + \frac{\varepsilon}{2}(h^2)'(\frac{z_1}{\varepsilon}) \left(\mathbf{n} \times \frac{\partial \mathbf{n}}{\partial z_2}\right) \\
& + \varepsilon h(\frac{z_1}{\varepsilon})\,[\mathbf{t}_1 \times \frac{\partial \mathbf{n}}{\partial z_2} + \frac{\partial \mathbf{n}}{\partial z_1} \times \mathbf{t}_2\,] + \varepsilon^2\, h^2(\frac{z_1}{\varepsilon}) \left(\frac{\partial \mathbf{n}}{\partial z_1} \times \frac{\partial \mathbf{n}}{\partial z_2}\right) = \\
& = \mathbf{t}_1 \times \mathbf{t}_2 + h'(\frac{z_1}{\varepsilon})\;(\mathbf{n} \times \mathbf{t}_2) + O(\varepsilon)\;,
\end{aligned}
$$

the vector of exterior (not necessarily unit) normal on Γ_2^ε.

For the wrinkled boundary we can now see by a direct computation that (see [13]):

Lemma 1. *Let dS_ε be the surface element on Γ_2^ε with $\varepsilon < \varepsilon_0$ and let dS be the surface element on Γ_2. Then*

$$dS_\varepsilon = \sqrt{1 + G(z, z_1/\varepsilon)}\;(1 + o(\varepsilon)\,)\;dS\;,$$

where $o(\varepsilon) \to 0$ as $\varepsilon \to 0$ uniformly with respect to $z \in \mathcal{G}$ and

$$G(z,\xi) = \frac{|\mathbf{t}_2(z)|^2\; h'(\xi)^2}{det[\,\mathbf{t}_i(z)\;\cdot\;\mathbf{t}_j(z)\,]} \quad . \tag{7}$$

Remark 2. *When the basis $\mathbf{t}_i$ is orthonormal (as for instance in [6], where Γ_2 is plain, parametrised by identity $\mathbf{x}(z) = z$), function $G(z,\xi) = G(\xi) = |h'(\xi)|^2$. Locally, parametrization can be chosen such that the metric tensor $[\mathbf{t}_i \cdot \mathbf{t}_j]$ is scalar, i.e. it has the form $g(z)\,\mathbf{I}$ (see, e.g. [9], Théorème 1, p 116). With such choice we get $G = |\nabla h|^2$.*

4. Existence, uniqueness, regularity and a priori estimates

We begin this section by stating some existence, uniqueness and regularity results for the problem (1)-(3). Since the proofs follow exactly the same lines as for the similar results proved in Temam [18] for the case of the Dirichlet's boundary conditions we leave them as an easy exercise to the reader.

Theorem 1. *Suppose that f^ε and g^ε satisfy the assumptions (i)-(iii). Then the problem (1)-(3) admits a solution $u^\varepsilon \in V_\varepsilon$, $p^\varepsilon \in L^2(\Omega_\varepsilon)/\mathbf{R}$. In addition, there exists a constant $M > 0$ (independent of ε but depending on κ_0 and μ) such that, if*

$$\|f^\varepsilon\|_{V_\varepsilon'} + \|g^\varepsilon\|_{L^2(\Gamma_2^\varepsilon)} < M\;, \tag{8}$$

then the solution is unique.

The following result was stated in lemma 4 in [13]:

Lemma 2. *Suppose that the assumptions (i), (ii), including the assumption (8) hold. Then there exists some $C > 0$, independent of ε, such that*

$$\|u^\varepsilon\|_{V_\varepsilon} \leq C \quad . \tag{9}$$

The pressure estimate in [13] was established only on Ω. Such estimate is insufficient for our purpose and we need stronger result. To prove such result the following technical lemma is needed:

Lemma 3. *Let $F \in L^2(\Omega_\varepsilon)^3$ and let $\int_{\Omega_\varepsilon} F = 0$. Then the problem*

$$div\, w = F \ \ in \ \ \Omega_\varepsilon, \qquad w = 0 \ \ on \ \ \partial\Omega_\varepsilon \tag{10}$$

has a solution $w \in H^1(\Omega_\varepsilon)^3$ satisfying the estimate

$$\|w\|_{H^1(\Omega_\varepsilon)} \leq C\|F\|_{L^2(\Omega_\varepsilon)} \tag{11}$$

with $C > 0$ independent of ε.

Proof. To start with we cut $\mathcal{G}$ in small, ε size parts by (for example) placing an ε-net on it. That way we divide $\mathcal{G}$ in $N(\varepsilon)$ parts, most of them being squares, except for the ones cutting the boundary (see Figure 2. below). Obviously the integer $N(\varepsilon)$ is of order ε^{-2}. We denote those ε-size subsets of $\mathcal{G}$ by $\mathcal{Z}_1^\varepsilon, \mathcal{Z}_2^\varepsilon, \ldots, \mathcal{Z}_{N(\varepsilon)}^\varepsilon$. We now define sets

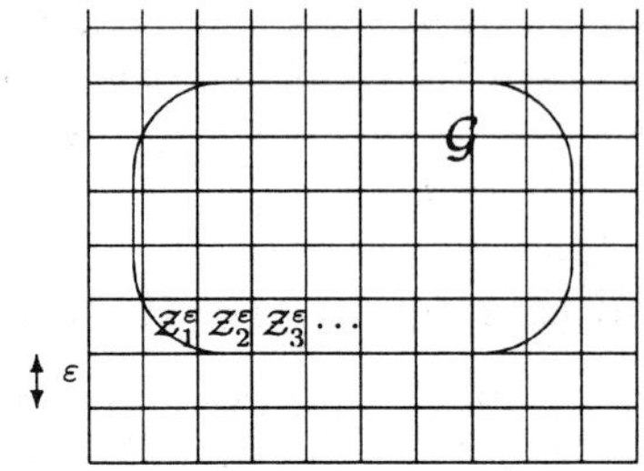

Figure 2. ε-net on $\mathcal{G}$

$$\mathcal{Y}_i^\varepsilon = \{\mathbf{x} = \mathbf{y}(z,t) \in \Omega_\varepsilon \ ; \ z \in \mathcal{Z}_i^\varepsilon \ , \ -\varepsilon < t < \varepsilon\, h(\frac{z_1}{\varepsilon}) \ \}$$
$$F_i^\varepsilon = \mathcal{Y}_i^\varepsilon \cap \Omega = \{\mathbf{x} = \mathbf{y}(z,t) \in \Omega_\varepsilon \ ; \ z \in \mathcal{Z}_i^\varepsilon \ , \ -\varepsilon < t < 0 \ \} \ .$$

Remark 3. *We notice that for each $\eta \in L^2(\mathcal{Y}_k^\varepsilon)$ such that $\int_{\mathcal{Y}_k^\varepsilon} \eta = 0$ there exists a solution of the problem*

$$div\, \psi = \eta \ \ in \ \ \mathcal{Y}_k^\varepsilon, \quad \psi = 0 \ \ on \ \ \partial\mathcal{Y}_k^\varepsilon \ ,$$

such that

$$\|\nabla\psi\|_{L^2(\mathcal{Y}_k^\varepsilon)} \le C_1\|\eta\|_{L^2(\mathcal{Y}_k^\varepsilon)}$$

with C_1 independent of ε. Indeed such claim follows by simple change of variables or from Lemma III.3.1 in [10].

Because of the remark 3 the idea is to split F in parts, each of them being defined on $\mathcal{Y}_k^\varepsilon$, having a zero mean value. To do so we use an explicite construction based on ideas from the proof of Theorem III.3.1 from [10].

We denote by χ_i^ε the characteristic function of the set F_i^ε. Now we construct the sequence of functions f_i in the following manner:

$$f_k(x) = \begin{cases} F(x) - \frac{\chi_k^\varepsilon(x)}{|F_k^\varepsilon|}\int_{\mathcal{Y}_k^\varepsilon} F, & x \in \mathcal{Y}_k^\varepsilon \\ 0, & x \in \Omega_\varepsilon\setminus\mathcal{Y}_1^\varepsilon \end{cases}, \quad k = 1,\dots,N(\varepsilon)$$

$$f_{N(\varepsilon)+1}(x) = \begin{cases} (1-\chi_1^\varepsilon(x) - \dots - \chi_{N(\varepsilon)}^\varepsilon(x))\,F(x) + \\ +\frac{\chi_1^\varepsilon(x)}{|F_1^\varepsilon|}\int_{\mathcal{Y}_1^\varepsilon} F + \dots + \frac{\chi_{N(\varepsilon)}^\varepsilon(x)}{|F_{N(\varepsilon)}^\varepsilon|}\int_{\mathcal{Y}_{N(\varepsilon)}^\varepsilon} F, & x \in \Omega \\ 0, & x \in \Omega_\varepsilon\setminus\Omega \end{cases}.$$

Such functions have the following properties:

$$\operatorname{supp} f_k \subset \overline{\mathcal{Y}_k^\varepsilon}, \qquad \int_{\mathcal{Y}_k^\varepsilon} f_k = 0 \tag{12}$$

$$F = \sum_{k=1}^{N(\varepsilon)+1} f_k, \quad \|f_k\|_{L^2(\mathcal{Y}_k^\varepsilon)} \le C\,\|F\|_{L^2(\mathcal{Y}_k^\varepsilon)},$$

with C independent of ε and k. We can now pose a sequence of problems

$$\begin{aligned} \operatorname{div} w^k &= f_k \text{ in } \mathcal{Y}_k^\varepsilon, \quad w^k = 0 \text{ on } \partial\mathcal{Y}_k^\varepsilon, \quad k = 1,\dots,N(\varepsilon)\,, \\ \operatorname{div} w^{N(\varepsilon)+1} &= f_{N(\varepsilon)+1} \text{ in } \Omega, \quad w^{N(\varepsilon)+1} = 0 \text{ on } \partial\Omega\,, \end{aligned}$$

Since (12) holds those problems are solvable and each of them admits a solution $w^k \in H_0^1(\mathcal{Y}_k^\varepsilon)^3$ such that

$$\|\nabla w^k\|_{L^2(\mathcal{Y}_k^\varepsilon)} \le C_1\|f_k\|_{L^2(\mathcal{Y}_k^\varepsilon)} \le C_2\,\|F\|_{L^2(\mathcal{Y}_k^\varepsilon)}, \quad k = 1,\dots,N(\varepsilon)$$

and

$$\|\nabla w^{N(\varepsilon)+1}\|_{L^2(\Omega)} \le C_3\|F\|_{L^2(\Omega)}\,.$$

Constants C_1, C_3 can be chosen independent of ε (see remark 3). Due to (12) we know that C_2 is also independent of k and our claim follows by summation with respect to k, with $w = \sum_{k=1}^{N(\varepsilon)+1} w^k$. □

Lemma 4. *Let the assumptions (i)-(iii) and (8) hold. Then there exists a constant $C > 0$ independent of ε, such that*

$$\|p^\varepsilon\|_{L^2(\Omega_\varepsilon)/\mathbf{R}} \leq C \ . \tag{13}$$

Proof. The proof relies on lemma 3. Indeed, it gives the existence of function ϕ, a solution of the problem

$$\begin{cases} \operatorname{div}\phi = \tilde{p}^\varepsilon & \text{in } \Omega_\varepsilon \\ \phi = 0 & \text{on } \partial\Omega_\varepsilon \ , \end{cases}$$

such that

$$\|\nabla\phi\|_{L^2(\Omega_\varepsilon)^3} \leq C\|\tilde{p}^\varepsilon\|_{L^2(\Omega_\varepsilon)} \ ,$$

where $\tilde{p}^\varepsilon = p^\varepsilon - \frac{1}{|\Omega_\varepsilon|}\int_{\Omega_\varepsilon} p^\varepsilon dx$. Using ϕ as a test function in problem (1) we get

$$\mu\int_{\Omega_\varepsilon} \nabla u^\varepsilon \nabla\phi + \int_{\Omega_\varepsilon} (u^\varepsilon\nabla)u^\varepsilon\phi =_{V_\varepsilon'} \langle f^\varepsilon \mid \phi\rangle_{V_\varepsilon} + \|\tilde{p}^\varepsilon\|^2_{L^2(\Omega_\varepsilon)} \ .$$

It leads to

$$\|\tilde{p}^\varepsilon\|^2_{L^2(\Omega_\varepsilon)} \leq C\{\mu\|u^\varepsilon\|_{V_\varepsilon} + \|f^\varepsilon\|_{V_\varepsilon'} + \|\nabla u^\varepsilon\|^2_{L^2(\Omega)}\}\|\tilde{p}^\varepsilon\|_{L^2(\Omega_\varepsilon)} \ ,$$

giving (13). □

5. The convergence

The estimates (9), (13) are independent of ε so that the restrictions of $(u^\varepsilon, p^\varepsilon)$ on Ω satisfy

$$\|u^\varepsilon\|_{H^1(\Omega)^3} \ , \ \|p^\varepsilon\|_{L^2(\Omega)/\mathbf{R}} \leq C \ . \tag{14}$$

with C independent of ε. Thus there exist a subsequence $\{u^\varepsilon, p^\varepsilon\}_{\varepsilon>0}$ and $(u \in H^1(\Omega)^3, p \in L^2(\Omega)/\mathbf{R})$ such that, as $\varepsilon \to 0$,

$$\begin{cases} u^\varepsilon \rightharpoonup u & \text{weakly in } H^1(\Omega)^3 \\ p^\varepsilon \rightharpoonup p & \text{weakly in } L^2(\Omega)/\mathbf{R} \ . \end{cases} \tag{15}$$

In analogy with section 2 we introduce the functional spaces

$$V_0 = \{v \in H^1(\Omega)^3 \ , \ \operatorname{div} v = 0 \ , \ v = 0 \text{ on } \Gamma_1 \ , \ v \parallel \mathbf{t}_2 \text{ on } \Gamma_2 \ \}$$
$$V_{\Gamma_1} = \{v \in H^1(\Omega)^3 \ , \ \operatorname{div} v = 0 \ , \ v = 0 \text{ on } \Gamma_1 \ \}$$

equipped by the norm

$$\|v\|_{V_0} = \|v\|_{V_{\Gamma_1}} = \|\nabla v\|_{L^2(\Omega)^9} \ .$$

In order to pass to the limit, as $\varepsilon \to 0$, in (4), we assume (in addition) that the restrictions of f^ε on Ω and g^ε satisfy

$$f^\varepsilon \to f^0 \text{ strongly in } V'_{\Gamma_1} \tag{16}$$

$$\overline{g}^\varepsilon \to g^0 \text{ strongly in } L^2(\Gamma_2)^3\,, \tag{17}$$

where $\overline{g}^\varepsilon[\mathbf{y}(z,0)] = g^\varepsilon\{\mathbf{y}[z,h_\varepsilon(z)]\}$ for $z \in \mathcal{G}$, i.e. for $\mathbf{x} = \mathbf{y}(z,0) \in \Gamma_2$ and $\mathbf{y}_\varepsilon = \mathbf{y}[z,h_\varepsilon(z)] \in \Gamma_2^\varepsilon$. As $\operatorname{div} u^\varepsilon \rightharpoonup \operatorname{div} u$ and for any $\phi \in C^1(\mathbf{R}^3)$ one has

$$0 = \int_{\Omega_\varepsilon} u^\varepsilon \nabla\phi \to \int_\Omega u\nabla\phi = \int_{\Gamma_2} (u\cdot\mathbf{n})\phi$$

obviously $u \cdot \mathbf{n} = 0$ on Γ_2.

We also have:

Lemma 5. *Let $\{v^\varepsilon\}$ be a sequence such that*

$$\|v^\varepsilon\|_{H^1(\Omega_\varepsilon)} \le C, \quad v^\varepsilon \cdot \mathbf{n}^\varepsilon = 0 \;\; on \;\; \Gamma_2^\varepsilon$$

and let v be the weak limit of the restriction $v^\varepsilon|_\Omega$. Then $v \in V_0$.

Proof. It remains to prove that $v \parallel \mathbf{t}_2 = 0$. First of all, a simple computation yields that for the function

$$\overline{v}^\varepsilon(\mathbf{x}) = v^\varepsilon(\mathbf{y}(z,\varepsilon h(z_1/\varepsilon))), \quad \mathbf{x} = \mathbf{y}(z,0) \in \Gamma_2\,,$$

we have

$$\|\overline{v}^\varepsilon - v^\varepsilon\|_{L^2(\Gamma_2)} \le C\varepsilon^{1/2}\;.$$

Now, since $v^\varepsilon \cdot \mathbf{n}^\varepsilon = 0$ on Γ_2^ε we have $\overline{v}^\varepsilon \cdot [\mathbf{t}_1 \times \mathbf{t}_2 + h'(z_1/\varepsilon)\,(\mathbf{n}\times\mathbf{t}_2) + O(\varepsilon)] = 0$ on Γ_2 or

$$v^\varepsilon \cdot \mathbf{n} = |\mathbf{t}_1 \times \mathbf{t}_2|^{-1} h'(z_1/\varepsilon)\; v^\varepsilon \cdot (\mathbf{n} \times \mathbf{t}_2) + E_\varepsilon \;\text{ on }\; \Gamma_2\,,$$

where $\|E_\varepsilon\|_{L^2(\Gamma_2)} \le C\,\varepsilon^{1/2}$. But then, multiplying by $h'(z_1/\varepsilon)$, we get

$$h'(z_1/\varepsilon)^2\; v^\varepsilon \cdot (\mathbf{n}\times\mathbf{t}_2) = |\mathbf{t}_1 \times \mathbf{t}_2|\; (h'(z_1/\varepsilon)\; v^\varepsilon \cdot \mathbf{n} - E_\varepsilon)\;. \tag{18}$$

Due to the compactness of the trace operator from $H^1(\Omega_\varepsilon)$ to $L^2(\Gamma_2)$ we have that $v^\varepsilon \to v$ strongly in $L^2(\Gamma_2)$ so that the left-hand side in (18) converges weakly in $L^2(\Gamma_2)$ to $(\int_0^1 (h')^2)\; v$. The right hand side weakly converges to $(\int_0^1 h')\; v \cdot \mathbf{n}$ which is equal to 0. Now $v \perp \mathbf{n}$ and $v \perp \mathbf{n} \times \mathbf{t}_2$ implies that $v \parallel \mathbf{t}_2$. □

We can now state our main result:

Theorem 2. *The functions (u,p) are the solution of the problem*

$$(u\nabla)u - \mu\Delta u + \nabla p = f^0, \quad div\, u = 0 \ in \ \Omega \tag{19}$$

$$u = 0 \ on\ \Gamma_1, \quad u \parallel \mathbf{t}_2 \ on\ \Gamma_2 \tag{20}$$

$$\mathbf{t}_2 \cdot (\mu \frac{\partial u}{\partial \mathbf{n}} + \kappa\, \lambda\, u) = \lambda\, g^0 \cdot \mathbf{t}_2 \ on\ \Gamma_2 \tag{21}$$

where $\lambda(z) = \int_0^1 \sqrt{1+|G(z,\xi)|}d\xi$, (note that $\lambda > 1$ and that in cases described in remark 2 λ is equal to the length of one wrinkle). Moreover, if the assumptions (i)-(ii) and (8) with $f^0, \lambda g^0, \lambda\kappa$ in place of $f^\varepsilon, g^\varepsilon, \kappa$ hold, then (19)-(21) has a unique solution and the whole sequence $\{(u^\varepsilon, p^\varepsilon)\}_{\varepsilon>0}$ converges to (u,p).

Proof. We begin by choosing the test functions. The tangential plane on the wavy boundary is spanned by vectors $\mathbf{t}_1^\varepsilon$, $\mathbf{t}_2^\varepsilon$ and its normal is $\mathbf{n}^\varepsilon$. Let $\phi \in V_0$. We denote by ϕ_i, $i = 1,2,3$ its components in $\mathbf{t}_1, \mathbf{t}_2, \mathbf{n}$ basis. Since $\phi \in V_0$ we have $\phi_1 = \phi_3 = 0$ on Γ_2. We extend ϕ_1, ϕ_3 by zero to $\Omega_\varepsilon \backslash \Omega$. We extend the remaining component ϕ_2 to $\Omega_\varepsilon \backslash \Omega$ by continuity. Such extension, denoted by the same symbol, now verifies $\phi \in H^1(\Omega_\varepsilon)^3$,

$$\|\phi\|_{H^1(\Omega_\varepsilon)} \le C$$

with C independent of ε. Unfortunately, due to the curvedness of the boundary Γ_2, our test function does not, necessarily, belong to V_ε since, in general

$$\phi \cdot \mathbf{n}^\varepsilon = \varepsilon\, h'(z_1/\varepsilon)\, h(z_1/\varepsilon)\, (\mathbf{n} \times \frac{\partial \mathbf{n}}{\partial z_2}) \ . \tag{22}$$

However, by direct computation, using (22), we obtain

$$\|\phi \cdot \mathbf{n}^\varepsilon\|_{L^2(\Gamma_2^\varepsilon)} \le C\, \varepsilon, \quad \|\phi \cdot \mathbf{n}^\varepsilon\|_{H^{1/2}(\Gamma_2^\varepsilon)} \le C\, \varepsilon^{\frac{1}{2}} \ . \tag{23}$$

For such ϕ we get from (1)

$$\int_{\Omega_\varepsilon} (u^\varepsilon\nabla)u^\varepsilon\phi + \mu \int_{\Omega_\varepsilon} \nabla u^\varepsilon \nabla\phi =$$

$$= {}_{H^{-1/2}(\Gamma_2^\varepsilon)} \langle -\mu\frac{\partial u^\varepsilon}{\partial \mathbf{n}} + p^\varepsilon\, \mathbf{n}^\varepsilon \mid (\phi \cdot \mathbf{n}^\varepsilon)\, \mathbf{n}^\varepsilon \rangle_{H^{1/2}(\Gamma_2^\varepsilon)} +$$

$$+ {}_{V_\varepsilon'}\langle f^\varepsilon \mid \phi \rangle_{V_\varepsilon} + \int_{\Omega_\varepsilon\backslash\Omega} p^\varepsilon \operatorname{div}\phi + \int_{\Gamma_2^\varepsilon} (g^\varepsilon - \kappa\, u^\varepsilon)\, \phi \ .$$

Using (15)-(17) and, if necessary, extracting the subsequence we pass to the limit as $\varepsilon \to 0$. Passage through all the terms is trivial, except for

$$M_\varepsilon = {}_{H^{-1/2}(\Gamma_2^\varepsilon)} \langle -\mu\frac{\partial u^\varepsilon}{\partial \mathbf{n}} + p^\varepsilon\, \mathbf{n}^\varepsilon \mid (\phi \cdot \mathbf{n}^\varepsilon)\, \mathbf{n}^\varepsilon \rangle_{H^{1/2}(\Gamma_2^\varepsilon)}$$

and the last one

$$N_\varepsilon = \int_{\Gamma_2^\varepsilon} (g^\varepsilon - \kappa u^\varepsilon)\,\phi \ .$$

There we have

$$\begin{aligned}
N_\varepsilon = & \int_{\Gamma_2^\varepsilon} g^\varepsilon \phi = \\
= & \int_{\mathcal{G}} g^\varepsilon[\mathbf{y}(z, h_\varepsilon(z))]\phi[\mathbf{y}(z, h_\varepsilon(z))]\sqrt{1 + |G(z, z_1/\varepsilon)|}(1 + o(\varepsilon))dS = \\
= & \int_{\mathcal{G}} \overline{g}^\varepsilon \phi[\mathbf{y}(z,0)]\ \lambda_\varepsilon\ dS + \\
& + \int_{\mathcal{G}} \overline{g}^\varepsilon \int_0^{h_\varepsilon(z)} \frac{\partial}{\partial t}\phi[\mathbf{y}(z,t)]dt\ \lambda_\varepsilon\ dS + o(\varepsilon) = I^1 + I^2 + o(\varepsilon)\ , \qquad (24)
\end{aligned}$$

where $\lambda_\varepsilon(z) = \sqrt{1 + |G(z, z_1/\varepsilon)|}$,

$$I^1 = \int_{\mathcal{G}} \overline{g}^\varepsilon \phi[\mathbf{y}(z,0)]\ \lambda_\varepsilon\ dS$$

and

$$I^2 = \int_{\mathcal{G}} \overline{g}^\varepsilon \int_0^{\varepsilon h(z_1/\varepsilon)} \frac{\partial}{\partial t}\phi[\mathbf{y}(z,t)]dt\ \lambda_\varepsilon\ dS \ .$$

Since the Jacobian $J^t = \det([\mathbf{t}_i \cdot \mathbf{t}_j]_{i,j=1,2,3})$, $\mathbf{t}_3 = \mathbf{n}$, satisfies the uniform (with respect to t) estimate $c_0 \le |J^t| \le c_1$, we have

$$\begin{aligned}
|I^2| \le & C \int_{\mathcal{G}} \int_0^{\varepsilon h(z_1/\varepsilon)} |\nabla\phi|\ |J^t| dt\, dS = \\
= & C \int_{\Omega_\varepsilon \setminus \Omega} |\nabla \phi| \to 0\ .
\end{aligned}$$

The periodicity lemma (see for example Bensoussan, Lions and Papanicolaou [5]) gives

$$\lambda_\varepsilon \rightharpoonup \lambda \ \text{ weak* in } L^\infty(\mathcal{G})\ ,$$

where

$$\lambda(z) = \int_0^1 (1 + G(z,\xi))^{1/2}\, d\xi \ .$$

Thus we have

$$I^1 = \int_{\Gamma_2} \phi\, \overline{g}^\varepsilon\ \lambda_\varepsilon \to \int_{\Gamma_2} \lambda\, g^0\, \phi\ .$$

We treat analogously the integral

$$\int_{\Gamma_2^\varepsilon} u^\varepsilon\ \phi$$

since $u^\varepsilon \to u$ strongly in $L^2(\Gamma_2)^3$ due to the compactness of the trace operator $\gamma_0 \in \mathcal{L}(H^1(\Omega_\varepsilon), L^2(\Gamma_2))$. For the remaining term M_ε we proceed as follows: Due to the Green's formula

$$\begin{aligned} M_\varepsilon &= {}_{H^1(\Omega_\varepsilon)'} \langle \mathrm{div}\{-\mu \nabla u^\varepsilon + p^\varepsilon \mathbf{I}\} \,|\, \psi \rangle_{H^1(\Omega_\varepsilon)} - \\ &- \int_{\Omega_\varepsilon} (-\mu \nabla u^\varepsilon \nabla \psi + p^\varepsilon \,\mathrm{div}\, \psi) = M_1^\varepsilon + M_2^\varepsilon \,, \end{aligned}$$

where $\psi \in H^1(\Omega_\varepsilon)^3$ is such that

$$\psi = (\phi \cdot \mathbf{n}^\varepsilon)\, \mathbf{n}^\varepsilon \text{ on } \Gamma_2^\varepsilon, \quad \psi = 0 \text{ on } \Gamma_1^\varepsilon$$

and

$$\|\psi\|_{H^1(\Omega_\varepsilon)} \le C \|(\phi \cdot \mathbf{n}^\varepsilon)\mathbf{n}^\varepsilon\|_{H^{1/2}(\Gamma_2^\varepsilon)} \le C\,\varepsilon^{\frac{1}{2}} \;.$$

Now

$$\begin{aligned} |\, M_1^\varepsilon \,| &= \left| {}_{V_\varepsilon'}\langle f^\varepsilon | \psi \rangle_{V_\varepsilon} - \int_{\Omega_\varepsilon} (u^\varepsilon \nabla) u^\varepsilon \,\phi \right| \le C(\|u^\varepsilon\|_{V_\varepsilon}^2 + \\ &+ \|f^\varepsilon\|_{V_\varepsilon'}) \, \|\psi\|_{H^1(\Omega_\varepsilon)} \le C\,\varepsilon^{\frac{1}{2}} \end{aligned}$$

and

$$|\, M_2^\varepsilon | \le C(\|u^\varepsilon\|_{V_\varepsilon} + \|p\|_{L^2(\Omega_\varepsilon)}) \|\psi\|_{H^1(\Omega_\varepsilon)} \le C\,\varepsilon^{\frac{1}{2}} \;.$$

Thus $M_\varepsilon \to 0$ as $\varepsilon \to 0$. It remains to prove the convergence of the pressure. That part is trivial, since for $\varphi \in C_0^\infty(\Omega)^3$ we have

$$\int_\Omega (u^\varepsilon \nabla) u^\varepsilon \varphi + \mu \int_\Omega \nabla u^\varepsilon \nabla \varphi = {}_{V_\varepsilon'} \langle f^\varepsilon \,|\, \varphi \rangle_{V_\varepsilon} + \int_\Omega p^\varepsilon \,\mathrm{div}\, \varphi \;.$$

Passing directly to the limit as $\varepsilon \to 0$ we find that (u, p) satisfy

$$\int_\Omega (u \nabla) u \varphi + \mu \int_\Omega \nabla u \nabla \varphi = {}_{V_{\Gamma_1}'} \langle f^0 \,|\, \varphi \rangle_{V_{\Gamma_1}} + \int_\Omega p \,\mathrm{div}\, \varphi \,,$$

i.e.

$$\begin{aligned} &(u\nabla)u - \mu \Delta u + \nabla p = f^0, \quad \mathrm{div}\, u = 0 \text{ in } \Omega \\ &u = 0 \text{ on } \Gamma_1, \quad u \parallel \mathbf{t}_2 \text{ on } \Gamma_2 \\ &\mathbf{t}_2 \cdot \left(\mu \frac{\partial u}{\partial \mathbf{n}} - \lambda (g^0 - \kappa\, u) \right) = 0 \text{ on } \Gamma_2 \;. \end{aligned}$$

□

References

[1] Achdou Y., Pironneau O., Domain decomposition and wall laws, C.R.Acad.Sci.Paris, Série I, 320 (1995), 541-547.

[2] Achdou Y., Pironneau O., Valentin F., Effective boundary conditions for laminar flows over periodic rough boundaries, J.Comp.Phys., 147 (1998), 187-218.

[3] Amirat Y., Simon J., Influence de la rugosité en hydrodinamique laminaire, C. R. Acad. Sci. Paris, Série I, 322 (1996), 313-318.

[4] Belyaev A.G., Average for the third boundary-value problem for the Poisson equation in a domain with rapidly oscillating boundary (in Russian), Vestnik Moskov. Univ. (Ser. I, Math. Mech.) 6 (1988), 63-66.

[5] Bensoussan A., Lions J.L., Papanicolaou G.C., Asymptotic analysis for periodic structures, North Holland, 1978.

[6] Checkin G., Friedman A., Piatnitski A., The boundary-value problem in domains with very rapidly oscillating boundary, J.Math.Anal.Appl., 231 (1999) 1, 213-234.

[7] Cioranescu D., Donato P., Ene H.I., Homogenization of the Stokes problem with non homogeneous slip boundary conditions, Math.Methods Appl.Sci., 19 (1996) 11, 857-881.

[8] Cioranescu D., Donato P., Ene H.I., Fluid flows in dielectric porous media, Optimization, optimal control and PDE, Proceedings of the 1 st. Franco-Romanian Conference, Iasi Romania 1992, ISNM 107, 3-10, 1992.

[9] Dubrovine B., Novikov S., Fomenko A., Géométrie contemporaine, Première partie, Mir, Moscou, 1982.

[10] Galdi G.P., An introduction to the mathematical theory of the Navier-Stokes equations, Vol. I,II, Springer, Berlin, 1992.

[11] Jäger W., Mikelić A., On the interface conditions of Beavers, Joseph and Safman, SIAM J. Appl.Math., Vol 60, No 4 (2000), 1111-1127.

[12] Jäger W., Mikelić A., On the roughness-induced effective boundary condition for an incompressible viscous flow, J.Differential Equations, 170 (2001), 96-122.

[13] Marušić-Paloka E., Average of the Navier's law on the rapidly oscillating boundary, J.Math.Anal.Appl., 259 (2001), 685-701.

[14] Marušić-Paloka E., Errata "Average of the Navier's law on the rapidly oscillating boundary", submitted to J.Math.Anal.Appl.

[15] Navier C.L.M.H., Sur les lois de l'équilibre et du mouvement des corps élastiques, Mem. Acad. R. Sci. Inst. France, 6 (1827) 369.

[16] Schatzman M., On the eigenvalues and eigenvectors of the Laplace operator in a thin set with Neumann boundary condition, Appl.Anal., 61 (1996), 293-306.

[17] Sanchez-Palencia E., Non-Homogeneous Media and Vibration Theory, Springer Lecture Notes in Physics 127, Springer-Verlag, Berlin, 1980.

[18] Temam R., Navier–Stokes Equations, Theory and Numerical Analysis, North–Holland, Amsterdam, 1984.

[19] Tuck E.O., Kozubov A., A laminar roughness boundary condition, J.Fluid Mech., 300 (1995), 59-70.

RELAXATION OF SOME ENERGY FUNCTIONALS RELATED TO THE FORMATION OF MICROSTRUCTURE*

Andrija Raguž
Department of Mathematics, Faculty of Science
Bijenička 30, 10 000 Zagreb, Croatia
andrija@math.hr

Abstract A variant of the energy studied in [1] perturbed by highly oscillatory term $a(\varepsilon^{-\beta}s)$ is considered, where $\beta > 0$ is a given parameter. The original functional is replaced by

$$J_\varepsilon(v) := \int_\Omega (\varepsilon^2 v''(s)^2 + W(v'(s)) + a(\varepsilon^{-\beta}s)v(s)^2)ds \, .$$

Expected oscillatory behavior is discussed, relaxation of the rescaled functional J_ε, the associated Γ-limit and its minimizer are proposed by an averaging procedure adjusted to capture multiscale behavior of the minimizing sequence (v^ε) as $\varepsilon \longrightarrow 0$.

Keywords: Relaxation, microstructure, Γ-convergence, Young measures

1. Introduction

The purpose of this paper is to discuss one of the recent advances concerning the development of mathematical tools for studying multiscale variational problems.

Typically, the limit of functionals whose minimizers exhibit oscillations on several fast scales should describe complex behavior of materials able to form *a microstructure*, that is to say, a fine hierarchy of structures appearing on the microscale. Increasing number of experimental and theoretical efforts in recent years yielded the basis for the further research in the matter, but rigorous mathematical analysis has been developed rather slowly.

*This work is supported in part by the Croatian Ministry of Science and Technology through project 037015–Oscillatory solutions of partial differential equations.

Classical result, based on De Giorgi's notion of Γ-convergence, the famous Theorem of Modica and Mortola from 1977, was one of the first results to highlight the treatment of variational problems changing their type in the limit. However, multiscale problems can not be modeled by such means alone.

Apart from the *two-scale convergence* approach of Allaire and Nguetseng, and the notion of H-measure of Tartar [11] and Gérard [5] and its variants, until recently there was no tool available examining multiscale behavior of minimizing sequences coming from different problems in PDEs and continuum mechanics. In particular, singularly perturbed problems with small parameter ε accounting for singular effects due to the introduction of higher-order gradient regularization have remained beyond the scope of classical methods, like formal asymptotic expansion, since such a procedure can only predict oscillatory behavior of the minimizers, but can not offer an insight into what exactly happens when ε tends to zero.

We follow an approach due to G. Alberti and S. Müller having the advantage of proposing an effective procedure, based on the notion of *Young measure*, which enables us to capture the oscillations in the limit, therefore offering the mathematical model of formation of the microstructures within some physical systems of interest, such as alloys etc.

2. Formulation of the problem

Our attention is being focused on one-dimensional singularly perturbed functionals of Ginzburg-Landau type, with some penalizing terms attached. In view of the results in [1], we expect that such a class should represent the starting point for further research in a sense that we can anticipate what happens in more realistic three-dimensional models.

The main feature of this approach is that it leads to well-posed minimization problems for the relaxed functionals, and, as a consequence, a unique minimizer is selected in the limit. An example studied in detail in [1] is the following one:

$$I_\varepsilon(v) := \int_\Omega \left(\varepsilon^2 v''(s)^2 + W(v'(s)) + a(s)v(s)^2\right) ds \, ,$$

where $\Omega \subseteq \mathbf{R}$ is a bounded open interval, $v \in \mathrm{H}^2_{per}(\Omega)$ and a is a Borel measurable function satisfying $0 < \alpha_1 \leq a(s) \leq \alpha_2 < +\infty$ for almost every $s \in \Omega$. W is chosen to be a typical non-convex function of the form $W(t) = (t^2 - 1)^2$, but any non-negative locally Lipschitz-continuous function with precisely two zeros and with superlinear growth at infinity may be chosen as well. The main steps in the approach are:

1 Determine the characteristic length scale of minimizing sequences of I_ε, and based on that, propose the blowup transformation $R^\varepsilon_s v$ capturing the

limiting behavior. Finally, characterize the class of all Young measures $(\nu_s)_{s\in\Omega}$ generated by the sequence of 'ε-blowups' of functions v^ε.

2 Rewrite the rescaled functionals $\varepsilon^{-\frac{2}{3}}I_\varepsilon(v)$ as $\int_\Omega f_s^\varepsilon(R_s^\varepsilon v)ds$ for a suitable choice of $R_s^\varepsilon v$ and f_s^ε.

3 Identify the Γ-limit f_s of f_s^ε for almost every $s \in \Omega$.

4 Identify the Γ-limit of naturally defined relaxed functionals on the space of Young measures with values in the 'pattern' space

5 Determine the minimizer for the relaxed functional in the limit and prove its uniqueness.

After rescaling, which is the first step in insuring that the limit is non-trivial, we ought to choose an admissible relaxation over the space of measures, namely the space of *Young measures on micro-patterns* or *Two-scale Young measures* expecting the convergence of minimizers to take place now. Here *the pattern space* is the space of all Borel measurable mappings modulo equivalence almost everywhere, and in fact it is generated by the map $t \mapsto R_s^\varepsilon v(t)$. This space can be chosen compact and metrizable.

Such a program was originally carried out by Alberti and Müller in [1], and it turns out that a characteristic length scale is $\varepsilon^{\frac{1}{3}}$, the admissible blowup is defined as

$$R_s^\varepsilon v(t) := \varepsilon^{-\frac{1}{3}}v(s+\varepsilon^{\frac{1}{3}}t)\ , \quad t \in \mathbf{R}\,,$$

while the unique minimizer for the functional in the limit is a non-homogeneous Young measure which is at almost every point equal to the unique translation-invariant measure supported on the translational orbit

$$\mathcal{O}(y_h) := \{T_\tau y_h : \tau \in \mathbf{R}\}$$

of the h-periodic sawtooth function (depending on $s \in \Omega$) $y_h : \mathbf{R} \longrightarrow \mathbf{R}$ defined by the duality pairing

$$\langle \varepsilon_{y_h}, g\rangle := \frac{1}{h}\int_0^h g(T_\tau y_h)d\tau\ ,$$

where, for $s \in \Omega$,

$$h(s) := L_0 \cdot a(s)^{-\frac{1}{3}}\ , \quad L_0 := \left(96\int_{-1}^{1}\sqrt{W}\right)^{\frac{1}{3}}$$

and

$$y_h(t) := |t| - \frac{h}{4}\ , \quad |t| \le \frac{h}{2}\ . \tag{1}$$

3. Some results

Our objective is to consider the functional

$$J_\varepsilon(v) := \int_\Omega (\varepsilon^2 v''(s)^2 + W(v'(s)) + a(\varepsilon^{-\beta}s)v(s)^2)ds ,$$

where $\beta > 0$. As shown by comprehensive analysis in [8], the characteristic scale length associated to the one-dimensional variants of I_ε is $\varepsilon^{\frac{1}{3}}$, and the shorter scale ε^β is expected not to change the quantities describing macroscopic energy. Longer scales ε^β, however, may significantly change features of the minimizers, and, as a consequence, involvement of longer scales puts us in the situation to look for the admissible relaxation which in the limit keeps track of oscillations not only over characteristic length scale, but over longer scales as well.

Indeed, all minimizers of J_ε form the structure of several scales in the sense above. When trying to ignore the penalizing effect, we end up with the same blowup $R_s^\varepsilon v$, and we consider the cases $\beta > \frac{1}{3}$ and $\beta < \frac{1}{3}$, while the critical case $\beta = \frac{1}{3}$ is expected to be more delicate.

Firstly, if $\beta > \frac{1}{3}$, i.e. if the externally imposed scale ε^β is shorter than the fast scale $\varepsilon^{\frac{1}{3}}$ appearing in the definition of $R_s^\varepsilon v$, a relaxation proposed in this case is exactly the one studied in [1], since the step 3, namely the Γ-convergence of the integrands

$$f_s^\varepsilon(x) := \frac{1}{2r}\int_{-r}^{r} \left(\varepsilon^{\frac{2}{3}} x''(t)^2 + \varepsilon^{-\frac{2}{3}} W(x'(t)) + a(\varepsilon^{-\beta}(s+\varepsilon^{\frac{1}{3}}t))x(t)^2\right)dt ,$$

if $x \in \mathrm{H}^2_{per}\langle -r, r\rangle$ (and $+\infty$ otherwise) still holds, where the limit f is now independent of $s \in \Omega$, and reads

$$f(x) := \begin{cases} \frac{A_0}{2r}\mathrm{card}(Sx' \cap \langle -r, r\rangle) + \overline{a}\frac{1}{2r}\int_{-r}^{r} x(t)^2 dt, & \text{if } x \in \mathcal{S} \\ +\infty, & \text{otherwise,} \end{cases}$$

where

$$A_0 := 2\int_{-1}^{1}\sqrt{W} , \quad \overline{a} := \frac{1}{\mu(\Omega)}\int_\Omega a(s)ds ,$$

$\mathcal{S}$ stands for the class of continuous sawtooth functions having derivative ± 1, such that derivative has only finitely many jumps, Sx' stands for the set of all discontinuities of x',

Apart from an application of the Theorem of Modica and Mortola (cf. [7]), the proof of Γ-convergence relies on the property that a convergence

$$a_s^\varepsilon \xrightarrow{\ *\ } \overline{a} \quad \text{in } \mathrm{L}^\infty(\mathbf{R}) \quad (\text{for almost every } s \in \Omega) ,$$

where

$$a_s^\varepsilon(t) := a(\varepsilon^{-\beta}s + \varepsilon^{\frac{1}{3}-\beta}t) ,$$

implies that the integrands ϕ_s^ε and ϕ defined by

$$\phi_s^\varepsilon(x) := \frac{1}{2r}\int_{-r}^{r} a_s^\varepsilon(t)x(t)^2 dt ,$$

$$\phi(x) := \frac{1}{2r}\int_{-r}^{r} \overline{a}x(t)^2 dt$$

satisfy

$$\phi_s^\varepsilon(x_\varepsilon) \longrightarrow \phi(x) ,$$

whenever $x_\varepsilon \xrightarrow{L^2} x$. It is not difficult to see that, following the sequence of results in [1] (see Proposition 3.1–Corollary 3.13) naturally defined relaxed functionals F^ε Γ-converge to the limit F in the topology of Young measures on micropatterns. Moreover, F admits a unique minimizer in such a space, which is the unique homogeneous Young measure associated to the symmetric functions y_h as in the case $\beta = 0$, but now the coarserest period h does not depend on the chosen point $s \in \Omega$:

$$h = L_0 \cdot \overline{a}^{-\frac{1}{3}} .$$

This concludes our discussion of the case $\beta > \frac{1}{3}$.

If $0 < \beta < \frac{1}{3}$, the sequence (a_s^ε) does not converge and the best one could get is

$$\lim_{\varepsilon \longrightarrow 0} d(a_s^\varepsilon, \mathcal{O}(a)) = 0 \quad \text{(for almost every } s \in \Omega) ,$$

where d is a metric consistent with the weak* topology in $L^\infty(\mathbf{R}; [-1, 1])$, and

$$\mathcal{O}(a) := \{T_\tau a : \tau \in \mathbf{R}\}$$

is the set of all functional translations associated to Ω-periodic function a. Since such weaker form of convergence does not suffices to conclude that f_s^ε Γ-converge to some limit (in the weak* topology of L^∞), we are compelled to look for other effective procedure in accordance with steps 1–5, that is, for different blowup functions, and different integrands f_s^ε, having the advantage of leading to convergence of the relaxed functionals F^ε and eventually to the limiting relaxed functional F with some unique minimizer. An additional averaging procedure was proposed in the ending lines of [1], and it is our intention to discuss the proposal in some detail.

It is reasonable to expect that due to the similar results of [8] the minimizers of the ε-problem are locally well approximated by the continuous periodic sawtooth function having the period

$$P^\varepsilon = L_0\varepsilon^{\frac{1}{3}} \cdot a(\varepsilon^{-\beta}s)^{-\frac{1}{3}} .$$

Thus the sequence of Dirac masses associated to ε-blowups of the minimizers converges to the homogeneous two-scale Young measure

$$\nu_s = \frac{1}{\mu(\Omega)} \int_\Omega \varepsilon_{x_t} dt \ ,$$

where x_t is the sawtooth function having the period $h_t = L_0 a(t)^{-\frac{1}{3}}$ defined as in (1).

It turns out that such a minimizing measure does not give good representation in the limit, since the limiting functional F in this setting is affine and minimal value of the penalizer does not enter the formula for the rescaled limiting energy (cf. [1]). Therefore, the question of possibility of getting a more natural limiting energy arises. This is done by considering a more convenient blow-up, namely

$$\mathcal{R}_s^\varepsilon w(t,\tau) := \varepsilon^{-\frac{1}{3}} w(s + \varepsilon^\beta t + \varepsilon^{\frac{1}{3}}\tau) \ .$$

Such a hierarchical blowup with two small scales is hoped to capture oscillations over both scales ε^β and $\varepsilon^{\frac{1}{3}}$, due to the introduction of the fast variable t weighted by ε^β. As before, the main difficulty in this setting is the lack of convergence of the terms

$$a_s^\varepsilon(t,\tau) := a(\varepsilon^{-\beta} s + t + \varepsilon^{\frac{1}{3}-\beta}\tau) \ .$$

In order to overcome this, we introduce a probability Radon measure λ_∞ on $\mathbf{R}$ defined as an asymptotic renormalization of one-dimensional Lebesgue measure λ: λ_∞ is the weak* limit of the measures λ_R as $R \longrightarrow +\infty$, where

$$\lambda_R(A) := \frac{1}{2R} \lambda(A \cap \langle -R, R\rangle)$$

holds for every Borel measurable set $A \subseteq \mathbf{R}$.

In the next step, we consider the pattern space

$$K_2 := \left\{ w : \mathbf{R}^2 \longrightarrow [-\infty, +\infty] \ , w \text{ is Borel measurable} \right\} ,$$

and define the functionals $f_s^\varepsilon : K_2 \longrightarrow [0, +\infty]$ by

$$f_s^\varepsilon(w) := \int_{\mathbf{R}} \frac{1}{2r} \int_{-r}^{r} \left(\varepsilon^{\frac{2}{3}} \left(\frac{d^2}{d\tau^2} w(t,\tau) \right)^2 + \varepsilon^{-\frac{2}{3}} W\left(\frac{d}{d\tau} w(t,\tau) \right) \right) d\tau d\lambda_\infty(t)$$
$$+ \int_{\mathbf{R}} \frac{1}{2r} \int_{-r}^{r} a(\varepsilon^{-\beta} s + t + \varepsilon^{\frac{1}{3}-\beta}\tau) w(t,\tau)^2 d\tau d\lambda_\infty(t) \ ,$$

if $w(t,\cdot) \in \mathrm{H}^2\langle -r, r\rangle$ for almost every $t \in \mathbf{R}$, and by $+\infty$ otherwise. We expect that it is possible to perform step 3 of the program now, so that for almost every $s \in \Omega$

$$f_s^\varepsilon \xrightarrow{\ \Gamma\ } f \ ,$$

as $\varepsilon \longrightarrow 0$ in the topology of K_2, where $f : K_2 \longrightarrow [0, +\infty]$ is now defined by the formula

$$f(w) := \int_{\mathbf{R}} \left(\frac{A_0}{2r} \operatorname{card}\Big(Sw'(t, \cdot) \cap \langle -r, r \rangle \Big) + a(t) \frac{1}{2r} \int_{-r}^{r} w(t, \tau)^2 d\tau \right) d\lambda_\infty(t) ,$$

if $w(t, \cdot) \in S$ for almost every $t \in \mathbf{R}$, and by $+\infty$ otherwise. Indeed, due to the lower semicontinuity of $\mathrm{L}^1(\mathbf{R}; \lambda_\infty)$-norm on K_2 it is enough to show that

$$\int_{\mathbf{R}} \frac{1}{2r} \int_{-r}^{r} a(\varepsilon^{-\beta} s + t + \varepsilon^{\frac{1}{3} - \beta} \tau) w_\varepsilon(t, \tau)^2 d\tau d\lambda_\infty(t)$$

converges to

$$\int_{\mathbf{R}} \frac{1}{2r} \int_{-r}^{r} a(t) w(t, \tau)^2 d\tau d\lambda_\infty(t) ,$$

provided that $w_\varepsilon \xrightarrow{\mathrm{L}^2} w$. For the rest of the proof we can argue by the Modica-Mortola Theorem (cf. [7]). Taking into account the definition of λ_∞ we anticipate that such convergence holds indeed. Once this step is completed, we are able to propose the reasonable relaxation of the functional J_ε (cf. Theorem 3.4 and subsection 3.5 in [1]). Relaxation is carried out over the space of K_2-valued Young measures, denoted by $YM(\Omega; K_2)$, where an element $\boldsymbol{\nu} \in YM(\Omega; K_2)$ is weakly$*$ measurable mapping $\boldsymbol{\nu} : \Omega \longrightarrow \mathcal{M}(K_2)$, $s \mapsto \nu_s$, usually written as $(\nu_s)_{s \in \Omega}$. Following [1], we expect that the sequences of Dirac masses $(\delta_{\mathcal{R}_s^\varepsilon v^\varepsilon})$, where $v^\varepsilon \in H^2_{per}(\Omega)$, generate a subset of probability Radon measures as $\varepsilon \longrightarrow 0$, denoted by $\mathcal{I}(K_2)$ and characterized by the relation

$$\langle \nu_s, g \circ T_{(\sigma_1, \sigma_2)} \rangle = \langle \nu_s, g \rangle , \quad g \in \mathrm{C}(K_2) , \quad (\sigma_1, \sigma_2) \in \mathbf{R}^2 ,$$

where $T_{(\sigma_1, \sigma_2)} : K_2 \longrightarrow K_2$ is the functional translation on K_2 associated to the point $(\sigma_1, \sigma_2) \in \mathbf{R}^2$. Thus, recalling that for $v \in \mathrm{H}^2_{per}\langle -r, r \rangle$ we have

$$\varepsilon^{-\frac{2}{3}} J_\varepsilon(v) = \int_\Omega \langle \delta_{\mathcal{R}_s^\varepsilon v}, f_s^\varepsilon \rangle ds ,$$

we define the functionals $F^\varepsilon, F : YM(\Omega; K_2) \longrightarrow [0, +\infty]$ as follows:

$$F^\varepsilon(\boldsymbol{\nu}) := \begin{cases} \int_\Omega \langle \nu_s, f_s^\varepsilon \rangle ds, & \text{if } \nu_s = \delta_{\mathcal{R}_s^\varepsilon w} \text{ for some } w \in \mathrm{H}^2_{per}(\Omega) \\ +\infty, & \text{otherwise,} \end{cases}$$

$$F(\boldsymbol{\nu}) := \begin{cases} \int_\Omega \langle \nu_s, f \rangle ds, & \text{if } \nu_s \in \mathcal{I}(K_2) \text{ for a.e. } s \in \Omega \\ +\infty, & \text{otherwise.} \end{cases}$$

Now, since $f_s^\varepsilon \xrightarrow{\Gamma} f$ for almost every $s \in \Omega$, we expect that $F^\varepsilon \xrightarrow{\Gamma} F$ holds, and that the unique minimizer for the functional F is given by

$$\eta_{\overline{w}} := \int_{\mathbf{R}} \int_{\mathbf{R}} \delta_{T_{(\sigma_1, \sigma_2)} \overline{w}} d\lambda_\infty(\sigma_2) d\lambda_\infty(\sigma_1) ,$$

where $\overline{w}$ satisfies

- $t \longrightarrow \overline{w}(t,\cdot)$ is Ω-periodic
- for almost every $t \in \mathbf{R}$ $\overline{w}(t,\cdot)$ is $\overline{h}(t)$-periodic continuous sawtooth function defined by

$$\overline{w}(t,\tau) := |\tau| - \frac{\overline{h}(t)}{4}\,, |\tau| \leq \frac{\overline{h}(t)}{2}\,,$$

where

$$\overline{h}(t) := L_0 \cdot a(t)^{-\frac{1}{3}}\,.$$

We point out that the original paper [1] indicates which results need to be obtained in order to prove that integral functionals (F^ε) converge (compare sections 4 and 5 in [1], in particular, Corollary 5.11), but the results themselves require more work and precise formulation of all necessary steps goes beyond the purpose of this paper. It is also worth mentioning that, in accordance with remarks in the ending lines of [1], as well as calculations sketched in this note, the approach of Alberti and Müller (outlined through steps 1–5) seems to be applicable to other similar problems of interest.

Acknowledgments

The author wishes to thank Prof. Stefan Müller for his constant encouragement and support which enabled the author to visit the Max Planck Institute for Mathematics in the Sciences, Leipzig, Germany (in February and June 2001). Many thanks goes to the staff of Max Planck Institute for the efforts to make the author's stay as pleasant as possible.

References

[1] G. Alberti, S. Müller: *A new approach to variational problems with multiple scales, Comm. Pure Appl. Math.* **54** (2001), 761-825.

[2] J.M. Ball: *A version of the fundamental theorem for Young measures*, in *PDE's and Continuum Models of Phase Transitions (M. Rascle and al., eds.), pp. 207-215 Lecture Notes in Physics* **344**, Springer, 1989.

[3] J.M. Ball, R.D. James: *Fine phase mixtures as minimizers of energy, Arch. Rational Mech. Anal.* **100** (1987), 13–52.

[4] A. DeSimone, R. V. Kohn, S. Müller, F. Otto: *Magnetic microstructures - a paradigm of multiscale problems, Proc. ICIAM 99 (Edinburgh, 1999), (Eds. J.M. Ball, J.C.R. Hunt), pp. 175-190,* Oxford University Press, 2000.

[5] P. Gérard: *Microlocal defect measures, Comm. PDE* **16** (1991), 1761–1794.

[6] P. Gérard: *Measures semi-classiques et ondes de Bloch*, in *Seminaire sur les Equations aux Derivées Partielles*, 1990-1991, École Polytechnique, Palaiseau 1991.

[7] L. Modica, S. Mortola: *Un esempio di* Γ*-convergenca, Bull. Un. Mat. Ital. (5)*, **14-B** (1977), 285-299.

[8] S. Müller: *Singular perturbations as a selection criterion for minimizing sequences, Calc. Var.* **1** (1993), 169–204.

[9] S. Müller: Private communication.

[10] L. Tartar: *Beyond Young measures, Meccanica* **30** (1995), 505–526.

[11] L. Tartar: *H-measures, a new approach for studying homogenisation, oscillations and concentration effects in partial differential equations, Proceedings of the Royal Society of Edinburgh* **A-115** (1990), 193-230.

A COARSE SPACE FOR ELASTICITY

Partition of Unity Rigid Body Motions Coarse Space

Marcus Sarkis
Instituto de Matemática Pura e Aplicada,
Est. Dona Castorina, 110, Rio de Janeiro, RJ, CEP 22420-320, Brazil
msarkis@impa.br
and
Mathematical Sciences Department, Worcester Polytechnic Institute,
Worcester, MA 01609, USA

Abstract In this paper, we develop overlapping Schwarz methods for a finite element discretization on linear elasticity. In order to make Schwarz methods scalable with respect to the number of subdomains, we add an appropriate coarse problem to the algorithms. The main purpose of this paper is to introduce and analyze a new coarse space for these methods. The proposed coarse space is based on a partition of unity (PU) and the local rigid body motions (RM) of the overlapping subdomains. We prove that the condition number of the algorithms grows only linearly with respect to the relative size of the overlap.

Keywords: Schwarz preconditioner, domain decomposition, coarse spaces, partition of unity, elliptic equations, finite elements

1. Introduction

In this paper, we introduce and analyze two-level overlapping Schwarz methods for unstructured meshes for a finite element discretization of linear elasticity. The novel part of the methods is the coarse space. Works on two-level methods and coarse spaces on unstructured meshes is an active area of research. Several different approaches have been introduced and some can be found in [1, 4, 5, 7, 8, 10, 11, 12, 13, 14, 15, 16, 18, 23] and papers cited therein. Related works to ours, based on two-level agglomeration techniques, can be found in [4, 14]. Their analysis use a class of partition of unity coarse space based on agglomeration smoothing techniques and they have proven an upper bound for the condition number which depends quadratically on the relative overlap. Inspired by some of the numerical results and analysis of [1, 6, 7, 17, 21], the author [19] introduced new coarse basis functions for the discrete Laplacian based on the kind of partition of unity used in the theoretical analysis of Schwarz methods. This kind of partition of unity has a controlled decaying property on the overlapping region and we have proved that the condition number of the algorithms grows only linearly with respect to the relative size of the overlap.

In this paper, we extend some of the results in [19] to linear elasticity. We combine the partition of unity and rigid body motions associated to the (overlapping) subdomains in order to design a new coarse space for elasticity. The partition of unity is used: 1) to localize the coarse basis functions to the subdomains, and 2) to force the coarse basis functions to have a smooth decaying to zero near the boundary of the subdomains. The rigid body motions associated to the subdomains allow us to use some of the theoretical arguments of Neumann-Neumann algorithms [15, 12]. The proposed coarse space, named PU-RM coarse space, has several advantages over traditional coarse spaces: 1) it is applicable to discretizations on unstructured meshes, 2) it is algebraic (see below), 3) the associated algorithms do not require that the subdomains be connected or that the boundary of the subdomains be smooth, 4) the coarse basis functions of the PU-RM coarse space are constructed explicitly and without the use of exact local solvers, 5) the stencil of the coarse matrix is sparser than the traditional ones, and 6) the support of the coarse basis functions is localized on the subdomains, and therefore easy in communication if implemented in a distributed memory parallel machine.

The preconditioners to be considered here are algebraic in the sense that the preconditioners are built in terms of the graph of the sparse matrix and the mesh partition. In this paper we provide a complete and simple mathematical analysis to a finite element problem, the linear elasticity with zero displacement (Dirichlet) boundary condition.

2. Linear Elasticity Discretization

We consider an isotropic elastic material in the configuration region $\Omega \subset \Re^2$. Let us denote $u^* = (u_1^*, u_2^*)^t$ to be the displacement and $f = (f_1, f_2)^t$ the body force. Let the region Σ be Ω or a subregion of Ω, and the spaces $\vec{H}^1(\Sigma)$, $\vec{L}_2(\Sigma)$, and $\vec{H}_0^1(\Sigma)$ to be the spaces $(H^1(\Sigma), H^1(\Sigma))^t$, $(L_2(\Sigma), L_2(\Sigma))^t$, and $(H_0^1(\Sigma), H_0^1(\Sigma))^t$, respectively. The weak formulation of the static theory of linear elasticity with zero boundary displacement boundary condition is given as follows.

Find $u^* \in \vec{H}_0^1(\Omega)$ such that

$$a(u^*, v) = f(v), \;\; \forall v \in \vec{H}_0^1(\Omega), \tag{1}$$

where

$$a(u^*, v) = \int_\Omega (\mu E(u^*) : E(v) + \lambda \text{div}(u^*)\, \text{div}(v))\; dx \;,$$

$$f(v) = \int_{\Omega} f \cdot v \, dx \ \text{ for } f \in \vec{L}^2(\Omega),$$

and

$$E(v) = \frac{1}{2}\left(\nabla v + (\nabla v)^t\right).$$

The positive constants μ and λ are called the Lamé constants. It is well-known that $a(\cdot,\cdot)$ is elliptic (using the First Korn Inequality; see Theorem 9.2.25 in [3]) and bounded, and therefore the system (1) is well posed.

For simplicity, let Ω be a bounded polygonal region in $\Re^2$ with a diameter of size $O(1)$. The extension of the results to $\Re^3$ can also be carried out using similar ideas. Let $T^h(\Omega)$ be a shape regular, quasi-uniform triangulation of grid size $O(h)$ of Ω, and $V \subset \vec{H}_0^1(\Omega)$ be the finite element space consisting of continuous piecewise linear functions associated with the triangulation. The extension of the theory for the case of local quasi-uniform triangulation is also straightforward.

We are interested in solving the discrete problem associated to (1): Find $u \in V$ such that

$$a(u,v) = f(v), \quad \forall\, v \in V. \tag{2}$$

Since $V \subset \vec{H}_0^1(\Omega)$, the discrete version is also well-posed.

Throughout this paper, C and C_0, are positive generic constants that do not depend of any of the mesh parameters, the number of subdomains, and the parameters λ and μ. All the domains and subdomains are assumed to be open; i.e., boundaries are not included in their definitions.

3. Algebraic Subregions

Given the domain Ω and triangulation $T^h(\Omega)$, we assume that a domain partition has been applied and resulted in N substructures (non-overlapping subregions) $\Omega_i, i = 1, \dots N$, of size $O(H)$, such that

$$\overline{\Omega} = \cup_{i=1}^N \overline{\Omega}_i, \qquad \Omega_i \cap \Omega_j = \emptyset, \quad \text{for} \quad j \neq i.$$

We define the overlapping subdomains Ω_i^δ as follows. Let Ω_i^1 be the one-overlap element extension of Ω_i, where $\Omega_i^1 \supset \Omega_i$ is obtained by including all the immediate neighboring elements $\tau_h \in T^h(\Omega)$ of Ω_i such that $\overline{\tau}_h \cap \overline{\Omega}_i \neq \emptyset$. Using the idea recursively, we can define a δ-extension overlapping subdomains Ω_i^δ

$$\Omega_i \subset \Omega_i^1 \subset \cdots \subset \Omega_i^\delta.$$

Here the integer $\delta \geq 1$ indicates the level of element extension and δh is the approximate length of the extension. We note that this extension can be coded easily through the knowledge of the adjacent matrix associated to the mesh and the partition.

In this paper we consider the linear elasticity problem for the case of Dirichlet boundary condition on the whole $\partial\Omega$. We want to design coarse spaces based on a class of partition of unit for which has been used as a very powerful tool in the theoretical analysis of Schwarz type domain decomposition methods. We note however that the partition of unity functions on this class do not necessarily vanish on $\partial\Omega$. Hence, they cannot be used straightforwardly as coarse basis functions since they should satisfy zero Dirichlet boundary conditions for Dirichlet type boundary problems. Hence, for the coarse basis functions that touch $\partial\Omega$, we modify them so that they have a controlled decaying to zero near $\partial\Omega$. To obtain such coarse basis functions, we next introduce a Dirichlet boundary treatment. Let Ω_B^1 be one layer of elements near the Dirichlet boundary $\partial\Omega$ and then define recursively,

$$\Omega_B^1 \subset \Omega_B^2 \subset \cdots \subset \Omega_B^\delta$$

with δ levels of extension by adding recursively neighboring elements.

To define and analyze the new methods, we subdivide Ω_i^δ as follow. Let $\gamma_i^\delta = \partial\Omega_i^\delta \backslash \partial\Omega, i = 1, \cdots, N$; i.e., the part of the boundary of Ω_i^δ that does not belong to the physical boundary of Ω, and let $\gamma_B^\delta = \partial\Omega_B^\delta \backslash \partial\Omega$. We define the interface overlapping boundary Γ^δ as the union of all the γ_i^δ and γ_B^δ; i.e., $\Gamma^\delta = \cup_{i=B,1}^N \gamma_i^\delta$. We also need the following subsets of Ω_i^δ

- $\Gamma_i^\delta = \Gamma^\delta \cap \Omega_i^\delta$ (local interface)
- $N_i^\delta = \Omega_i^\delta \backslash (\cup_{j \neq i} \Omega_j^\delta \cup \Omega_B^\delta \cup \Gamma_i^\delta)$ (non-overlapping region)
- $O_i^\delta = \Omega_i^\delta \backslash (N_i^\delta \cup \Gamma_i^\delta)$ (overlapping region)

We note that $\Omega_i^\delta = N_i^\delta \cup \Gamma_i^\delta \cup O_i^\delta$. The region O_i^δ is the overlapping region of Ω_i^δ excluding $\Gamma_i^\delta \cap \Omega_i^\delta$. The region N_i^δ is the subregion of Ω_i^δ which does not overlap any neighboring extended subdomain $\overline{\Omega}_j^\delta$ and $\overline{\Omega}_B^\delta$. We recall that the regions O_i^δ and N_i^δ are open sets.

4. Overlapping Schwarz Methods

We next describe the PU-RM coarse space and introduce the corresponding overlapping additive Schwarz method and a hybrid Schwarz method. We first consider the local spaces and local problems.

4.1 The Local Spaces

We introduce the local spaces as

$$V_i^\delta = V \cap \vec{H}_0^1(\Omega_i^\delta), \quad i = 1, 2, \cdots, N,$$

extended by zero to $\Omega\backslash\Omega_i^\delta$. It is easy to verify that

$$V = V_1^\delta + V_2^\delta + \cdots + V_N^\delta. \tag{3}$$

The property (3) gives robustness for the preconditioners defined in this paper in the sense that a convergence is always attained independently of the quality of the partitioning. This is an advantage over some iterative substructuring methods in which are based on the requirement that all the substructures Ω_i must be connected. We point out that the space decomposition given by (3) is not a direct sum if $\delta > 1$. This increases robustness for the methods when the Ω_i have rough (*zigzag*) boundaries. By extending the substructures to Ω_i^δ, we allow the possibility of decomposing a function of V as a sum of functions of V_i^δ without the *zigzag* behavior. So it is possible to obtain low energy decompositions, and hence better lower bounds for the condition number of the preconditioners are obtained.

We define the local projections (local problems) $P_i^\delta : V \to V_i^\delta$ as follows:

$$a(P_i^\delta u, v) = a(u, v), \quad \forall v \in V_i^\delta. \tag{4}$$

We next introduce the PU-RM coarse space V_0^δ.

4.2 A PU-RM Coarse Space

We next construct a partition of unity θ_i^δ such that $\theta_i^\delta \in V_i^\delta$, $0 \le \theta_i^\delta(x) \le 1$, $|\nabla\theta_i^\delta(x)| \le C/(\delta h)$ in the interior of the elements, and $\sum_{i=B,1}^N \theta_i^\delta \equiv 1$. Such construction is natural, algebraic and easy to implement. We first construct the function $\hat\theta_B^\delta \in V_i^\delta$ as follows. We let $\hat\theta_B^\delta(x) = 1$ for nodes x on $\partial\Omega$. For the first layer of neighboring nodes x of $\partial\Omega$ we let $\hat\theta_B^\delta(x) = (\delta - 1)/\delta$, For the second layer of neighboring nodes x of $\partial\Omega$ we let $\hat\theta_B^\delta(x) = (\delta - 2)/\delta$, and recursively until $k = \delta - 1$, we let $\hat\theta_B^\delta(x) = (\delta - k)/\delta$ for the $(k)st$ layer of neighboring nodes x of $\partial\Omega$. For the reminaing nodes x of $\overline\Omega$ we let $\hat\theta_B^\delta(x) = 0$. Similarly, for $i = 1, \cdots, N$, we let $\hat\theta_i^\delta(x) = 1$ for nodes x of $\overline\Omega_i\backslash\Omega_B^\delta$. For the first layer of neighboring nodes x of $\overline\Omega_i\backslash\Omega_B^\delta$ we let $\hat\theta_i^\delta(x) = (\delta - 1)/\delta$, and recursively until $k = \delta - 1$, we let $\hat\theta_i^\delta(x) = (\delta - k)/\delta$ for the $(k)st$ layer of neighboring nodes x of $\overline\Omega_i\backslash\Omega_B^\delta$. For the remaining nodes x of $\overline\Omega$ we let $\hat\theta_i^\delta(x) = 0$. It is easy to verify that $0 \le \hat\theta_i^\delta(x) \le 1$, and for quasi-uniform triangulation $|\nabla\hat\theta_i^\delta(x)| \le C/(\delta h)$ in the interior of the elements. The partition of unity θ_i^δ is defined as

$$\theta_i^\delta = I_h\left(\frac{\hat\theta_i^\delta}{\sum_{j=B,1}^N \hat\theta_j^\delta}\right).$$

Here I_h is the regular pointwise linear interpolation operator from the continuous functions to piecewise linear and continuous functions. It is easy to verify that $\sum_{i=B,1}^N \theta_i^\delta(x) = 1$, $0 \le \theta_i^\delta(x) \le 1$, and $|\nabla\theta_i^\delta(x)| \le C/(\delta h), \forall x \in \overline\Omega$.

We next consider the key ingredient for designing the coarse space for linear elasticity: the local rigid body motions. We let

$$\vec{RM}(\Sigma) = \{v \in \vec{L}_2(\Sigma) : v = c + b(x_2, -x_1)^t, c \in \Re^2, b \in \Re\}$$

be the space of rigid body motions functions on Σ. An important property of the space $\vec{RM}(\Sigma)$, and which will play an important role in the later analysis of our methods, is given as follows. If $v \in \vec{RM}(\Sigma)$ then $a_\Sigma(v,v) \equiv 0$. In addition, in certain extent the converse is also true; if $a_\Sigma(v,v) \equiv 0$ and Σ is connected, then $v \in \vec{RM}(\Sigma)$. Here, the bilinear form $a_\Sigma(\cdot,\cdot)$ on $\vec{H}^1(\Sigma) \times \vec{H}^1(\Sigma)$ is given by

$$a_\Sigma(u,v) = \int_\Sigma (2\mu E(u) : E(v) + \lambda \text{div}(u)\, \text{div}(v))\, dx\, .$$

The PU-RM coarse space V_0^δ is defined as

$$V_0^\delta = \{\sum_{i=1}^{N} \vec{I}_h \left([c_i + b_i(x_2, -x_1)^t]\theta_i^\delta \right), \forall c_i \in \Re^2, \forall b_i \in \Re\}. \tag{5}$$

Here, the interpolator $\vec{I}_h = (I_h, I_h)^t$ is the regular componentwise pointwise linear interpolation operator. We note that we do not include the θ_B^δ in the sum of (5), and therefore the number of degrees of freedom of V_0^δ is $3N$. The function θ_B^δ is needed only to define the others functions $\theta_i^\delta, i = 1, \cdots, N$.

We define the global projection (global problem) $P_0^\delta : V \to V_0^\delta$ as follows:

$$a(P_0^\delta u, v) = a(u,v), \quad \forall v \in V_0^\delta.$$

4.3 Preconditioners

We consider two preconditioners:

- The two-level overlapping additive Schwarz operator [9] given by

$$P_{as}^\delta = \sum_{i=0}^{N} P_i^\delta,$$

- The hybrid Schwarz operator [15, 16] given by

$$P_{hyb}^\delta = P_0^\delta + (I - P_0^\delta)(\sum_{i=1}^{N} P_i^\delta)(I - P_0^\delta).$$

5. Theoretical Analysis

It is possible to show that the solution of (2) is the solution of the preconditioned system $P_{as}^{\delta}u = g_{as}$ ($P_{hyb}^{\delta}u = g_{hyb}$), for an appropriate right hand side g_{as} (g_{hyb}); see [20]. These preconditioned systems are typically solved by the conjugate gradient method, without further preconditioning, using $a(\cdot,\cdot)$ as the inner product. The preconditioned systems presented in this paper are applicable to any unstructure mesh and partitioning. The notions of subdomains, the classification of the regions O_i^{δ} and N_i^{δ} and the interfaces Γ_i^{δ}, etc., can all be defined in terms of the graph of the sparse matrix. In this section we provide a sharp and optimal estimate for a case in which each Ω_i^{δ} has a square shape or is an image of a reference square under reasonably smooth mapping. The analysis below can then be carried out on the reference square. We note however, that results can also be extended for any connected Lipschitz subregions that are not too distorted. We next prove the main result of the paper.

Theorem 1. *There exists a constant $C > 0$ such that for any $u \in V$ we have*

$$\kappa(P_{hyb}^{\delta}) \leq \kappa(P_{as}^{\delta}) \leq C(1 + \lambda/\mu)(1 + \frac{H}{\delta h}). \tag{6}$$

The constant C does not depend on h, δ, H, λ, and μ.

Proof.
We estimate the condition numbers of the operators P_{as}^{δ} and P_{hyb}^{δ} in terms of the fine mesh size h, the subdomain size H, the overlapping factor δ, and the parameters λ and μ. We shall follow the abstract additive Schwarz theory [22, 20] to analyze the additive versions, where three assumptions have to be checked and three paramenters C_0, ω and $\rho(E)$ estimated; see Theorem 1 in [20]. Two assumptions are trivial to check: $\omega = 1$ since we use exact solvers, and $\rho(E) \leq C$, where C is the maximum number of subdomains Ω_i^{δ} overlapping a common point. Hence, if the overlap is of the order of H or less, then C will not depend of H and h. So our focus on the rest of the paper is in bounding C_0. The first inequality of (6) follows directly from the Lemma 3.2 in [16]; see also Theorem 4 in [20].

What remains to complete the proof is to derive a bound for C_0; i.e., to find C_0 such that for any given $u \in V$, there exist $u_i \in V_i^{\delta}$, such that

$$u = \sum_{i=0}^{N} u_i, \tag{7}$$

and

$$\sum_{i=0}^{N} a(u_i, u_i) \leq C_0^2 a(u, u). \tag{8}$$

We next concentrate on defining the decomposition $u = \sum_{i=0}^{N} u_i$. Let $V_{i,0}^{\delta}, i = 1, \cdots, N$, be the spaces with three degrees of freedom generated by the $[c_i^{\delta} + b_i^{\delta}(x_2, -x_1)^t]\theta_i^{\delta}$ coarse basis functions. We introduce the interpolation-like operator $I_0^{\delta} = \sum_{i=1}^{N} I_{i,0}^{\delta}$, where the $I_{i,0}^{\delta} : V \to V_{i,0}^{\delta}$ is defined as follows:

$$(I_{i,0}^{\delta} u)(x) := \vec{I}_h \left([\bar{c}_i^{\delta} + \bar{b}_i^{\delta}(x_2, -x_1)^t]\theta_i^{\delta}(x) \right),$$

where

$$\bar{b}_i^{\delta} = -\frac{1}{2|\Omega_i^{\delta}|} \int_{\Omega_i^{\delta}} \mathrm{rot}(u(x))\, dx$$

and

$$\bar{c}_i^{\delta} := \frac{1}{|\Omega_i^{\delta}|} \int_{\Omega_i^{\delta}} [u(x) - \bar{b}_i^{\delta}(x_2, -x_1)^t]\, dx.$$

Here $|\Omega_i^{\delta}|$ is the area of the region Ω_i^{δ}. Let us denote

$$w_i = u - [\bar{c}_i^{\delta} + \bar{b}_i^{\delta}(x_2, -x_1)^t].$$

We remark that the definitions of $\bar{b}_i^{\delta}$ and the $\bar{c}_i^{\delta}$ imply that $\mathrm{rot}(w_i)$ and the two components of w_i have average zero on Ω_i^{δ}. We define the u_i as follows. Let $u_0 \in V_0^{\delta}$ be defined as

$$u_0 = I_0^{\delta} u = \sum_{i=1}^{N} I_0^{i,\delta} u = \sum_{i=1}^{N} \vec{I}_h \left([\bar{c}_i^{\delta} + \bar{b}_i^{\delta}(x_2, -x_1)^t]\theta_i^{\delta} \right),$$

and the $u_i \in V_i^{\delta}, i = 1, \cdots, N$ as

$$u_i = \vec{I}_h(\vartheta_i^{\delta} u) - \vec{I}_h \left([\bar{c}_i^{\delta} + \bar{b}_i^{\delta}(x_2, -x_1)^t]\theta_i^{\delta} \right).$$

Here, the piecewise linear functions $\vartheta_i^{\delta} \in H^1(\Omega_i^{\delta})$ are modifications of the θ_i^{δ} and are defined so that they form a partition of unity $\sum_{i=1}^{N} \vartheta_i^{\delta} \equiv 1$ on the whole $\overline{\Omega}$, satisfy $0 \le \vartheta_i^{\delta} \le 1$, $|\nabla \vartheta_i^{\delta}| \le C/(\delta h)$, and $\vartheta_i^{\delta}(x) = \theta_i^{\delta}(x), \forall x \in \Omega \backslash \overline{\Omega}_B^{\delta}$. It is therefore easy to see that (7) holds. We postpone the construction of the ϑ_i^{δ} to later in this paper. We note that the ϑ_i^{δ} are not used in the algorithms; they are used only as a tool for the analysis.

The strategy of the proof is to first obtain a bound for $\sum_{i=1}^{N} a(u_i, u_i)$. Once done, we obtain a bound for $\sum_{i=0}^{N} a(u_i, u_i)$ by noticing that $u_0 = u - \sum_{i=1}^{N} u_i$, and therefore with triangular inequalities and properties of $\rho(\mathcal{E})$ we obtain

$$a(u_0, u_0) \le 2 \left(a(u, u) + a(\sum_{i=1}^{N} u_i, \sum_{i=1}^{N} u_i) \right)$$

$$\leq 2\left(a(u,u)+\rho(\mathcal{E})\sum_{i=1}^{N}a(u_i,u_i)\right).$$

It remains to obtain a bound for $\sum_{i=1}^{N} a(u_i, u_i)$. To do so, we first decompose the $u_i, i = 1, \cdots, N$ as $u_i = u_i^0 + u_i^B$ where

$$u_i^0 = \vec{I}_h(\theta_i^\delta w_i) \text{ and } \ u_i^B = \vec{I}_h((\vartheta_i^\delta - \theta_i^\delta)u)), \tag{9}$$

and use

$$\sum_{i=1}^{N} a(u_i,u_i) \leq 2\sum_{i=1}^{N} a(u_i^0,u_i^0) + 2\sum_{i=1}^{N} a(u_i^B,u_i^B), i = 1,\cdots,N.$$

The next steps concentrate in obtaining a bound for each $a(u_i^0, u_i^0)$. The estimation of $a(u_i^0, u_i^0)$ is done as in standard additive Schwarz methods [9]. Let K be an element of Ω_i^δ, and $\bar{\theta}_i^{K,\delta}$ be the average of θ_i^δ on K. We have

$$\begin{aligned} a_K(u_i^0,u_i^0) &= a_K(\vec{I}_h(\theta_i^\delta w_i), \vec{I}_h(\theta_i^\delta w_i)) \\ &\leq 2a_K(\vec{I}_h(\bar{\theta}_i^{K,\delta} w_i), \vec{I}_h(\bar{\theta}_i^{K,\delta} w_i)) \quad (10)\\ &+ 2a_K(\vec{I}_h([\theta_i^\delta - \bar{\theta}_i^{K,\delta}]w_i), \vec{I}_h([\theta_i^\delta - \bar{\theta}_i^{K,\delta}]w_i)). \quad (11) \end{aligned}$$

For the term (10), we use that $|\bar{\theta}_i^{K,\delta}| \leq 1$ (it follows from $0 \leq \theta_i^\delta \leq 1$) to obtain

$$a_K(\vec{I}_h(\bar{\theta}_i^{K,\delta} w_i), \vec{I}_h(\bar{\theta}_i^{K,\delta} w_i)) \leq a_K(w_i, w_i).$$

For obtaining a bound for the term (11) we do the following. If $K \subset N_i^\delta$ then $\theta_i^\delta \equiv 1$, and hence $\|\theta_i^\delta - \bar{\theta}_i^{K,\delta}\|_{L^\infty(K)} = 0$ and (11) vanishes. It remains to consider only the case $K \subset O_i^\delta$. To do so, we first use an inverse inequality; i.e., any linear function v defined on an element K of size $O(h)$ satisfies $|v|_{H^1(K)} \leq C/h\|v\|_{L^2(K)}$. We obtain

$$\begin{aligned} &a_K(\vec{I}_h([\theta_i^\delta - \bar{\theta}_i^{K,\delta}]w_i), \vec{I}_h([\theta_i^\delta - \bar{\theta}_i^{K,\delta}]w_i)) \\ &\leq C\max\{\mu,\lambda\}h^{-2}\|(\vec{I}_h([\theta_i^\delta - \bar{\theta}_i^{K,\delta}]w_i)\|^2_{\vec{L}^2(K)}. \end{aligned}$$

We then use that $\|\theta_i^\delta - \bar{\theta}_i^{K,\delta}\|_{L^\infty(K)} \leq C/\delta$ (it follows from $|\nabla\theta_i^\delta| \leq C/(\delta h)$) to have

$$\begin{aligned} &C\max\{\mu,\lambda\}h^{-2}\|(\vec{I}_h([\theta_i^\delta - \bar{\theta}_i^{K,\delta}]w_i)\|^2_{\vec{L}^2(K)} \\ &\leq C\max\{\mu,\lambda\}(\delta h)^{-2}\|w_i\|^2_{\vec{L}^2(K)}. \end{aligned}$$

Hence, using that the support u_i^0 is contained in $\overline{\Omega}_i^\delta$, and the previous calculations, we have

$$a(u_i^0, u_i^0) = \sum_{K \in \Omega_i^\delta} a_K(u_i^0, u_i^0)$$

$$\leq C \left(a_{\Omega_i^\delta}(w_i, w_i) + \max\{\mu, \lambda\} \frac{1}{(\delta h)^2} \|w_i\|^2_{\vec{L}^2(O_i^\delta)} \right).$$

We next use similar arguments as in Lemma 3 in [11]. By using simple manipulations of the Fundamental Theorem of Calculus and the Cauchy-Schwarz inequality, we have

$$\|w_i\|^2_{\vec{L}_2(O_i^\delta)} \leq C \left(\delta h \|w_i\|^2_{\vec{L}_2(\partial\Omega_i^\delta)} + (\delta h)^2 |w_i|^2_{\vec{H}^1(O_i^\delta)} \right),$$

and with a trace theorem for domains (here Ω_i^δ) with size of $O(H)$ we obtain

$$\|w_i\|^2_{\vec{L}_2(\partial\Omega_i^\delta)} \leq C \left(H^{-1} \|w_i\|^2_{\vec{L}^2(\Omega_i^\delta)} + H |w_i|^2_{\vec{H}^1(\Omega_i^\delta)} \right).$$

Hence,

$$(\delta h)^{-2} \|w_i\|^2_{\vec{L}^2(O_i^\delta)} \leq C \left((1 + \frac{H}{\delta h}) |w_i|^2_{\vec{H}^1(\Omega_i^\delta)} + \frac{1}{H(\delta h)} \|w_i\|^2_{\vec{L}^2(\Omega_i^\delta)} \right),$$

and therefore,

$$a(u_i^0, u_i^0) \leq C \max\{\mu, \lambda\} \left((1 + \frac{H}{\delta h}) |w_i|^2_{\vec{H}^1(\Omega_i^\delta)} + \frac{1}{H(\delta h)} \|w_i\|^2_{\vec{L}^2(\Omega_i^\delta)} \right).$$

Using that $\mathrm{rot}(w_i)$ and the two components of w_i have average zero on Ω_i^δ, we can apply the Second Korn inequality (see Theorem 9.2.12 in [3]) to obtain

$$|w_i|^2_{\vec{H}^1(\Omega_i^\delta)} + \frac{1}{H^2} \|w_i\|^2_{\vec{L}_2(\Omega_i^\delta)} \leq C \int_{\Omega_i^\delta} E(w_i) : E(w_i)\, dx,$$

and since

$$\int_{\Omega_i^\delta} \mathrm{div}(w_i)\, \mathrm{div}(w_i)\, dx \leq C \int_{\Omega_i^\delta} E(w_i) : E(w_i)\, dx,$$

we obtain

$$a(u_i^0, u_i^0) \leq C(1 + \lambda/\mu)(1 + \frac{H}{\delta h}) a_{\Omega_i^\delta}(w_i, w_i).$$

We then use that $u - w_i \in \vec{RM}(\Omega_i^\delta)$ (kernel of $a_{\Omega_i^\delta}(\cdot,\cdot)$) to obtain

$$a_{\Omega_i^\delta}(w_i, w_i) = a_{\Omega_i^\delta}(u, u).$$

The next step is to obtain a bound for $a(u_i^B, u_i^B)$. We note that for $i = 1, \cdots, N$

$$\vartheta_i^\delta(x) = \theta_i^\delta(x), \forall x \in \Omega \backslash \overline{\Omega}_B^\delta,$$

and therefore the support of u_i^B is on $\overline{\Omega}_B^\delta \cap \overline{\Omega}_i^\delta$. On $\overline{\Omega}_B^\delta$, we have $|\vartheta_i^\delta - \theta_i^\delta| \leq 1$ and $|\nabla \vartheta_i^\delta - \nabla \theta_i^\delta| \leq C/(\delta h)$. So we can use the similar arguments as before and a Friedrichs inequality (u vanishes on $\partial\Omega$) to obtain,

$$a(u_i^B, u_i^B) \leq C(1 + \lambda/\mu)(1 + \frac{H}{\delta h})a_{\Omega_i^\delta}(u, u).$$

We now sum all the contributions and use a coloring argument to obtain

$$\sum_{i=1}^{N} a(u_i, u_i) \leq C(1 + \lambda/\mu)a(u, u).$$

We now give the construction of the ϑ_i^δ. We next modify the coarse basis functions θ_i^δ on $(\overline{\Omega}_i^\delta \cap \overline{\Omega}_B^\delta)$ to define the partition of unity ϑ_i^δ. We first construct the function $\hat{\vartheta}_i^\delta \in H^1(\Omega_i^\delta)$. Let $\hat{\vartheta}_i^\delta(x) = 1$ and $\hat{\vartheta}_i^\delta(x) = 0$ for nodes x of $\overline{\Omega}_i$ and $\overline{\Omega} \backslash \Omega_i^\delta$, respectively. For the first layer of neighboring nodes x of $\overline{\Omega}_i$ we let $\hat{\vartheta}_i^\delta(x) = (\delta - 1)/\delta$, and recursively until $k = \delta - 1$, we let $\hat{\vartheta}_i^\delta(x) = (\delta - k)/\delta$ for the $(k)st$ layer of neighboring nodes x of $\overline{\Omega}_i$. The partition of unity ϑ_i^δ is defined as

$$\vartheta_i^\delta = I_h(\frac{\hat{\vartheta}_i^\delta}{\sum_{j=1}^{N} \hat{\vartheta}_j^\delta}).$$

It is easy to verify that $\sum_{i=1}^{N} \vartheta_i^\delta(x) = 1, \;\; 0 \leq \vartheta_i^\delta(x) \leq 1$, and $|\nabla \vartheta_i^\delta(x)| \leq C/(\delta h)$ in the interior of the elements. Also, $\vartheta_i^\delta(x) = \theta_i^\delta(x), i = 1, \cdots, N$, when $x \in \Omega \backslash \Omega_B^\delta$. □

We note that the discretization considered in this paper gives satisfactory (second order accurate) convergent finite element approximation to the elasticity problem when λ/μ is not large. It can be shown [2, 3] that the apriori error estimate of this finite element method deteriorates as $\lambda \gg \mu$; this phenomenon is called *locking effect* or *volume locking*. We note that the upper bound estimate of the preconditioners presented here also follows the similar patterns. Here also, we cannot remove the λ/μ dependence on the upper bound estimates for the conditioning number of the preconditioned systems. To see this we use the following arguments. If $\mathrm{div}(u) = 0$ and λ is close to ∞, the only way to obtain a decomposition stable with respect to λ is to have the all the $\mathrm{div}(u_i) = 0$. However, it is easy to see that $\mathrm{div}(u_0) = 0$ implies that u_0 vanishes. Hence, there is no global communication and therefore the condition number must have a H dependence on the upper bound estimation. For incompressible ($\lambda = \infty$) or almost incompressible materials, other discretizations based on hybrid or non-conforming finite elements approximations [2, 3] are more appropriate and they will not be considered here.

Acknowledgements: A special thanks goes to Zlatko Drmač and local organizers of the conference for the hospitality and support. The work was supported in part also by the NSF grant CCR-9984404.

References

[1] Bjørstad P., Dryja M., and Vainikko E. (1996). *Additive Schwarz methods without subdomain overlap and with new coarse spaces.* Applied Parallel Computing in Industrial Problems and Optimization, Springer, Lecture Notes in Computer Science, Vol. 1184.

[2] Braess D. (1997). *Finite elements: Theory, fast solvers, and applications in solid mechanics.* Cambridge University Press, Cambridge.

[3] Brenner S. C, and Scoot L. R. (1994). *The mathematical theory of finite element methods.* Springer-Verlag, Vol. 15, Texts in Applied Mathematics, New York.

[4] Brezina M., and Vaněk P. (1999). *A black-box iterative solvers based on a two-level Schwarz method.* Computing. Vol. 63, pp. 233–363.

[5] Cai X.-C. (1993). *An optimal two-level overlapping domain decomposition method for elliptic problems in two and three dimensions.* SIAM J. Sci. Comp., Vol. 14, pp. 239–247.

[6] Cai X.-C. and Sarkis M. (1999) A Restricted Additive Schwarz Preconditioner for General Sparse Linear systems. SIAM J. Sci. Comput. **21**, 792–797.

[7] Cai X.-C., Dryja M., and Sarkis M. (2001). *A restricted additive Schwarz preconditioner with harmonic overlap for symmetric positive definite linear systems.* SIAM J. Sci. Comp. (Submitted).

[8] Chan T., Smith B., and Zou J. (1996). *Overlapping Schwarz methods on unstructured meshes using non-matching coarse grids.* Numer. Math. Vol. 73, pp. 149–167.

[9] Dryja M., and Widlund O. (1987). *An additive variant of the Schwarz alternating method for the case of many subregions.* Department of Computer Science, Courant Institute, TR:339, also Ultracomputer Note 131.

[10] Dryja M., Smith B., and Widlund O. (1994). *Schwarz analysis of iterative substructuring algorithms for elliptic problems in three dimensions.* SIAM J. Numer. Anal., Vol. 31(6), pp. 1662–1694.

[11] Dryja M., and Widlund O. (1994). *Domain decomposition algorithms with small overlap.* SIAM J. Sci. Comp., Vol.15, pp. 604–620.

[12] Dryja M., and Widlund O. (1995). *Schwarz methods of Neumann-Neumann type for three-dimensional elliptic finite elements problems.* Comm. Pure Appl. Math., Vol. 48, pp. 121–155.

[13] Farhat C., and Roux F. (1991) *A Method of finite element tearing and interconnecting and its parallel solution algorithm.* Int. J. Numer. Mech. Engrg., Vol. 32, pp. 1205–1227.

[14] Jenkins E., Kees C., Kelley C., and Miller C. (2001). *An aggregation-based domain decomposition preconditioner for groundwater flow.* SIAM J. Sci. Comp., Vol. 25, pp. 430–441.

[15] Mandel J. (1993). *Balancing domain decomposition.* Communications in Numerical Methods in Engineerings, Vol 9, pp. 233–241.

[16] Mandel J. (1994). *Hybrid domain decomposition with unstructured subdomains.* Contemporary Mathematics, Vol. 157, pp. 103–112.

[17] M. Lesoinne, M. Sarkis, U. Hetmaniu, and C. Farhat. (2001). *A linearized method for the frequency analysis of three-dimensional fluid/structure interaction problem in all flow regimes.* Comp. Meth. Appl. Mech. Eng., Vol. 190, pp. 3121–3146.

[18] Sarkis M. (1997). *Nonstandard coarse spaces and Schwarz methods for elliptic problems with discontinuous coefficients using nonconforming elements.* Numer. Math., Vol 77, pp. 383–406.

[19] Sarkis M. (2001). *Partition of unity coarse spaces and Schwarz methods with harmonic overlap.* In the Proceedings of the Workshop in Domain Decomposition, ETH Zurich, June, 2001. To be published by Springer-Verlag in the LNCSE series.

[20] Sarkis M. (2002). *Domain decomposition methods.* In the Proceedings of the Applied Mathematics and Scientific Computing, Dubrovnik, Croacia, June, 2001. To be published by Kluwer Academic Press.

[21] Sarkis M., and Koobus B. (2000). *A scaled and minimum overlap restricted additive Schwarz method with application on aerodynamics.* Comp. Meth. Appl. Mech. and Eng., 184, pp. 391–400.

[22] B. Smith, Bjørstad P., and Gropp W. (1995). *Domain decomposition: Parallel multilevel methods for elliptic partial differential equations.* Cambridge University Press.

[23] Tezaur R. Vaněk P., and Brezina M. (1995). *Two-level method for solids on unstructured meshes.* Center for Computational Mathematics Report CCM TR 73, University of Colorado at Denver.

NUMERICAL APPROXIMATIONS OF THE SEDIMENT TRANSPORT EQUATIONS

Luka Sopta
University of Rijeka
51000 Rijeka, Vukovarska 58, Croatia
luka.sopta@rijeka.riteh.hr

Nelida Črnjarić-Žic
University of Rijeka
51000 Rijeka, Vukovarska 58, Croatia
nelida@rijeka.riteh.hr

Senka Vuković
University of Rijeka
51000 Rijeka, Vukovarska 58, Croatia
senka.vukovic@ri.hinet.hr

Abstract The mathematical model describing the physical phenomenon of one-dimensional bed-load sediment transport in channels and rivers consists of three equations. Two of them represent conservation laws for one-dimensional shallow water equations, and third is the conservation law governing bed-load sediment transport. Here we consider one possible type of the sediment flux proposed by Hudson and Sweby [7]. We compare numerical results for test problems using different numerical schemes: Q-scheme, Hubbard's scheme, ENO Roe and ENO locally Lax-Friedrichs scheme. The obtained results illustrate good properties of ENO schemes with the source term decomposition, developed by authors. We also prove that these schemes have the exact C-property when applied to the sediment transport equations.

Keywords: conservation laws, source term decomposition, exact C-property, sediment transport

Introduction

The sediment transport model consist of the one-dimensional shallow water equations

$$\partial_t h + \partial_x (vh) = 0 \tag{1}$$

$$\partial_t (vh) + \partial_x (hv^2 + \frac{1}{2} gh^2) = -gh\partial_x H \tag{2}$$

and the equation that describes the moving of the sediment on the bottom of the river, i.e. the bed-updating equation

$$\partial_t H + \xi \partial_x q = 0. \tag{3}$$

Here $\xi = \frac{1}{1-\epsilon}$, ϵ is porosity of the riverbed which depends of the type of the sediment and $q(v, h)$ is the total volumetric sediment transport rate in the x direction i.e. the sediment transport flux (see Figure 1). We consider the case

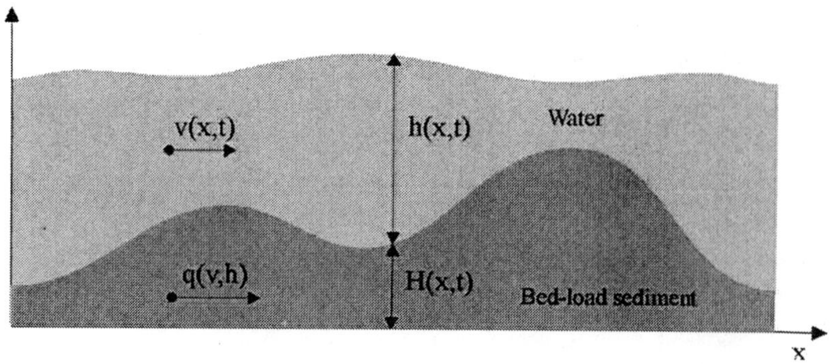

Figure 1. Sediment Transport.

when the sediment flux has the form

$$q(v) = Av^m, 1 \leq m \leq 4$$

([7]). There are other possible choices for the sediment flux (see [9]). The values of A depend of the elevation and can be obtained experimentally.

In the first section we present formulation of the sediment transport equations used in the numerical approximations. In the second section we give a brief overview of all the schemes that are used in this work and we emphasize the schemes with the decomposed source term evaluation. In the third section we prove the main result: for the observed formulation of the system, Q-scheme, Hubbard's scheme and new versions of the ENO schemes with the source term decomposition, developed by authors, have the exact C-property. Finally, in the fourth section we present numerical results of the test problem proposed in [7].

1. Numerical treatment of the sediment transport equations

For the system of equations (1), (2) and (3), rewritten in the homogeneous form it can be proved that the Jacobian matrix is, under the assumptions $h(x,t) > 0$ when $v(x,t) > 0$ and $v(x,t) \geq 0$ on all the domain, nonsingular with real and different eigenvalues ([7]), so the system is hyperbolic. Because in the homogeneous form the variables are non-conservative, for numerical treatment, the system of equations (1), (2) and (3), is rewritten in the vector form

$$\partial_t \mathbf{u} + \partial_x \mathbf{f} = \mathbf{g} \tag{4}$$

where

$$\mathbf{u} = \begin{pmatrix} h \\ hv \\ H \end{pmatrix}, \mathbf{f} = \begin{pmatrix} hv \\ hv^2 + gh(\frac{1}{2}h + H) \\ \xi q \end{pmatrix}, \mathbf{g} = \begin{pmatrix} 0 \\ gH\partial_x h \\ 0 \end{pmatrix}. \tag{5}$$

The Jacobian matrix $\mathbf{A} = \frac{\partial \mathbf{f}}{\partial \mathbf{u}}$ is, under the same assumptions as for the homogeneous form of the system, nonsingular with real and different eigenvalues. Because of that, for the numerical approximations of the system (4), (5) we use numerical schemes that are otherwise developed for the hyperbolic systems. The fact that the term $\mathbf{g}$ is not a real source term, because it includes the derivative of the variable, will not cause any difficulty for the numerical approximations presented in the paper.

2. Numerical schemes

For the numerical approximations a mesh with uniform width Δx and time step Δt is chosen. Cells $[x_{i-\frac{1}{2}}, x_{i+\frac{1}{2}}], i = 0, \ldots, N$, where $x_{i\pm\frac{1}{2}} = x_i \pm \frac{\Delta x}{2}$ and points $x_i = i\Delta x$ as the ith cell center are defined.

If the solution $\mathbf{u}$ at time t is known, from the differential form of the conservation law using the standard finite volume approach follows, that the numerical approximation $\mathbf{u}_i(t)$, which represents the average of $\mathbf{u}$ over the ith cell, evolves according to

$$\frac{d}{dt}\mathbf{u}_i(t) = -\frac{1}{\Delta x}(\mathbf{f}_{i+\frac{1}{2}} - \mathbf{f}_{i-\frac{1}{2}}) + \mathbf{g}_i. \tag{6}$$

Here $\mathbf{f}_{i+\frac{1}{2}}$ represents numerical flux at the right ith cell boundary, while $\mathbf{g}_i$ represents numerical source term over the ith cell. The left side of (6) is the time operator that belongs to the system, while the right side is the numerical approximation for

$$\mathbf{L}(\mathbf{u}, x, t) = -\frac{\partial \mathbf{f}(\mathbf{u})}{\partial x} + \mathbf{g}(\mathbf{u}, x, t).$$

Q-scheme and Hubbard's scheme. In the Q-scheme the simple forward Euler method with time step Δt is used for the time integration. The Hubbard's scheme includes also the limited higher order correction terms that ensure second order accuracy in space and time in the smooth regions of the flow, while enforcing a TVD property. For both schemes, the equation (6) can be written in the same form

$$\mathbf{u}_i^{n+1} = \mathbf{u}_i^{n} - \frac{\Delta t}{\Delta x}(\mathbf{f}_{i+\frac{1}{2}} - \mathbf{f}_{i-\frac{1}{2}}) + \mathbf{g}_i \Delta t$$

with the numerical flux defined as

$$\begin{aligned} \mathbf{f}_{i+\frac{1}{2}} &= \frac{1}{2}(\mathbf{f}_i + \mathbf{f}_{i+1}) \\ &\quad -\frac{1}{2}\sum_{p=1}^{m}(|\lambda^{(p)}_{i+\frac{1}{2}}|(1-\phi(\theta^{(p)}_{i+\frac{1}{2}})(1-|\nu^{(p)}_{i+\frac{1}{2}}|))\mathbf{l}^{(p)}_{i+\frac{1}{2}}\Delta\mathbf{u}_{i+\frac{1}{2}})\mathbf{r}^{(p)}_{i+\frac{1}{2}} \end{aligned} \tag{7}$$

where $\Delta \mathbf{u}_{i+\frac{1}{2}} = \mathbf{u}_{i+1} - \mathbf{u}_i$ and

$$\nu^{(p)}_{i+\frac{1}{2}} = \frac{\Delta t}{\Delta x}\lambda^{(p)}_{i+\frac{1}{2}}, \quad \theta^{(p)}_{i+\frac{1}{2}} = \frac{\mathbf{l}^{(p)}_{i+\frac{1}{2}}\Delta\mathbf{u}_{I+\frac{1}{2}}}{\mathbf{l}^{(p)}_{i+\frac{1}{2}}\Delta\mathbf{u}_{i+\frac{1}{2}}}, \quad I = i - sgn(\lambda^{(p)}_{i+\frac{1}{2}}). \tag{8}$$

The function ϕ that appears in (7) is taken to be identically equal 0 in the Q-scheme, while in the Hubbard's scheme ϕ is one of the possible flux limiters (see [8]). It can be noticed that the second order correction terms for Hubbard's scheme are actually hidden in the flux evaluation.

Eigenvalues, left and right eigenvectors, $\lambda^{(p)}_{i+\frac{1}{2}}$, $\mathbf{l}^{(p)}_{i+\frac{1}{2}}$ and $\mathbf{r}^{(p)}_{i+\frac{1}{2}}$, are of the linearized approximation of the Jacobian matrix on the $(i+\frac{1}{2})$th cell boundary $\mathbf{A}_{i+\frac{1}{2}} = \frac{\partial \mathbf{f}}{\partial \mathbf{u}}(\frac{\mathbf{u}_i + \mathbf{u}_{i+1}}{2})$.

It remains to define the numerical approximation for the source term. In an effort to keep the same order of accuracy of the schemes when the source term is included and to avoid the numerical errors produced by the pointwise evaluation, the decomposed approach is used, so the numerical source term over the ith cell is given by

$$\mathbf{g}_i = \mathbf{g}_{i+\frac{1}{2},+} + \mathbf{g}_{i-\frac{1}{2},-}. \tag{9}$$

Here

$$\mathbf{g}_{i+\frac{1}{2},\pm} = \frac{1}{2}\sum_{p=1}^{m}(1 \mp sgn(\lambda^{(p)}_{i+\frac{1}{2}})(1-\phi(\theta^{(p)}_{i+\frac{1}{2}})(1-|\nu^{(p)}_{i+\frac{1}{2}}|))\mathbf{l}^{(p)}_{i+\frac{1}{2}}\mathbf{g}_{i,i+1})\mathbf{r}^{(p)}_{i+\frac{1}{2}} \tag{10}$$

and $\mathbf{g}_{i,i+1}$ is the approximation of the source term dependent on the conservation law on the obtained position. The parameters that appear in (10) are used the same as in (7).

When approximating the standard hyperbolic conservation law with the dominant source term, except the described upwind treatment of the source term, there are other possible ways to obtain better accuracy, for example a semi-implicit source term evaluation (see [11]). Because in that method the numerical approximation of the Jacobian matrix $\frac{\partial \mathbf{g}}{\partial \mathbf{u}}$ is needed, it is not exactly clear how this term can be approximated numerically for system (5), so we avoid this approach in the paper.

Essentially non-oscillatory (ENO) schemes. In the ENO schemes the Runge-Kutta method of *r*th order is used for time integration. Depending to the type of the numerical approximation for the flux we can distinguish ENO Roe with entropy fix (ENORF) and ENO locally Lax-Friedrichs scheme (ENOLLF) (see [3, 4, 10]).

In both cases the components of the flux in the *p*th characteristic field are evaluated using the ENO reconstruction via primitive function for functions $\frac{1}{2}(\mathbf{f}+\alpha^{(p)}_{i+\frac{1}{2}}\mathbf{u})\cdot\mathbf{l}^{(p)}_{i+\frac{1}{2}}$ and $\frac{1}{2}(\mathbf{f}-\alpha^{(p)}_{i+\frac{1}{2}}\mathbf{u})\cdot\mathbf{l}^{(p)}_{i+\frac{1}{2}}, p = 1,\dots,m$. In the Roe approach $\alpha^{(p)}_{i+\frac{1}{2}} = 0$ while in the ENOLLF formulation $\alpha^{(p)}_{i+\frac{1}{2}} = \max_{[x_i,x_{i+1}]}|\lambda^{(p)}|$. The final result in the ENO flux reconstruction can be expressed in the form

$$\mathbf{f}_{i+\frac{1}{2}} = \sum_{p=1}^{m} f^{(p)}_{i+\frac{1}{2}}\mathbf{r}^{(p)}_{i+\frac{1}{2}}, \quad \text{where } f^{(p)}_{i+\frac{1}{2}} = f^{(p)}_{i+\frac{1}{2},+} + f^{(p)}_{i+\frac{1}{2},-}$$

$$\mathbf{f}^{(p)}_{i+\frac{1}{2},\pm} = \frac{1}{2}\left(\mathbf{f}_{I^{(p)}_{\pm}(i,0)} \pm \alpha^{(p)}_{i+\frac{1}{2}}\mathbf{u}_{I^{(p)}_{\pm}(i,0)}\right)\cdot\mathbf{l}^{(p)}_{i+\frac{1}{2}} + P^{(p)}_{i+\frac{1}{2},\pm} \tag{11}$$

$$P^{(p)}_{i+\frac{1}{2},\pm} = \sum_{l=1}^{r}\left(\prod_{j=I^{(p)}_{\pm}(i,l-1),j\neq i+1}^{I^{(p)}_{\pm}(i,l-1)+l}(x_{i+\frac{1}{2}} - x_{j-\frac{1}{2}})\right)a^{(p)}_{I^{(p)}_{\pm}(i,l),l,\pm}. \tag{12}$$

The coefficients $a^{(p)}_{j,l,\pm}$ are computed as

$$a^{(p)}_{j,l,\pm} = \frac{1}{2(l+1)}\left([\mathbf{f}_j,\dots,\mathbf{f}_{j+l}] \pm \alpha^{(p)}_{i+\frac{1}{2}}[\mathbf{u}_j,\dots,\mathbf{u}_{j+l}]\right)\cdot\mathbf{l}^{(p)}_{i+\frac{1}{2}}$$

where $[\mathbf{f}_j,\dots,\mathbf{f}_{j+l}]$ (resp. $[\mathbf{u}_j,\dots,\mathbf{u}_{j+l}]$) stands for the *l*th divided difference of $\mathbf{f}$ (resp. $\mathbf{u}$).

Here the essence of the ENO polynomial interpolation is hidden in the choice of the indices $I^{(p)}_{\pm}(i,l), l = 0,\dots,r$. They are generated through the following recursive algorithm

$$I_{\pm}^{(p)}(i,0) = \begin{cases} i, & \text{if } \lambda_{i+\frac{1}{2}}^{(p)} > 0 \\ i+1, & \text{otherwise} \end{cases} \quad \text{for ENORF formulation,}$$

$$I_{+}^{(p)}(i,0) = i, I_{-}^{(p)}(i,0) = i+1 \text{ for ENOLLF formulation,}$$

and

$$I_{\pm}^{(p)}(i,l) = \begin{cases} I_{\pm}^{(p)}(i,l-1)-1, \text{ if } |a_{I_{\pm}^{(p)}(i,l-1)-1,l,\pm}^{(p)}| < |a_{I_{\pm}^{(p)}(i,l-1),l,\pm}^{(p)}| \\ I_{\pm}^{(p)}(i,l-1), \qquad \text{otherwise} \end{cases}. \tag{13}$$

For numerical approximations of the source term in the standard versions of ENO schemes a pointwise evaluation is used and we will refer to them as to the ENO schemes with the source term added. The new versions of the ENO schemes, developed by authors, use a decomposed approach for the evaluation in the similar way as it was done in the Q-scheme. We call the modified schemes ENO schemes with the source term decomposed. Procedure of evaluation is the following: first, the numerical source term $\mathbf{g}_i$ is split as in (9). The pth characteristic field components of these decomposed parts of the source terms at the $(i+\frac{1}{2})$th cell boundary are

$$\mathbf{g}_{i+\frac{1}{2},\pm}^{(p)} = \frac{1}{2}(1 \mp sgn(\lambda_{i+\frac{1}{2}}^{(p)}))\mathbf{G}_{i,i+1} \cdot \mathbf{l}_{i+\frac{1}{2}}^{(p)} + Q_{i+\frac{1}{2},+}^{(p)} + Q_{i+\frac{1}{2},-}^{(p)}. \tag{14}$$

The definition of the term $\mathbf{G}_{j,k} = \mathbf{G}(\mathbf{u}_j, \mathbf{u}_k, x_j, x_k)$ is related to the source term function. The exact expression depends on the particular system of conservation laws and it will be defined for the case of one-dimensional sediment transport equations in the next section. Polynomial parts $Q_{i+\frac{1}{2},+}^{(p)}$ and $Q_{i+\frac{1}{2},-}^{(p)}$ depend on the scheme and they are needed for balancing polynomial corrections of the numerical flux. Now it just remains to construct the polynomial corrections $Q_{i+\frac{1}{2},\pm}^{(p)}$.

Obviously balancing cannot be maintained unless the stencils in ENO reconstruction are equal, so the construction begins with the form

$$Q_{i+\frac{1}{2},\pm}^{(p)} = \frac{1}{\Delta x}\sum_{l=1}^{r}\left(\prod_{j=I_{\pm}^{(p)}(i,l-1),j\neq i+1}^{I_{\pm}^{(p)}(i,l-1)+l}(x_{i+\frac{1}{2}} - x_{j-\frac{1}{2}})\right) b_{I_{\pm}^{(p)}(i,l),l,\pm}^{(p)} \tag{15}$$

where the $I_{\pm}^{(p)}(i,l), l = 0,\ldots,r$ are the same as for the flux. The flux gradient - source term balancing now depends only on the choice of the coefficients $b_{j,l,\pm}^{(p)}$.

The source term polynomial coefficients are defined as

$$b^{(p)}_{j,l,\pm} = \frac{1}{2(l+1)}\left([\mathbf{g}_j,\ldots,\mathbf{g}_{j+l}] \pm sgn(\lambda^{(p)}_{i+\frac{1}{2}})\beta_{i+\frac{1}{2}}[\mathbf{z}_j,\ldots,\mathbf{z}_{j+l}]\right)\cdot \mathbf{l}^{(p)}_{i+\frac{1}{2}}.$$

The terms $[\mathbf{g}_j,\ldots,\mathbf{g}_{j+k}]$ and $[\mathbf{z}_j,\ldots,\mathbf{z}_{j+k}], k = 2,\ldots,r$ are evaluated exactly as divided differences, just for $k = 1$ the expressions are defined with $[\mathbf{g}_j,\mathbf{g}_{j+1}] = \mathbf{G}_{j,j+1}$ and $[\mathbf{z}_j,\mathbf{z}_{j+1}] = \mathbf{Z}_{j,j+1}$. The exact definitions of $\mathbf{Z}_{j,k}$ and $\beta_{i+\frac{1}{2}}$ are related to the source term function and their formulation must be solved differently for every particular system of conservation laws.

In the schemes we also propose a modified way of chousing stencils in the expressions (13): instead of comparing coefficients $|a^{(p)}_{I^{(p)}_{\pm}(i,l-1)-1}|$ and $|a^{(p)}_{I^{(p)}_{\pm}(i,l-1)}|$ values $|a^{(p)}_{I^{(p)}_{\pm}(i,l-1)-1} - b^{(p)}_{I^{(p)}_{\pm}(i,l-1)-1}|$ and $|a^{(p)}_{I^{(p)}_{\pm}(i,l-1)} - b^{(p)}_{I^{(p)}_{\pm}(i,l-1)}|$ should be compared.

Using the presented schemes a high order of accuracy is obtained even in the approximation of non-homogeneous hyperbolic systems.

3. Exact C-property

Following the original idea of C-property for the shallow water equations given in [1], the numerical scheme for the sediment transport equations must be exactly compatible with the quiescent flow:

$$h + H = const.\text{ and } v = 0. \tag{16}$$

That means that the numerical solution must not change in time, i.e. the equation

$$-\frac{1}{\Delta x}(\mathbf{f}_{i+\frac{1}{2}} - \mathbf{f}_{i-\frac{1}{2}}) + \mathbf{g}_{i+\frac{1}{2},+} + \mathbf{g}_{i-\frac{1}{2},-} = 0 \tag{17}$$

has to be verified.

Eigenvalues and left and right eigenvectors in the quiescent flow case are $\lambda^{(p)}_{i+\frac{1}{2}} = (-1)^{(p)}c_{i,i+1}, p = 1,2,\ \lambda^{(3)}_{i+\frac{1}{2}} = 0,$

$$\mathbf{l}^{(p)}_{i+\frac{1}{2}} = \begin{pmatrix} c_{i,i+1} \\ (-1)^{(p)} \\ g\bar{h}_{i,i+1}/c_{i,i+1} \end{pmatrix}, p = 1,2 \text{ and } \mathbf{l}^{(3)}_{i+\frac{1}{2}} = \begin{pmatrix} 0 \\ 0 \\ -g\bar{h}_{i,i+1}/c_{i,i+1} \end{pmatrix} \tag{18}$$

where $\bar{h}_{i,i+1} = \frac{h_i+h_{i+1}}{2}$, $c_{i,i+1} = \sqrt{g(\bar{h}_{i,i+1} + \bar{H}_{i,i+1})}$ and $\bar{H}_{i,i+1} = \frac{H_i+H_{i+1}}{2}$.

Exact C-property for the Q-scheme. In the algorithm for Q-scheme the source term approximation is taken to be:

$$\mathbf{g}_{i,i+1} = \begin{pmatrix} 0 \\ g\bar{H}_{i,i+1}\frac{h_{i+1}-h_i}{\Delta x} \\ 0 \end{pmatrix}.$$

For the Q-scheme the equation (17) in the pth characteristic field for $p = 1, 2$ becomes

$$\frac{1}{2\Delta x}(-\mathbf{f}_i - \mathbf{f}_{i+1} + c_{i,i+1}(\mathbf{u}_{i+1} - \mathbf{u}_i) + (1 - (-1^{(p)}))\mathbf{g}_{i,i+1}\Delta x) \cdot \mathbf{l}^{(p)}_{i+\frac{1}{2}}$$
$$+\frac{1}{2\Delta x}(\mathbf{f}_{i-1} + \mathbf{f}_i - c_{i-1,i}(\mathbf{u}_i - \mathbf{u}_{i-1}) + (1 + (-1^{(p)}))\mathbf{g}_{i-1,i}\Delta x) \cdot \mathbf{l}^{(p)}_{i-\frac{1}{2}} = 0.$$

After taking into account the expressions for the eigenvalues and eigenvectors in the quiescent flow case (18) and some computations the above equation can be proved. In the third characteristic field the equation (17) become

$$\frac{1}{2\Delta x}(-\mathbf{f}_i - \mathbf{f}_{i+1} + \mathbf{g}_{i,i+1}\Delta x) \cdot \mathbf{l}^{(p)}_{i+\frac{1}{2}} + \frac{1}{2\Delta x}(\mathbf{f}_{i-1} + \mathbf{f}_i + \mathbf{g}_{i-1,i}\Delta x) \cdot \mathbf{l}^{(p)}_{i-\frac{1}{2}} = 0.$$

Note that both summands in the above expression are equal 0, because fluxes and source terms are orthogonal to the left eigenvector in the observed field. With this, the proof of the exact C-property for the Q-scheme on the sediment transport equations is finished.

Exact C-property for the Hubbard's scheme. The expressions for the flux and the source term in Hubbard's scheme, can be divided in two parts. The first part is the same as in the Q-scheme, while the second is the flux limited correction of the second order. Because the balancing of the first part is inherited from the Q-scheme, we must only prove the equation

$$\left(|\lambda^{(p)}_{i+\frac{1}{2}}|(\mathbf{u}_{i+1} - \mathbf{u}_i) - sgn(\lambda^{(p)}_{i+\frac{1}{2}})\mathbf{g}_{i,i+1}\Delta x\right) \cdot \mathbf{l}^{(p)}_{i+\frac{1}{2}} = 0, p = 1, 2, 3.$$

For $p = 1, 2$ the above equation is a consequence of the expressions for the eigenvalues and eigenvectors in the quiescent flow case, the definition of the term $\mathbf{g}_{i,i+1}$ and the condition (16). For $p = 3$ above relation is obvious, because $\lambda^{(p)}_{i+\frac{1}{2}} = 0$.

Exact C-property for the ENO schemes. In order to apply the ENO schemes with the source term decomposed, we must define terms $\mathbf{G}_{j,k}$, $\mathbf{Z}_{j,k}$ and $\beta_{i+\frac{1}{2}}$. The definitions that will finally lead to the exact C-property are

$$\mathbf{G}_{j,k} = \begin{pmatrix} 0 \\ g\bar{H}_{j,k}\frac{h_k - h_j}{x_k - x_j} \\ 0 \end{pmatrix}, \mathbf{Z}_{j,k} = \begin{pmatrix} 0 \\ \frac{h_k - h_j}{x_k - x_j} \\ 0 \end{pmatrix} \text{ and } \beta_{i+\frac{1}{2}} = g\bar{H}_{i,i+1}$$

where $\bar{H}_{j,k} = \frac{H_j + H_k}{2}$. Let us notice that in the expression (11) for the flux and in the expression (14) for the decomposed source term we can distinguish two parts, similarly as it was done for the Hubbard's scheme. The first is the result of the simple upwinding and the second is the ENO polynomial correction of r-th order. The balancing for these two parts will be proved separately. Similarly to the proof of balancing for the Q-schemes, simple computation leads to the relation

$$\begin{aligned} &-\tfrac{1}{2\Delta x}(\mathbf{f}_{I_+^{(p)}(i,0)} + \mathbf{f}_{I_-^{(p)}(i,0)} - \mathbf{f}_{I_+^{(p)}(i-1,0)} - \mathbf{f}_{I_-^{(p)}(i-1,0)})+ \\ &+\tfrac{1}{2}(1 - sgn(\lambda^{(p)}_{i+\frac{1}{2}}))\mathbf{G}_{i,i+1} + \tfrac{1}{2}(1 + sgn(\lambda^{(p)}_{i+\frac{1}{2}}))\mathbf{G}_{i-1,i} = 0 \end{aligned} \tag{19}$$

for $p = 1, 2, 3$ and all $i = 1, \ldots, N-1$. Proof of the relation (19) is very similar to the proof of the C-property for the Q-scheme. The balancing of the polynomial corrections depends on whether equations

$$-\frac{1}{\Delta x} P^{(p)}_{i+\frac{1}{2},\pm} + Q^{(p)}_{i+\frac{1}{2},\pm} = 0$$

for $p = 1, 2, 3$, $i = 1, \ldots, N-1$ holds. By examining expressions (12) and (15) it follows that this reduces to equations

$$a^{(p)}_{j,l,\pm} - b^{(p)}_{j,l,\pm} = 0 \tag{20}$$

for $l = 1, \ldots, r$ and $p = 1, 2, 3$, $i = 0, \ldots, N-1$. For the case $p = 1, 2$ and $l = 1$ is

$$\begin{aligned} &a^{(p)}_{j,1,\pm} - b^{(p)}_{j,1,\pm} = \tfrac{1}{4}\left([\mathbf{f}_j, \mathbf{f}_{j+1}] - [\mathbf{g}_j, \mathbf{g}_{j+1}]\right) \cdot \mathbf{l}^{(p)}_{i+\frac{1}{2}} \pm \\ &\pm\tfrac{1}{4}\left(\alpha^{(p)}_{i+\frac{1}{2}}[\mathbf{u}_j, \mathbf{u}_{j+1}] \mp sgn(\lambda^{(p)}_{i+\frac{1}{2}})\beta_{i+\frac{1}{2}}[h_j, h_{j+1}]\right) \cdot \mathbf{l}^{(p)}_{i+\frac{1}{2}} = 0 \end{aligned} \tag{21}$$

It is very simple to prove that the first summand is equal 0, while the proof that the second summand is equal 0 follows from the proof of the C-property for the Hubbard's scheme. For $p = 3$, the first summand is 0, because of the orthogonality of vectors and in the second one $\lambda^{(p)}_{i+\frac{1}{2}} = 0$. Because the coefficients $a^{(p)}_{j,l,\pm}$ and $b^{(p)}_{j,l,\pm}$ are defined recursively, the equations (20) follow immediately from (21).

4. Numerical results

The problem discussed in this section was proposed in [7]. In order to obtain realistic physical initial conditions with a moving riverbed, we suppose at first 'dummy' initial conditions with a fixed riverbed defined with

$$H(x,0) = \begin{cases} \sin^2\left(\frac{\pi(x-300)}{200}\right) & \text{, if } 300 \le x \le 500 \\ 0 & \text{, otherwise} \end{cases},$$

$$h(x,0) = 10 - H(x,0) \text{ and } v(x,0) = \frac{10}{h(x,0)}$$

for $x \in [0, 1000]$. Used expressions for the sediment transport flux and porosity are $q(v) = v^3$ and $\epsilon = 0.2$.

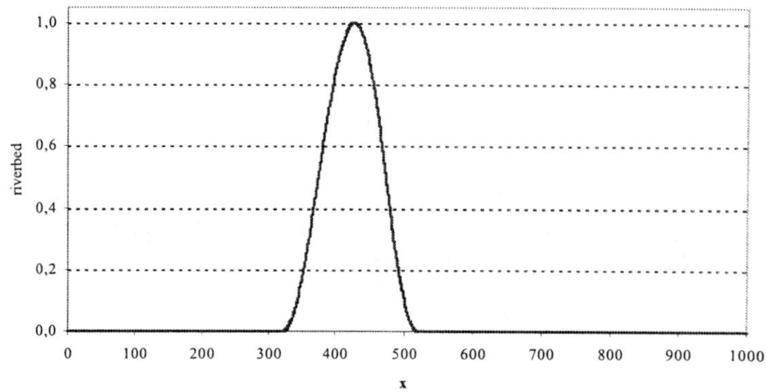

Figure 2. The initial riverbed.

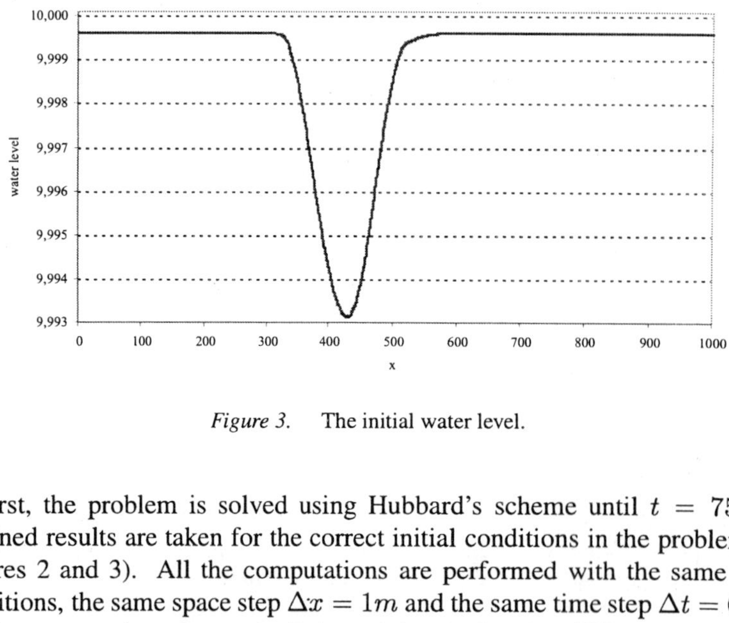

Figure 3. The initial water level.

First, the problem is solved using Hubbard's scheme until $t = 75s$ and obtained results are taken for the correct initial conditions in the problem (see Figures 2 and 3). All the computations are performed with the same initial conditions, the same space step $\Delta x = 1m$ and the same time step $\Delta t = 0,02s$. All the presented results are until the maximum time $t = 700s$.

The computations are performed using Q-scheme, Hubbard's scheme, the 2nd order ENORF and ENOLLF scheme with source term decomposed, 2nd order ENORF and ENOLLF scheme with source term added. In the Figures 4 and 5 a 'zoom in' on the details of the riverbed and the water level at different

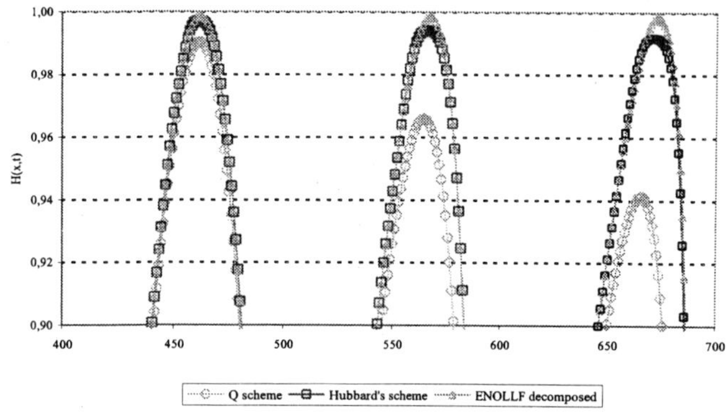

Figure 4. Comparison of the results obtained by different schemes, riverbed at $t = 100s, 400s, 700s$.

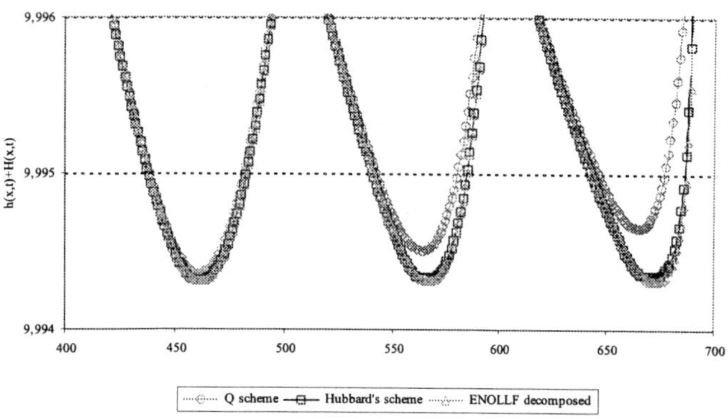

Figure 5. Comparison of the results obtained by different schemes, water level at $t = 100s, 400s, 700s$.

times are given. It can be noticed that numerical results computed with Q-scheme and Hubbard's scheme have tendency of decreasing, in the opposite with those obtained by ENO decomposed scheme.

The improvement obtained by the source term decomposition can be clearly seen in the Figures 6 and 7. Pointwise evaluation of the source term produce spurious numerical oscillations, while the source term decomposition strategy maintains the results with high order of accuracy. Presented results are obtained with the Roe version of the ENO scheme. The similar conclusion can be done for the ENOLLF scheme. The results have again quite a significant numerical errors when the source term is pointwise evaluated. In the Figure 8 we give

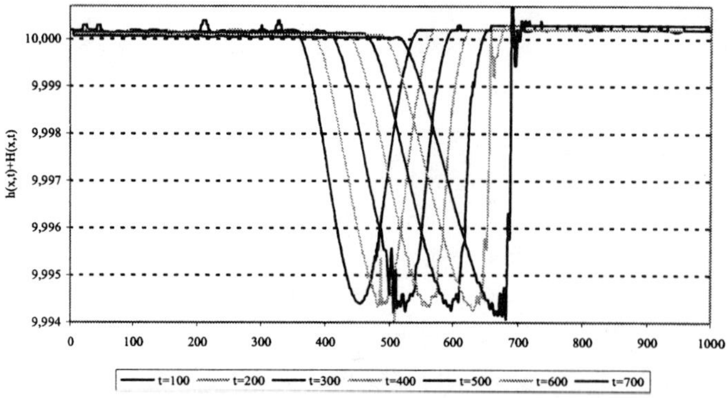

Figure 6. Results obtained with the standard ENORF scheme, water level at different times.

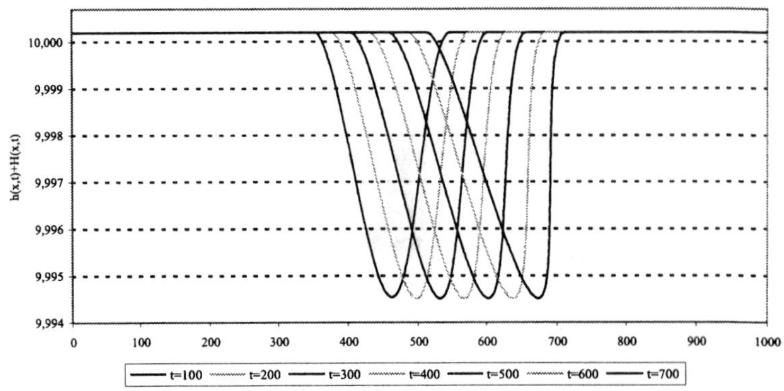

Figure 7. Results obtained with the ENORF decomposed scheme, water level at different times.

"zoom in" on the details of the water level in order to make clear the improvement obtained by the decomposed source term evaluation.

5. Conclusion remarks

In this paper we presented the system of equations that can be used to mathematically model the sediment transport in channels and rivers. The main result is the proof that Q-scheme, Hubbard's scheme and ENO schemes with the source term decomposition have the exact C-property when applied to these equations. So even in the case when the source term is stiff numerical errors could be avoided using a decomposed approach for the numerical evaluation

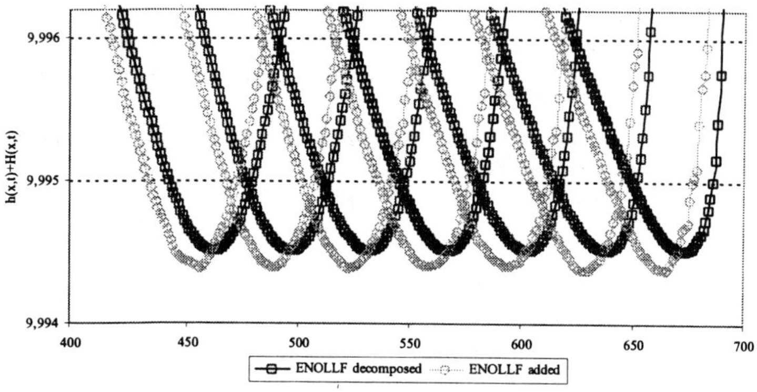

Figure 8. Comparison of the ENOLLF added scheme and ENOLLF decomposed scheme, details of the water level at $t = 100s, 200s, 300s, 400s, 500s, 600s, 700s$.

of the source term. Similar results can also be obtained for the weighted ENO schemes ([10]).

References

[1] A. Bermùdez, A. Dervieux, J. A. Désidéri and M. E. Vázquez, *Upwind schemes for the two-dimensional shallow water equations with variable depth using unstructured meshes*, Comput. Methods Appl. Mech. Eng. 155, 49 (1998).

[2] A.Bermùdez and M. E. Vázquez, *Upwind methods for hyperbolic conservation laws with source terms*, Comput.and Fluids 23(8), 1049 (1994).

[3] A.Harten and S.Osher, *Uniformly high-order accurate non-oscillatory schemes I*, SIAM Journal on Numerical Analysis 24, 279 (1987).

[4] A.Harten, B.Engquist, S.Osher and S.R.Chakravarthy, *Uniformly high-order accurate non-oscillatory schemes III*, Journal of Computational Physics 71, 231 (1987).

[5] M. E.Hubbard and P. Garcia-Navarro, *Flux difference splitting and the balancing of source terms and flux gradients*, Numerical Analysis Report, University of Reading, Department of Mathematics, (1999).

[6] J. Hudson, *Numerical techniques for the shallow water equations*, Numerical Analysis Report, University of Reading, Department of Mathematics, (1999).

[7] J. Hudson and P. K. Sweby, *Numerical Formulations for Approximating the Equations Governing Bed-Load Sediment Transport in Channels nad Rivers*, Numerical Analysis Report, University of Reading, Department of Mathematics, (2000).

[8] R. J. LeVeque, *Balancing source terms and flux gradients in high-resolution Godunov methods: the quasi-steady wave propagation algorithm*, Numerical Analysis Report, Department of Applied Mathematics and Department of Mathematics, University of Washington, (1998).

[9] L.C. van Rijn, *Sediment Transport, Part I: Bed-Load Transport*, Proc. ASCE Journal of Hydraulics Division, 110, HY 10, 1613 (1984)

[10] C.-W. Shu, *Essentially non-oscillatory and weighted essentially non-oscillatory shock-capturing schemes for hyperbolic conservation laws*, NASA/CR-97-206253, ICASE Report No. 97-65, Institute for Comp. App. in Science and Eng., NASA Langley Research Center, (1997).

[11] J. Burguete and P. Garcia-Navarro, *Efficient construction of high-resolution TVD conservative schemes for equations with the source terms: application to shallow water flows*, Int. Journal for Numerical Methods in Fluids 37, 209 (2001)

A MODEL OF IRREGULAR CURVED RODS*

Josip Tambača
Department of Mathematics, University of Zagreb
Bijenička cesta 30, 10000 Zagreb, Croatia
tambaca@math.hr

Abstract In this paper we derive a model of curved elastic rods with a piecewise smooth middle curve from the model of curved rods with smooth middle curves. The obtained model is the same as the model of junction of two straight rods derived directly from the three–dimensional linearized elasticity.

Keywords: linearized elasticity, curved rod model, irregular curved rods, junction model.

1. Introduction

The equilibrium problem for curved elastic rods is modelled by the one–dimensional differential equation for functions of arc–length parameter of the middle curve of the rod, see Jurak and Tambača (2001), Sanchez–Hubert and Sanchez–Palencia (1999). The model is derived in the form of a variational equation for the components of the displacement of the middle curve in the local basis (tangent, normal and binormal at each point of the middle curve) and an auxiliary function describing the torsion of the cross–section of the rod. The coefficients in the variational are expressed in terms of the Lamé elasticity coefficients, inertia moments of the cross–sections of the rod and the derivatives of the curvature and the torsion of the middle curve, so smoothness of the middle curve is necessary for the variational formulation.

In this paper we derive the model of elastic curved rods with irregular middle curves from the model of curved rods with smooth middle curves. By a irregular curve we mean a piecewise smooth curve i.e. continuous union of finitely many smooth curves. The method we use is the following. The irregular curve is approximated by a family of smooth curves. Then the family of associated solutions of the curved rod models is considered. The Frenet basis of the irregular curve is discontinuous while the equilibrium displacement should be continu-

*This work is supported by grant 037004 by Croatian Ministry of Science and Technology.

ous. Therefore the components of the displacement in the local basis should be discontinuous. Therefore the curved rod model for a family of smooth curves is expressed in terms of the displacement through the differential formulation of the model. Then it is shown that the family of solutions of the model for smooth curves converges to the unique solution of the problem of the same form, called the model of irregular curved rods. The explicit formula for the solution of both, the model for smooth and irregular rods, is used to give the error estimate. Then it is shown that the obtained model is the same as the model of junction of two straight rods connected at right angle derived by Le Dret (1989) directly from the three–dimensional linearized elasticity.

The 3-dimensional Euclidean space is identified with $\mathbb{R}^3$; $\cdot$ and $|\cdot|$ denote the Euclidean inner product and associated norm. The canonical basis in $\mathbb{R}^3$ is denoted by $(\boldsymbol{e}_1, \boldsymbol{e}_2, \boldsymbol{e}_3)$. Vectors, matrices, vector-valued and matrix-valued functions are denoted by bold-faced letters. The derivative is denoted by $'$. $\|\cdot\|_2$ and $\|\cdot\|_\infty$ denote usual norm of Lebesgue spaces $L^2(0,\ell)$ and $L^\infty(0,\ell)$. $\|\cdot\|$ without a subscript denotes the operator norm.

2. Curved rod model

Let $(\mathcal{C}_\varepsilon, \varepsilon > 0)$ be a family of simple curves. Let for each $\varepsilon > 0$ the curve $\mathcal{C}_\varepsilon$ be parameterized by the natural parameterization $\boldsymbol{\Phi}_\varepsilon \in C^3([0,\ell];\mathbb{R}^3)$. Let $\mathcal{C}$ be parameterized by a piecewise smooth function $\boldsymbol{\Phi}$. More precisely, let there be $n \in \mathbb{N}$ and $0 = s_0 < s_1 < \cdots < s_n = \ell$ such that

$$\boldsymbol{\Phi} \in C^1([s_{k-1}, s_k];\mathbb{R}^3), \ k = 1, \dots n, \qquad \boldsymbol{\Phi} \in C([0,\ell];\mathbb{R}^3).$$

We assume

$$\boldsymbol{\Phi}_\varepsilon \to \boldsymbol{\Phi} \text{ strongly in } H^1(0,\ell)^3 \tag{1}$$

when ε tends to 0. The tangent vector on the curve $\mathcal{C}_\varepsilon$, $\varepsilon > 0$ at $\boldsymbol{\Phi}_\varepsilon(s)$ and the tangent vector on the curve $\mathcal{C}$ at $\boldsymbol{\Phi}(s)$ are given by

$$\boldsymbol{t}_\varepsilon(s) = \boldsymbol{\Phi}_\varepsilon'(s), \ s \in [0,\ell], \qquad \boldsymbol{t}(s) = \boldsymbol{\Phi}'(s), \ s \in (s_{k-1}, s_k), \ k = 1, \dots n.$$

For $s_0, \dots, s_n$ the tangents are

$$\boldsymbol{t}(0+), \qquad \boldsymbol{t}(s_k-), \ \boldsymbol{t}(s_k+), \ k = 1, \dots, n-1, \qquad \boldsymbol{t}(\ell-),$$

where $f(s\pm) = \lim_{\substack{h\to 0\\ h>0}} f(s \pm h)$. A finite number of continuous curves of finite length cannot cover the unit sphere, see e. g. Tambača (2000). Therefore there is a unit vector $\boldsymbol{k}$ such that

$$\begin{aligned} &\pm\boldsymbol{k} \neq \boldsymbol{t}_\varepsilon(s), \ s \in [0,\ell], \qquad \varepsilon > 0, \\ &\pm\boldsymbol{k} \neq \boldsymbol{t}(s), \ s \in (s_{k-1}, s_k), \ \pm\boldsymbol{k} \neq \boldsymbol{t}(s_k-), \ \pm\boldsymbol{k} \neq \boldsymbol{t}(s_{k-1}+), \ k = 1, \dots, n. \end{aligned}$$

Then at each point $\mathbf{\Phi}_\varepsilon(s)$ of $\mathcal{C}_\varepsilon, \varepsilon > 0$ the local basis can be defined by

$$\boldsymbol{t}_\varepsilon(s),\ \boldsymbol{n}_\varepsilon(s) = \frac{\boldsymbol{k} \times \boldsymbol{t}_\varepsilon(s)}{|\boldsymbol{k} \times \boldsymbol{t}_\varepsilon(s)|},\ \boldsymbol{b}_\varepsilon(s) = \boldsymbol{t}_\varepsilon(s) \times \boldsymbol{n}_\varepsilon(s), \quad s \in [0, \ell].$$

Let at $\mathbf{\Phi}(s)$ the local basis be defined by

$$\boldsymbol{t}(s), \boldsymbol{n}(s) = \frac{\boldsymbol{k} \times \boldsymbol{t}(s)}{|\boldsymbol{k} \times \boldsymbol{t}(s)|}, \boldsymbol{b}(s) = \boldsymbol{t}(s) \times \boldsymbol{n}(s),\ s \in (s_{k-1}, s_k),\ k = 1, \ldots, n.$$

Then $\mathbf{Q}_\varepsilon = \begin{bmatrix} \boldsymbol{t}_\varepsilon & \boldsymbol{n}_\varepsilon & \boldsymbol{b}_\varepsilon \end{bmatrix} \in C^2([0,\ell]; \mathrm{SO}(3))$, $\mathbf{Q} = \begin{bmatrix} \boldsymbol{t} & \boldsymbol{n} & \boldsymbol{b} \end{bmatrix} \in L^2(0, \ell; \mathrm{SO}(3))$ and

$$\mathbf{Q}_\varepsilon \to \mathbf{Q} \text{ strongly in } L^2(0,\ell)^9. \tag{2}$$

We define $\mathbf{R}^\varepsilon(s) = \mathbf{Q}_\varepsilon'(s)^T \mathbf{Q}_\varepsilon(s), s \in [0, \ell]$.

Let $S \subset \mathbb{R}^2$ be a bounded simply connected domain with the Lipschitz boundary that satisfies (points in $\mathbb{R}^2$ are denoted by (z^2, z^3))

$$\int_S z^2 dz^2 dz^3 = \int_S z^3 dz^2 dz^3 = \int_S z^2 z^3 dz^2 dz^3 = 0.$$

$\mathcal{C}_\varepsilon$ is assumed to be the middle curve of the elastic curved rod, with the Lamé coefficients λ and μ and with cross–sections S. Let us define the constant matrix

$$\mathbf{H} = \begin{bmatrix} \mu K & 0 & 0 \\ 0 & EI_n & 0 \\ 0 & 0 & EI_b \end{bmatrix},$$

where E is the Young modulus, I_n, I_b are the inertia moments with respect to the center of mass, while μK is the torsional rigidity of the cross–section S. Therefore $\mathbf{H}$ describes elastic properties of the material the rod is made of and geometrical properties of its cross–section.

The rod is assumed to be clamped at $\mathbf{\Phi}_\varepsilon(0)$, subjected to a body force with the line density $\tilde{\boldsymbol{f}}_\varepsilon \in L^2(0,\ell)^3$ and the contact force $\tilde{\boldsymbol{g}}_\varepsilon$ at the end $\mathbf{\Phi}_\varepsilon(\ell)$. We denote $\boldsymbol{f}_\varepsilon = \mathbf{Q}_\varepsilon^T \tilde{\boldsymbol{f}}_\varepsilon, \boldsymbol{g}_\varepsilon = \mathbf{Q}_\varepsilon^T \tilde{\boldsymbol{g}}_\varepsilon$. Let us define the function space

$$\begin{aligned} \mathcal{W}^\varepsilon &= \Big\{ (v_t, v_n, v_b, \psi) \in H^1(0,\ell) \times H^2(0,\ell) \times H^2(0,\ell) \times H^1(0,\ell) : \\ &\qquad v_t' - R_{12}^\varepsilon v_n - R_{13}^\varepsilon v_b = 0, \boldsymbol{v}(0) = 0, \psi(0) = v_n'(0) = v_b'(0) = 0 \Big\} \end{aligned}$$

and linear operators on $\mathcal{W}^\varepsilon$ with values in $L^2(0,\ell)$:

$$\begin{aligned} E_2^\varepsilon(\boldsymbol{v}, \psi) &= (v_n' + R_{12}^\varepsilon v_t - R_{23}^\varepsilon v_b)' + R_{13}^\varepsilon \psi - R_{23}^\varepsilon (v_b' + R_{13}^\varepsilon v_t + R_{23}^\varepsilon v_n), \\ E_3^\varepsilon(\boldsymbol{v}, \psi) &= -(v_b' + R_{13}^\varepsilon v_t + R_{23}^\varepsilon v_n)' + R_{12}^\varepsilon \psi - R_{23}^\varepsilon (v_n' + R_{12}^\varepsilon v_t - R_{23}^\varepsilon v_b), \\ E_4^\varepsilon(\boldsymbol{v}, \psi) &= \psi' + R_{12}^\varepsilon (v_b' + R_{13}^\varepsilon v_t + R_{23}^\varepsilon v_n) - R_{13}^\varepsilon (v_n' + R_{12}^\varepsilon v_t - R_{23}^\varepsilon v_b); \end{aligned}$$

here we use notations $\boldsymbol{v} = (v_t, v_n, v_b)$ and $\tilde{\boldsymbol{v}} = v_t \boldsymbol{t}_\varepsilon + v_n \boldsymbol{n}_\varepsilon + v_b \boldsymbol{b}_\varepsilon = \mathbf{Q}_\varepsilon \boldsymbol{v}$, so v_t, v_n, v_b stand for components of $\tilde{\boldsymbol{v}}$ in the local basis of the middle curve. The condition $v_t' - R_{12}^\varepsilon v_n - R_{13}^\varepsilon v_b = 0$ is the condition of approximate inextensibility of the middle curve of the rod, so $\mathcal{W}^\varepsilon$ is a function space of inextensible displacements of the curved rod.

Then the equilibrium state of the curved rod is unique solution of the following one–dimensional linear curved rod model (see Jurak and Tambača (2001), Sanchez–Hubert and Sanchez–Palencia (1999)): find $(\boldsymbol{u}_\varepsilon, \phi_\varepsilon) \in \mathcal{W}^\varepsilon$ such that

$$\int_0^\ell \mathbf{H} \begin{bmatrix} E_4^\varepsilon(\boldsymbol{u}_\varepsilon, \phi_\varepsilon) \\ E_3^\varepsilon(\boldsymbol{u}_\varepsilon, \phi_\varepsilon) \\ E_2^\varepsilon(\boldsymbol{u}_\varepsilon, \phi_\varepsilon) \end{bmatrix} \cdot \begin{bmatrix} E_4^\varepsilon(\boldsymbol{v}, \psi) \\ E_3^\varepsilon(\boldsymbol{v}, \psi) \\ E_2^\varepsilon(\boldsymbol{v}, \psi) \end{bmatrix} = \boldsymbol{v}(\ell) \cdot \boldsymbol{g}_\varepsilon + \int_0^\ell \boldsymbol{f}_\varepsilon \cdot \boldsymbol{v}, \qquad (3)$$

for all $(\boldsymbol{v}, \psi) \in \mathcal{W}^\varepsilon$. The function $\tilde{\boldsymbol{u}}_\varepsilon = \mathbf{Q}_\varepsilon \boldsymbol{u}_\varepsilon$ is the displacement of the middle curve, while ϕ_ε is the angle of the torsion of the cross–section of the rod.

It is easy to see that for $\mathcal{C}$ such equation can not be written. Namely, the matrix function $\mathbf{R}$, the analogue of $\mathbf{R}_\varepsilon$, can not be defined, even if the middle curve is piecewise C^3. Still, we will consider the behavior of the solution of (3) when ε tends to zero. Therefore in what follows we assume

$$\tilde{\boldsymbol{f}}_\varepsilon \to \tilde{\boldsymbol{f}} \text{ strongly in } L^2(0,\ell)^3, \quad \text{and} \quad \tilde{\boldsymbol{g}}_\varepsilon \to \tilde{\boldsymbol{g}} \text{ in } \mathbb{R}^3. \qquad (4)$$

3. Curved rod model of irregular rods

In this section we consider the behavior of the solution of (3) when ε tends to zero, i.e. when $\mathcal{C}_\varepsilon$ tends to $\mathcal{C}$. The limit function is identified as the unique solution of the equation we call the curved rod model of irregular rods.

For irregular rods the displacement of the middle curve is expected to be continuous, in difference with the components in the local basis. Therefore for $\varepsilon > 0$ we use the differential formulation (see Jurak and Tambača (2001), Sanchez–Hubert and Sanchez–Palencia (1999)) which is formally equivalent to the variational formulation (3): find a quadruple $(\tilde{\boldsymbol{p}}_\varepsilon, \tilde{\boldsymbol{q}}_\varepsilon, \tilde{\boldsymbol{\omega}}_\varepsilon, \tilde{\boldsymbol{u}}_\varepsilon)$ that satisfies

$$\begin{aligned} &\tilde{\boldsymbol{p}}_\varepsilon' + \tilde{\boldsymbol{f}}_\varepsilon = 0 \text{ on } (0,\ell), \quad \tilde{\boldsymbol{p}}_\varepsilon(\ell) = \tilde{\boldsymbol{g}}_\varepsilon, \\ &\tilde{\boldsymbol{q}}_\varepsilon' + \boldsymbol{t}_\varepsilon \times \tilde{\boldsymbol{p}}_\varepsilon = 0 \text{ on } (0,\ell), \quad \tilde{\boldsymbol{q}}_\varepsilon(\ell) = 0, \\ &\tilde{\boldsymbol{\omega}}_\varepsilon' - \mathbf{Q}_\varepsilon \mathbf{H}^{-1} \mathbf{Q}_\varepsilon^T \tilde{\boldsymbol{q}}_\varepsilon = 0 \text{ on } (0,\ell), \quad \tilde{\boldsymbol{\omega}}_\varepsilon(0) = 0, \\ &\tilde{\boldsymbol{u}}_\varepsilon' + \boldsymbol{t}_\varepsilon \times \tilde{\boldsymbol{\omega}}_\varepsilon = 0 \text{ on } (0,\ell), \quad \tilde{\boldsymbol{u}}_\varepsilon(0) = 0. \end{aligned} \qquad (5)$$

$\tilde{\boldsymbol{p}}_\varepsilon$ is the contact force, $\tilde{\boldsymbol{q}}_\varepsilon$ contact couple, $\tilde{\boldsymbol{\omega}}_\varepsilon$ describes the rotation of the cross–sections, while $\tilde{\boldsymbol{u}}_\varepsilon$ is the displacement of the middle curve of the rod. The first two equations are equilibrium equations, while the last two are constitutive equations. In this model the rod is inextensible and unshearable.

This system of ordinary differential equations can be integrated step by step. Immediate consequence of $\tilde{\boldsymbol{f}}_\varepsilon \in L^2(0,\ell)^3$, $\mathbf{Q}_\varepsilon \in C^2([0,\ell];\mathrm{SO}(3))$ is the existence and uniqueness of the solution

$$\tilde{\boldsymbol{p}}_\varepsilon \in H^1(0,\ell)^3, \quad \tilde{\boldsymbol{q}}_\varepsilon \in H^2(0,\ell)^3, \quad \tilde{\boldsymbol{\omega}}_\varepsilon \in H^3(0,\ell)^3, \quad \tilde{\boldsymbol{u}}_\varepsilon \in H^3(0,\ell)^3.$$

Theorem 1. *When ε tends to zero the solution $(\tilde{\boldsymbol{p}}_\varepsilon, \tilde{\boldsymbol{q}}_\varepsilon, \tilde{\boldsymbol{\omega}}_\varepsilon, \tilde{\boldsymbol{u}}_\varepsilon)$ of (5) converges strongly in $H^1(0,\ell)^{12}$ to the unique solution $(\tilde{\boldsymbol{p}}, \tilde{\boldsymbol{q}}, \tilde{\boldsymbol{\omega}}, \tilde{\boldsymbol{u}}) \in H^1(0,\ell)^{12}$ of the system*

$$\begin{aligned} &\tilde{\boldsymbol{p}}' + \tilde{\boldsymbol{f}} = 0 \text{ on } (0,\ell), \quad \tilde{\boldsymbol{p}}(\ell) = \tilde{\boldsymbol{g}}, \\ &\tilde{\boldsymbol{q}}' + \boldsymbol{t} \times \tilde{\boldsymbol{p}} = 0, \quad \tilde{\boldsymbol{q}}(\ell) = 0, \\ &\tilde{\boldsymbol{\omega}}' - \mathbf{Q}\mathbf{H}^{-1}\mathbf{Q}^T\tilde{\boldsymbol{q}} = 0, \quad \tilde{\boldsymbol{\omega}}(0) = 0, \\ &\tilde{\boldsymbol{u}}' + \boldsymbol{t} \times \tilde{\boldsymbol{\omega}} = 0, \quad \tilde{\boldsymbol{u}}(0) = 0, \end{aligned} \tag{6}$$

where the equations are satisfied in $L^2(0,\ell)^3$.

Proof. The assumption (4) and $(5)_1$ imply that there is $\tilde{\boldsymbol{p}} \in H^1(0,\ell)^3$ such that $\tilde{\boldsymbol{p}}_\varepsilon \to \tilde{\boldsymbol{p}}$ strongly in $H^1(0,\ell)^3$. Therefore (1) implies $\boldsymbol{t}_\varepsilon \times \tilde{\boldsymbol{p}}_\varepsilon \to \boldsymbol{t} \times \tilde{\boldsymbol{p}}$ strongly in $L^2(0,\ell)^3$. $(5)_2$ then implies that there is $\tilde{\boldsymbol{q}} \in H^1(0,\ell)^3$ such that $\tilde{\boldsymbol{q}}_\varepsilon \to \tilde{\boldsymbol{q}}$ strongly in $H^1(0,\ell)^3$. Now we estimate

$$\begin{aligned} &\|\mathbf{Q}_\varepsilon\mathbf{H}^{-1}\mathbf{Q}_\varepsilon^T\tilde{\boldsymbol{q}}_\varepsilon - \mathbf{Q}\mathbf{H}^{-1}\mathbf{Q}^T\tilde{\boldsymbol{q}}\|_2 \\ &\quad \le \|\mathbf{Q}_\varepsilon\mathbf{H}^{-1}\mathbf{Q}_\varepsilon^T(\tilde{\boldsymbol{q}}_\varepsilon - \tilde{\boldsymbol{q}})\|_2 + \|\mathbf{Q}_\varepsilon\mathbf{H}^{-1}(\mathbf{Q}_\varepsilon - \mathbf{Q})^T\tilde{\boldsymbol{q}}\|_2 \\ &\qquad + \|(\mathbf{Q}_\varepsilon - \mathbf{Q})\mathbf{H}^{-1}\mathbf{Q}^T\tilde{\boldsymbol{q}}\|_2 \\ &\quad \le \|\mathbf{H}^{-1}\|\|\tilde{\boldsymbol{q}}_\varepsilon - \tilde{\boldsymbol{q}}\|_2 + 2\|\mathbf{H}^{-1}\|\|\mathbf{Q}_\varepsilon - \mathbf{Q}\|_2 \sup_{t\in[0,\ell]} |\tilde{\boldsymbol{q}}(t)|. \end{aligned} \tag{7}$$

The continuous imbedding of $H^1(0,\ell)$ in $C([0,\ell])$ now implies the convergence $\mathbf{Q}_\varepsilon\mathbf{H}^{-1}\mathbf{Q}_\varepsilon^T\tilde{\boldsymbol{q}}_\varepsilon \to \mathbf{Q}\mathbf{H}^{-1}\mathbf{Q}^T\tilde{\boldsymbol{q}}$ strongly in $L^2(0,\ell)^3$. Hence $(5)_3$ implies that there is $\tilde{\boldsymbol{\omega}} \in H^1(0,\ell)^3$ such that $\tilde{\boldsymbol{\omega}}_\varepsilon \to \tilde{\boldsymbol{\omega}}$ strongly in $H^1(0,\ell)^3$. The same argument as in the case of $\tilde{\boldsymbol{q}}_\varepsilon$ applied on $(5)_4$ implies that there is $\tilde{\boldsymbol{u}} \in H^1(0,\ell)^3$ such that $\tilde{\boldsymbol{u}}_\varepsilon \to \tilde{\boldsymbol{u}}$ strongly in $H^1(0,\ell)^3$. The continuous imbedding $H^1(0,\ell) \hookrightarrow C([0,\ell])$ implies that $\tilde{\boldsymbol{p}}, \tilde{\boldsymbol{q}}, \tilde{\boldsymbol{\omega}}, \tilde{\boldsymbol{u}}$ are continuous and that the boundary conditions are preserved in the limit. Therefore the limit quadruple $(\tilde{\boldsymbol{p}}, \tilde{\boldsymbol{q}}, \tilde{\boldsymbol{\omega}}, \tilde{\boldsymbol{u}})$ satisfies (6). □

Remark 1 (Transmission conditions). The equations (6) are fulfilled pointwisely in open intervals (s_{k-1}, s_k), $k = 1, \dots, n$. The first derivative of $\tilde{\boldsymbol{q}}, \tilde{\boldsymbol{\omega}}, \tilde{\boldsymbol{u}}$ has the first order jump at points $s_1, \dots, s_{n-1}$. The contact conditions (conditions of transmission) in these points are continuity of the functions $\tilde{\boldsymbol{p}}, \tilde{\boldsymbol{q}}, \tilde{\boldsymbol{\omega}}, \tilde{\boldsymbol{u}}$, i.e.

$$\tilde{\boldsymbol{p}}(s_k-) = \tilde{\boldsymbol{p}}(s_k+),\ \tilde{\boldsymbol{q}}(s_k-) = \tilde{\boldsymbol{q}}(s_k+),\ \tilde{\boldsymbol{\omega}}(s_k-) = \tilde{\boldsymbol{\omega}}(s_k+),\ \tilde{\boldsymbol{u}}(s_k-) = \tilde{\boldsymbol{u}}(s_k+),$$

for $k = 1, \dots, n-1$.

Remark 2 (Variational formulation). Let us assume that the limit curve $\mathcal{C}$ is more regular, namely $\boldsymbol{\Phi} \in C^3([s_{k-1}, s_k]; \mathbb{R}^3), k = 1, \dots, n$. Then

$$\mathbf{Q} \in C^2([s_{k-1}, s_k]; \mathrm{SO}(3)), \quad \mathbf{R} = (\mathbf{Q}^T)'\mathbf{Q} \in C^1([s_{k-1}, s_k]; \mathrm{Skw}(3)),$$

for $k = 1, \dots, n$. The quadruple $(\tilde{\boldsymbol{p}}, \tilde{\boldsymbol{q}}, \tilde{\boldsymbol{\omega}}, \tilde{\boldsymbol{u}})$ as well as the forces $\tilde{\boldsymbol{f}}, \tilde{\boldsymbol{g}}$ are rewritten in the local basis of the rod and new functions are defined

$$\boldsymbol{p} = \mathbf{Q}^T\tilde{\boldsymbol{p}},\ \boldsymbol{q} = \mathbf{Q}^T\tilde{\boldsymbol{q}},\ \boldsymbol{\omega} = \mathbf{Q}^T\tilde{\boldsymbol{\omega}},\ \boldsymbol{u} = \mathbf{Q}^T\tilde{\boldsymbol{u}},\ \boldsymbol{f} = \mathbf{Q}^T\tilde{\boldsymbol{f}},\ \boldsymbol{g} = \mathbf{Q}^T\tilde{\boldsymbol{g}}.$$

(6) now yield the problem: find a quadruple $(\boldsymbol{p}, \boldsymbol{q}, \boldsymbol{\omega}, \boldsymbol{u})$ such that

$$\begin{aligned} &\boldsymbol{p}' - \mathbf{R}\boldsymbol{p} + \boldsymbol{f} = 0, \quad \boldsymbol{p}(\ell) = \boldsymbol{g}, \\ &\boldsymbol{q}' - \mathbf{R}\boldsymbol{q} + \boldsymbol{e}_1 \times \boldsymbol{p} = 0, \quad \boldsymbol{q}(\ell) = 0, \\ &\boldsymbol{\omega}' - \mathbf{R}\boldsymbol{\omega} - \mathbf{H}^{-1}\boldsymbol{q} = 0, \quad \boldsymbol{\omega}(0) = 0, \\ &\boldsymbol{u}' - \mathbf{R}\boldsymbol{u} + \boldsymbol{e}_1 \times \boldsymbol{\omega} = 0, \quad \boldsymbol{u}(0) = 0, \end{aligned} \tag{8}$$

where the equations are fulfiled pointwise on $(s_{k-1}, s_k), k = 1, \dots, n$. The last equation in (8) implies the inextensibility condition

$$u_t' - R_{12}u_n - R_{13}u_b = 0, \tag{9}$$

which then appears in the function space of the variational equation. The associated Lagrange's multiplier is p_t. The second equation in (8) implies $q_t' - R_{12}q_n - R_{13}q_b = 0$. Therefore apart from the displacement $\boldsymbol{u}$ there is another independent function in the variational formulation $\phi = \omega_t$ the same as in (3). The transmission conditions imply

$$\mathbf{Q}(s_k+)\boldsymbol{u}(s_k+) = \mathbf{Q}(s_k-)\boldsymbol{u}(s_k-),\ \mathbf{Q}(s_k+)\boldsymbol{\omega}(s_k+) = \mathbf{Q}(s_k-)\boldsymbol{\omega}(s_k-), \tag{10}$$

$$\mathbf{Q}(s_k+)\boldsymbol{q}(s_k+) = \mathbf{Q}(s_k-)\boldsymbol{q}(s_k-),\ \mathbf{Q}(s_k+)\boldsymbol{p}(s_k+) = \mathbf{Q}(s_k-)\boldsymbol{p}(s_k-), \tag{11}$$

for $k = 1, \dots, n-1$. The conditions (9) and (10) have to be satisfied by the test function in the variational formulation. Let

$$\begin{aligned} \mathcal{W} = \big\{(\boldsymbol{v}, \psi) :\ &\boldsymbol{v}|_{(s_{k-1},s_k)} \in H^1(s_{k-1}, s_k) \times H^2(s_{k-1}, s_k) \times H^2(s_{k-1}, s_k), \\ &\psi|_{(s_{k-1},s_k)} \in H^1(s_{k-1}, s_k), k = 1, \dots, n, \\ &(\boldsymbol{v}, \psi) \text{ satisfies (10) and (9)}, \\ &\boldsymbol{v}(0) = 0, v_n'(0) = 0, v_b'(0) = 0, \psi(0) = 0\big\}. \end{aligned}$$

The variational formulation is given by: find $(\boldsymbol{u}, \phi) \in \mathcal{W}$ such that

$$\sum_{k=1}^{n} \int_{s_{k-1}}^{s_k} \mathbf{H} \begin{bmatrix} E_4(\boldsymbol{u}, \phi) \\ E_3(\boldsymbol{u}, \phi) \\ E_2(\boldsymbol{u}, \phi) \end{bmatrix} \cdot \begin{bmatrix} E_4(\boldsymbol{v}, \psi) \\ E_3(\boldsymbol{v}, \psi) \\ E_2(\boldsymbol{v}, \psi) \end{bmatrix} = \boldsymbol{v}(\ell) \cdot \boldsymbol{g} + \sum_{k=1}^{n} \int_{s_{k-1}}^{s_k} \boldsymbol{f} \cdot \boldsymbol{v},$$

for all $(\boldsymbol{v}, \psi) \in \mathcal{W}$.

4. Error estimate

In this section we derive the error estimate with explicitly calculated constants for the approximation from Theorem 1. It is obtained using an explicit formula for the solution of (5). Let us denote

$$\tilde{\boldsymbol{W}}_\varepsilon = \begin{bmatrix} \tilde{\boldsymbol{\omega}}_\varepsilon \\ \tilde{\boldsymbol{u}}_\varepsilon \end{bmatrix}, \quad \tilde{\boldsymbol{Z}}_\varepsilon = \begin{bmatrix} \tilde{\boldsymbol{p}}_\varepsilon \\ \tilde{\boldsymbol{q}}_\varepsilon \end{bmatrix}, \quad \tilde{\boldsymbol{F}}_\varepsilon = \begin{bmatrix} \tilde{\boldsymbol{f}}_\varepsilon \\ 0 \end{bmatrix}, \quad \tilde{\boldsymbol{G}}_\varepsilon = \begin{bmatrix} \tilde{\boldsymbol{g}}_\varepsilon \\ 0 \end{bmatrix}.$$

For $\boldsymbol{v} \in \mathbb{R}^3$ we define $\mathbf{A}_{\boldsymbol{v}}$ to be the antisymmetric matrix with associated axial vector $\boldsymbol{v}$ i.e. $\mathbf{A}_{\boldsymbol{v}}\boldsymbol{x} = \boldsymbol{v} \times \boldsymbol{x}, \boldsymbol{x} \in \mathbb{R}^3$. We also define block matrices

$$\mathbf{O}_{\boldsymbol{v}} = \begin{bmatrix} \mathbf{I} & \mathbf{0} \\ \mathbf{A}_{\boldsymbol{v}} & \mathbf{I} \end{bmatrix}, \quad \boldsymbol{v} \in \mathbb{R}^3 \qquad \tilde{\mathbf{D}}_\varepsilon(s) = \begin{bmatrix} \mathbf{0} & -\mathbf{Q}_\varepsilon(s)\mathbf{H}^{-1}\mathbf{Q}_\varepsilon(s)^T \\ \mathbf{0} & \mathbf{0} \end{bmatrix};$$

here $\mathbf{I}$ and $\mathbf{0}$ stand for the identity and null matrix of order 3. The matrix inverse to $\mathbf{O}_{\boldsymbol{v}}$ is $\mathbf{O}_{-\boldsymbol{v}}$. The matrix norm of $\mathbf{A}_{\boldsymbol{v}}$ and $\mathbf{O}_{\boldsymbol{v}}$ are estimated by

$$\|\mathbf{A}_{\boldsymbol{v}}\| \leq |\boldsymbol{v}|, \qquad \|\mathbf{O}_{\boldsymbol{v}}\| = \sup_{\boldsymbol{x},\boldsymbol{y} \in \mathbb{R}^3} \frac{\sqrt{|\boldsymbol{x}|^2 + |\boldsymbol{v} \times \boldsymbol{x} + \boldsymbol{y}|^2}}{\sqrt{|\boldsymbol{x}|^2 + |\boldsymbol{y}|^2}} \leq 1 + |\boldsymbol{v}|.$$

Then using the same argument as in Jurak and Tambača (2001) the solution of (5) can be written in an explicit form

$$\tilde{\boldsymbol{Z}}_\varepsilon(s) = \mathbf{O}_{\boldsymbol{\Phi}_\varepsilon(\ell)-\boldsymbol{\Phi}_\varepsilon(s)} \left(\tilde{\boldsymbol{G}}_\varepsilon + \int_s^\ell \mathbf{O}^{-1}_{\boldsymbol{\Phi}_\varepsilon(\ell)-\boldsymbol{\Phi}_\varepsilon(t)} \tilde{\boldsymbol{F}}_\varepsilon(t)dt \right), \quad s \in [0,\ell], \tag{12}$$

$$\tilde{\boldsymbol{W}}_\varepsilon(s) = -\mathbf{O}_{\boldsymbol{\Phi}_\varepsilon(0)-\boldsymbol{\Phi}_\varepsilon(s)} \int_0^s \mathbf{O}^{-1}_{\boldsymbol{\Phi}_\varepsilon(0)-\boldsymbol{\Phi}_\varepsilon(t)} \tilde{\mathbf{D}}_\varepsilon(t) \tilde{\boldsymbol{Z}}_\varepsilon(t)dt, \; s \in [0,\ell]. \tag{13}$$

The same explicit formula (12), (13) holds for the continuous solution of (6) with just dropping the index ε.

Now we are able to estimate the difference of the solution of ε–problem (5) and the limit problem (6). The norm defined by

$$\|\boldsymbol{f}\|_* = \| \, |\boldsymbol{f}| \, \|_\infty = \sup_{t \in [0,\ell]} |\boldsymbol{f}(t)|,$$

is equivalent to the norm $\|\cdot\|_\infty$ on $C([0,\ell];\mathbb{R}^n), n \in \{3,6\}$.

Lemma 1. *The following estimates hold*

$$\begin{aligned}
|\tilde{\boldsymbol{W}}_\varepsilon(s) - \tilde{\boldsymbol{W}}(s)| \leq \|\mathbf{H}^{-1}\| \Big(& (1+\ell)^3 \ell |\tilde{\boldsymbol{g}}_\varepsilon - \tilde{\boldsymbol{g}}| + (1+\ell)^4 \ell^{3/2} \|\tilde{\boldsymbol{f}}_\varepsilon - \tilde{\boldsymbol{f}}\|_2 \\
& + 2(1+\ell)^3 \sqrt{\ell} \left(|\tilde{\boldsymbol{g}}| + 2(1+\ell)\sqrt{\ell} \|\tilde{\boldsymbol{f}}\|_2 \right) \|\mathbf{Q}_\varepsilon - \mathbf{Q}\|_2 \\
& + 6(1+\ell)^2 \ell \left(|\tilde{\boldsymbol{g}}| + 2(1+\ell)\sqrt{\ell} \|\tilde{\boldsymbol{f}}\|_2 \right) \|\boldsymbol{\Phi}_\varepsilon - \boldsymbol{\Phi}\|_* \Big), \qquad s \in [0,\ell]
\end{aligned}$$

$$
\begin{aligned}
\|\tilde{\boldsymbol{W}}'_\varepsilon - \tilde{\boldsymbol{W}}'\|_2 \le \|\mathbf{H}^{-1}\| \Big(& (\ell^3 + 2\ell^2 + 2\ell + 1)|\tilde{\boldsymbol{g}}_\varepsilon - \tilde{\boldsymbol{g}}| \\
& + (1+\ell)^2 (\ell^2 + \ell + 1)\sqrt{\ell}\|\tilde{\boldsymbol{f}}_\varepsilon - \tilde{\boldsymbol{f}}\|_2 \\
& + (3\ell^3 + 6\ell^2 + 5\ell + 2)\left(|\tilde{\boldsymbol{g}}| + 2(1+\ell)\sqrt{\ell}\|\tilde{\boldsymbol{f}}\|_2\right)\|\mathbf{Q}_\varepsilon - \mathbf{Q}\|_2 \\
& + 2(\ell^{5/2} + \ell^2 + \ell^{3/2} + \ell + 1)\left(|\tilde{\boldsymbol{g}}| + 2(1+\ell)\sqrt{\ell}\|\tilde{\boldsymbol{f}}\|_2\right)\|\boldsymbol{\Phi}_\varepsilon - \boldsymbol{\Phi}\|_* \Big).
\end{aligned}
$$

Proof. Using the fact that C is of length ℓ we estimate the function $\tilde{\boldsymbol{Z}}$:

$$
\begin{aligned}
|\tilde{\boldsymbol{Z}}(s)| &= \left| \mathbf{O}_{\boldsymbol{\Phi}(\ell)-\boldsymbol{\Phi}(s)} \left(\tilde{\boldsymbol{G}} + \int_s^\ell \mathbf{O}^{-1}_{\boldsymbol{\Phi}(\ell)-\boldsymbol{\Phi}(t)} \tilde{\boldsymbol{F}}(t)dt \right) \right| \\
&\le (1 + |\boldsymbol{\Phi}(\ell) - \boldsymbol{\Phi}(s)|) \left(|\tilde{\boldsymbol{G}}| + \int_0^\ell \left| \mathbf{O}^{-1}_{\boldsymbol{\Phi}(\ell)-\boldsymbol{\Phi}(t)} \tilde{\boldsymbol{F}}(t) \right| dt \right) \\
&\le (1+\ell) \left(|\tilde{\boldsymbol{G}}| + \int_0^\ell (1 + |\boldsymbol{\Phi}(\ell) - \boldsymbol{\Phi}(t)|) \left| \tilde{\boldsymbol{F}}(t) \right| dt \right) \\
&\le (1+\ell) \left(|\tilde{\boldsymbol{G}}| + (1+\ell)\sqrt{\ell}\|\tilde{\boldsymbol{F}}\|_2 \right).
\end{aligned}
$$

Therefore

$$
\|\tilde{\boldsymbol{Z}}\|_* \le (1+\ell)\left(|\tilde{\boldsymbol{G}}| + (1+\ell)\sqrt{\ell}\|\tilde{\boldsymbol{F}}\|_2 \right), \quad \|\tilde{\boldsymbol{Z}}\|_2 \le \sqrt{\ell}\|\tilde{\boldsymbol{Z}}\|_*. \tag{14}
$$

For each $s \in [0, \ell]$ in the same way we estimate the difference:

$$
\begin{aligned}
|\mathbf{Z}_\varepsilon(s) - \mathbf{Z}(s)| \le & \left| \mathbf{O}_{\boldsymbol{\Phi}_\varepsilon(\ell)-\boldsymbol{\Phi}_\varepsilon(s)} \left(\tilde{\boldsymbol{G}}_\varepsilon - \tilde{\boldsymbol{G}} \right) \right| \\
& + \left| \mathbf{O}_{\boldsymbol{\Phi}_\varepsilon(\ell)-\boldsymbol{\Phi}_\varepsilon(s)} \int_s^\ell \mathbf{O}^{-1}_{\boldsymbol{\Phi}_\varepsilon(\ell)-\boldsymbol{\Phi}_\varepsilon(t)} \left(\tilde{\boldsymbol{F}}_\varepsilon(t) - \tilde{\boldsymbol{F}}(t) \right) dt \right| \\
& + \left| \mathbf{O}_{\boldsymbol{\Phi}_\varepsilon(\ell)-\boldsymbol{\Phi}_\varepsilon(s)} \int_s^\ell \left(\mathbf{O}^{-1}_{\boldsymbol{\Phi}_\varepsilon(\ell)-\boldsymbol{\Phi}_\varepsilon(t)} - \mathbf{O}^{-1}_{\boldsymbol{\Phi}(\ell)-\boldsymbol{\Phi}(t)} \right) \tilde{\boldsymbol{F}}(t)dt \right| \\
& + \left| \left(\mathbf{O}_{\boldsymbol{\Phi}_\varepsilon(\ell)-\boldsymbol{\Phi}_\varepsilon(s)} - \mathbf{O}_{\boldsymbol{\Phi}(\ell)-\boldsymbol{\Phi}(s)} \right) \left(\tilde{\boldsymbol{G}} + \int_s^\ell \mathbf{O}^{-1}_{\boldsymbol{\Phi}(\ell)-\boldsymbol{\Phi}(t)} \tilde{\boldsymbol{F}}(t)dt \right) \right| \\
\le & (1+\ell)|\tilde{\boldsymbol{G}}_\varepsilon - \tilde{\boldsymbol{G}}| + (1+\ell)\int_0^\ell (1 + |\boldsymbol{\Phi}_\varepsilon(\ell) - \boldsymbol{\Phi}_\varepsilon(t)|)|\tilde{\boldsymbol{F}}_\varepsilon(t) - \tilde{\boldsymbol{F}}(t)|dt \\
& + (1+\ell)\int_0^\ell (|\boldsymbol{\Phi}_\varepsilon(\ell) - \boldsymbol{\Phi}(\ell)| + |\boldsymbol{\Phi}_\varepsilon(t) - \boldsymbol{\Phi}(t)|)|\tilde{\boldsymbol{F}}(t)|dt \\
& + 2\|\boldsymbol{\Phi}_\varepsilon - \boldsymbol{\Phi}\|_* \left(|\tilde{\boldsymbol{G}}| + \int_0^\ell (1 + |\boldsymbol{\Phi}(\ell) - \boldsymbol{\Phi}(t)|)|\tilde{\boldsymbol{F}}(t)|dt \right) \\
\le & (1+\ell)|\tilde{\boldsymbol{G}}_\varepsilon - \tilde{\boldsymbol{G}}| + (1+\ell)^2\sqrt{\ell}\|\tilde{\boldsymbol{F}}_\varepsilon - \tilde{\boldsymbol{F}}\|_2 \\
& + 2\left(|\tilde{\boldsymbol{G}}| + 2(1+\ell)\sqrt{\ell}\|\tilde{\boldsymbol{F}}\|_2 \right)\|\boldsymbol{\Phi}_\varepsilon - \boldsymbol{\Phi}\|_*.
\end{aligned}
\tag{15}
$$

Applying the same arguments as in (15) for $\tilde{\boldsymbol{W}}_\varepsilon(s) - \tilde{\boldsymbol{W}}(s)$ one gets

$$\begin{aligned}&|\tilde{\boldsymbol{W}}_\varepsilon(s) - \tilde{\boldsymbol{W}}(s)| \le \\ &\quad \le (1+\ell)^2\sqrt{\ell}\|\tilde{\mathbf{D}}_\varepsilon\tilde{\boldsymbol{Z}}_\varepsilon - \tilde{\mathbf{D}}\tilde{\boldsymbol{Z}}\|_2 + 4(1+\ell)\sqrt{\ell}\|\tilde{\mathbf{D}}\tilde{\boldsymbol{Z}}\|_2\|\boldsymbol{\Phi}_\varepsilon - \boldsymbol{\Phi}\|_*.\end{aligned}$$

Now, the estimate (7) implies

$$\begin{aligned}|\tilde{\boldsymbol{W}}_\varepsilon(s) - \tilde{\boldsymbol{W}}(s)| &\le \|\mathbf{H}^{-1}\|\Big((1+\ell)^2\sqrt{\ell}\|\tilde{\boldsymbol{Z}}_\varepsilon - \tilde{\boldsymbol{Z}}\|_2 \\ &+2(1+\ell)^2\sqrt{\ell}\|\tilde{\boldsymbol{Z}}\|_*\|\mathbf{Q}_\varepsilon - \mathbf{Q}\|_2 + 4(1+\ell)\sqrt{\ell}\|\tilde{\boldsymbol{Z}}\|_2\|\boldsymbol{\Phi}_\varepsilon - \boldsymbol{\Phi}\|_*\Big).\end{aligned}$$

The difference of the derivatives needs an L^2 estimate

$$\begin{aligned}&\|\tilde{\boldsymbol{W}}'_\varepsilon - \tilde{\boldsymbol{W}}'\|_2 \le \\ &\quad\le \|\mathbf{O}'_{\boldsymbol{\Phi}_\varepsilon(0)-\boldsymbol{\Phi}_\varepsilon(\cdot)}\int_0^\cdot \mathbf{O}^{-1}_{\boldsymbol{\Phi}_\varepsilon(0)-\boldsymbol{\Phi}_\varepsilon(t)}\Big(\tilde{\mathbf{D}}_\varepsilon(t)\tilde{\boldsymbol{Z}}_\varepsilon(t) - \tilde{\mathbf{D}}(t)\tilde{\boldsymbol{Z}}(t)\Big)\,dt\|_2 \\ &\quad+\|\mathbf{O}'_{\boldsymbol{\Phi}_\varepsilon(0)-\boldsymbol{\Phi}_\varepsilon(\cdot)}\int_0^\cdot\Big(\mathbf{O}^{-1}_{\boldsymbol{\Phi}_\varepsilon(0)-\boldsymbol{\Phi}_\varepsilon(t)} - \mathbf{O}^{-1}_{\boldsymbol{\Phi}(0)-\boldsymbol{\Phi}(t)}\Big)\tilde{\mathbf{D}}(t)\tilde{\boldsymbol{Z}}(t)dt\|_2 \\ &\quad+\|\Big(\mathbf{O}'_{\boldsymbol{\Phi}_\varepsilon(0)-\boldsymbol{\Phi}_\varepsilon(\cdot)} - \mathbf{O}'_{\boldsymbol{\Phi}(0)-\boldsymbol{\Phi}(\cdot)}\Big)\int_0^\cdot \mathbf{O}^{-1}_{\boldsymbol{\Phi}(0)-\boldsymbol{\Phi}(t)}\tilde{\mathbf{D}}(t)\tilde{\boldsymbol{Z}}(t)dt\|_2 \\ &\quad+\|\tilde{\mathbf{D}}_\varepsilon\tilde{\boldsymbol{Z}}_\varepsilon - \tilde{\mathbf{D}}\tilde{\boldsymbol{Z}}\|_2 \\ &\quad\le \|\int_0^\cdot \mathbf{O}^{-1}_{\boldsymbol{\Phi}_\varepsilon(0)-\boldsymbol{\Phi}_\varepsilon(t)}\Big(\tilde{\mathbf{D}}_\varepsilon(t)\tilde{\boldsymbol{Z}}_\varepsilon(t) - \tilde{\mathbf{D}}(t)\tilde{\boldsymbol{Z}}(t)\Big)\,dt\|_2 \\ &\quad+\|\int_0^\cdot\Big(\mathbf{O}^{-1}_{\boldsymbol{\Phi}_\varepsilon(0)-\boldsymbol{\Phi}_\varepsilon(t)} - \mathbf{O}^{-1}_{\boldsymbol{\Phi}(0)-\boldsymbol{\Phi}(t)}\Big)\tilde{\mathbf{D}}(t)\tilde{\boldsymbol{Z}}(t)dt\|_2 \\ &\quad+\|\boldsymbol{t}_\varepsilon - \boldsymbol{t}\|_2\int_0^\ell |\mathbf{O}^{-1}_{\boldsymbol{\Phi}(0)-\boldsymbol{\Phi}(t)}\tilde{\mathbf{D}}(t)\tilde{\boldsymbol{Z}}(t)|dt + \|\tilde{\mathbf{D}}_\varepsilon\tilde{\boldsymbol{Z}}_\varepsilon - \tilde{\mathbf{D}}\tilde{\boldsymbol{Z}}\|_2 \\ &\quad\le (\ell^2+\ell+1)\|\tilde{\mathbf{D}}_\varepsilon\tilde{\boldsymbol{Z}}_\varepsilon - \tilde{\mathbf{D}}\tilde{\boldsymbol{Z}}\|_2 + 2\ell\|\boldsymbol{\Phi}_\varepsilon - \boldsymbol{\Phi}\|_*\|\tilde{\mathbf{D}}\tilde{\boldsymbol{Z}}\|_2 \\ &\quad+(1+\ell)\sqrt{\ell}\|\mathbf{Q}_\varepsilon - \mathbf{Q}\|_2\|\tilde{\mathbf{D}}\tilde{\boldsymbol{Z}}\|_2.\end{aligned}$$

Using the estimate (7) we obtain

$$\begin{aligned}\|\tilde{\boldsymbol{W}}'_\varepsilon - \tilde{\boldsymbol{W}}'\|_2 &\le \|\mathbf{H}^{-1}\|\Big(\ell^2+\ell+1)\|\tilde{\boldsymbol{Z}}_\varepsilon - \tilde{\boldsymbol{Z}}\|_2 \\ &+2\ell\|\tilde{\boldsymbol{Z}}\|_2\|\boldsymbol{\Phi}_\varepsilon - \boldsymbol{\Phi}\|_* + (3\ell^2+3\ell+2)\|\tilde{\boldsymbol{Z}}\|_*\|\mathbf{Q}_\varepsilon - \mathbf{Q}\|_2\Big).\end{aligned}$$

The statement of the lemma follows using (14) and (15). □

5. Comparison with the junction model

In this section we derive the model of two straight rods connected at the right angle and compare it with the model derived by Le Dret (1989). Let $\boldsymbol{k} = \boldsymbol{e}_3$

and

$$\mathbf{\Phi}(s) = \begin{cases} (1-s,0,0), & s \in [0,1] \\ (0,s-1,0), & s \in [1,2] \end{cases}.$$

Then $\mathbf{Q}$ is a piecewise constant matrix function given by

$$\mathbf{Q}(s) = \begin{cases} \begin{bmatrix} -\boldsymbol{e}_1 & -\boldsymbol{e}_2 & \boldsymbol{e}_3 \end{bmatrix}, & s \in [0,1) \\ \begin{bmatrix} \boldsymbol{e}_2 & -\boldsymbol{e}_1 & \boldsymbol{e}_3 \end{bmatrix}, & s \in (1,2] \end{cases}.$$

For the straight rod is $\mathbf{R} = 0$, so the expressions for $\tilde{\boldsymbol{\omega}}, \tilde{\boldsymbol{q}}$ and $\tilde{\boldsymbol{p}}$ simplify to

$$\tilde{\boldsymbol{\omega}} = \mathbf{Q}\begin{pmatrix} \phi \\ -u_b' \\ u_n' \end{pmatrix}, \quad \tilde{\boldsymbol{q}} = \mathbf{QH}\begin{pmatrix} \phi' \\ -u_b'' \\ u_n'' \end{pmatrix}, \quad \tilde{\boldsymbol{p}} = \mathbf{Q}\begin{pmatrix} p_t \\ EI_2u_n''' \\ EI_3u_b''' \end{pmatrix},$$

so the equations in the local coordinates are given by

$$-p_t' = f_t, \quad -EI_2u_n'''' = f_n, \quad -EI_2u_b'''' = f_b, \quad u_t' = 0, \quad \phi'' = 0.$$

These equations written in the Cartesian coordinates are following

$$\begin{aligned} p_1' &= -f_1,\ \phi'' = 0,\ u_1' = 0,\ EI_2u_2'''' = f_2,\ EI_3u_3'''' = f_3, \quad \text{in } (0,1), \quad (16) \\ -p_2' &= f_2,\ \phi'' = 0,\ u_2' = 0,\ EI_2u_1'''' = f_1,\ EI_3u_3'''' = f_3, \quad \text{in } (1,2). \quad (17) \end{aligned}$$

The boundary conditions are

$$\begin{aligned} &u_1(0) = 0,\ u_2(0) = 0,\ u_3(0) = 0,\ \phi(0) = 0,\ u_3'(0) = 0,\ u_2'(0) = 0, \\ &p_2(2) = g_2, \quad -EI_2u_1'''(2) = g_1, \quad -EI_3u_3'''(2) = g_3, \qquad (18) \\ &\mu K\phi'(2) = 0, \quad EI_3u_3''(2) = 0, \quad EI_2u_1''(2) = 0. \end{aligned}$$

The transmission conditions are given by

$$\begin{aligned} &u_1(1+) = u_1(1-), \quad u_2(1+) = u_2(1-), \quad u_3(1+) = u_3(1-), \\ &-u_3'(1+) = \phi(1-), \quad \phi(1+) = u_3'(1-), \quad u_1'(1+) = u_2'(1-), \qquad (19) \\ &-EI_3u_3''(1+) = \mu K\phi'(1-),\ \mu K\phi'(1+) = EI_3u_3''(1-),\ u_1''(1+) = u_2''(1-), \\ &EI_2u_1'''(1+) = -p_1(1-),\ -p_2(1+) = EI_2u_2'''(1-),\ u_3'''(1+) = u_3'''(1-). \end{aligned}$$

The model following from (16), (17), (18) and (19) with $\boldsymbol{g} = 0$ was obtained by asymptotic techniques, taking the thickness of the rod to zero, directly from the three–dimensional linearized elasticity, see Le Dret (1989).

6. Perspectives

The model of irregular curved rods has been derived from the model of curved rods with a smooth middle curve by approximating the geometry of the middle curve of the rod.

Let us now do the opposite. Let a smooth curve be approximated by a piecewise linear curve of the same length. The error estimate from the Section 4 can be applied in this case too. Therefore, instead of solving the model of curved rod with a smooth curve one needs to solve the model for a rod with a piecewise linear curve. The error of such approximation depends on the approximation of the geometry and applied loads. The benefit of solving the model for a rod with piecewise linear middle curve is that on each linear piece the coefficients given by $\mathbf{R}$ are zero (see Section 5). That implies very simple ordinary differential equations and the general solution on each linear piece. The transmission conditions then determine the solution of the model by a regular system of linear equations. These considerations suggest a possible numerical procedure for solving the model of curved rods with arbitrary middle curve.

Acknowledgments

It is my pleasure to thank professor E. Marušić–Paloka for mentioning this problem to me and professor I. Aganović on his valuable comments.

References

Jurak, M. and Tambača, J. (2001). Linear curved rod model. General curve. *Math. Mod. Meth. Appl. Sci.* 11: 1237–1253.

Le Dret, H. (1989). Modeling of the junction between two rods. *J. Math pures et appl.*, 68: 365–397.

Sanchez–Hubert, J. and Sanchez–Palencia, E. (1999). Statics of curved rods on account of torsion and flexion. *Eur. J. Mech. A/Solids*, 18: 365–390.

Tambača, J. (2000). *Evolution model of curved rods.* PhD thesis, Department of Mathematics, University of Zagreb. (in Croatian).

EXISTENCE OF THE DENSITY OF STATES FOR SOME ALLOY TYPE MODELS WITH SINGLE SITE POTENTIALS THAT CHANGE SIGN

Ivan Veselić

Fakultät für Mathematik, Ruhr-Universität Bochum, Germany

www.ruhr-uni-bochum.de/mathphys/ivan

Abstract We study spectral properties of ergodic random Schrödinger operators on $L^2(\mathbb{R}^d)$. The density of states is shown to exist for a certain class of alloy type potentials with single site potentials of changing sign. The Wegner estimate we prove implies Anderson localization under certain additional assumptions. For some examples we discuss briefly some properties of the common and conditional densities of the random coupling constants used in the proof of the Wegner estimate.

Keywords: density of states, random Schrödinger operators, Wegner estimate, multiscale analysis, localization, indefinite single site potential

2000 AMS Mathematics Subject Classification 35J10, 35P20, 81Q10, 81Q15

1. Alloy type model and the integrated density of states

We consider Schrödinger operators with a potential which is a stochastic process ergodic with respect to translations from $\mathbb{Z}^d$. Such operators model quantum mechanical Hamiltonians which govern the motion of single electrons in disordered solids. The spectral properties of the Schrödinger operator are related to the dynamical behaviour of the electron wave packets and thus to the charge transport properties of the described solid, cf. e.g. [4, 12, 29]. The random potential considered in this note is of *alloy* or *continuous Anderson* type:

$$V : \Omega \times \mathbb{R}^d \to \mathbb{R}, \quad V_\omega(x) = \sum_{k \in \mathbb{Z}^d} u(x-k).$$

The function $u \in L^p$ ($p = 2$ for $d \leq 3$ and $p > \frac{d}{2}$ for $d \geq 4$) whose translates generate V_ω is called *single site potential*. The "strength" of the potentials

$u(\cdot - k)$ associated to different lattice sites $k \in \mathbb{Z}^d$ is a random variable $\omega_k : \Omega \to \mathbb{R}$. We assume that the sequence $(\omega_k)_{k \in \mathbb{Z}^d}$ is independent and identically distributed (iid), and moreover that each ω_k is distributed according to the probability measure μ which has a density f in the Sobolev space $W_c^{1,1}(\mathbb{R})$. We may identify the probability space $(\Omega, \mathbb{P})$ with the space $\times_{k \in \mathbb{Z}^d} \mathbb{R}$ equipped with the product measure $\otimes_{k \in \mathbb{Z}^d} \mu$. Thus an element $\omega \in \Omega$ is a random vector $(\omega_k)_{k \in \mathbb{Z}^d}$. The *alloy type model* is the random Schrödinger operator, or more precisely, the random family of Schrödinger operators, given by

$$H_\omega := H_0 + V_\omega, \quad \omega \in \Omega, \qquad H_0 := -\Delta + V_0 \tag{1}$$

where $V_0 \in L^p_{\text{loc}}(\mathbb{R}^d)$ is a $\mathbb{Z}^d$-periodic potential and p is chosen depending on the dimension, as above. A large part of the mathematical literature and results on random Schrödinger operators are devoted to the model just described, certain derivatives thereof, cf. e.g. [24, 6, 18] and its discrete version on $l^2(\mathbb{Z}^d)$ — actually the original model studied by Anderson [1]. An extensive list of references can be found in the textbooks [10, 20, 5, 30, 32]. The name alloy type model stems from the fact that one can consider $u(\cdot - k)$ as an atomic potential, say of screened Coulomb type, due to a nucleus sitting at the lattice site $k \in \mathbb{Z}^d$. While the lattice structure of the solid is fixed and all atoms are assumed to generate a potential of the same *shape*, the distribution of the different types of atoms on the lattice sites is random. This is then modelled by random nuclear charges ω_k which appear as *coupling constants* in front of the atomic potential $u(\cdot - k)$ and determine its "strength". Under the above assumptions on H_ω it is well known that the spectrum and its measure theoretic components are almost surely non-random. More precisely, there exist $\Omega' \subset \Omega$ of measure one and a subset of the reals Σ such that $\sigma(H_\omega) = \Sigma$ for all $\omega \in \Omega'$. The analog statement is true for the sets $\sigma_{disc}(H_\omega), \sigma_{ess}(H_\omega), \sigma_{ac}(H_\omega), \sigma_{sc}(H_\omega)$ and $\sigma_{pp}(H_\omega)$. (Here we denote by σ_{pp} the *closure* of the set of eigenvalues.) This property of the spectrum is called *selfaveraging*. It is due to the fact that on the infinite configuration space $\mathbb{R}^d$ the fluctuations of the potential V_ω, which are present in finite size cubes, cancel out thanks to the iid property of the random coupling constants. Part of the spectral properties of the family of Schrödinger operators H_ω is contained in the *integrated density of states* (IDS). It can be used to calculate the basic thermodynamic quantities of the corresponding non-interacting many particle system, cf. e.g. [17, 11]. In physical terms, the IDS counts the number of energy levels per unit volume, up to a given energy E. The possible existence of continuous spectrum makes a careful mathematical definition of the IDS necessary, which is as follows: Denote by $\Lambda = \Lambda_l$ the cube $[0, l[^d$ and with H^l_ω the restriction of H_ω to the interior of Λ_l with periodic boundary conditions. Then the normalized eigenvalue counting functions

$$N^l_\omega(E) = l^{-d} \#\{i | \ \lambda_i(H^l_\omega) < E\} = l^{-d} \operatorname{Tr} P^l_\omega(] - \infty, E[) \tag{2}$$

of H_ω^l converge for almost all ω to a limit $N(E) := \lim_{l\to\infty} N_\omega^l(E)$ which is ω-independent. This convergence holds true for all E which are continuity points of N. Here we denote with $P_\omega^l(I)$ the spectral projection of H_ω onto the interval I. Since the limit N is independent of the randomness, we encountered another selfaveraging quantity. Its set of points of increase $\mathcal{I} := \{E \in \mathbb{R} | \, N(E+\epsilon) - N(E-\epsilon) > 0\}$ coincides with the a.s. spectrum Σ.

This note is organized as follows. In the next section we state our main theorem and related results of other authors as well as the implications for Anderson localization, Section 3 contains a sketch of the proof of the main theorem and the last section is devoted to the discussion of the proof of localization using the multiscale analysis and to Wegner estimates for alloy type potentials with dependent random coupling constants. Furthermore we discuss some technical differences of the use of the common and the conditional density of the coupling constants.

2. Main theorem: A Wegner estimate for indefinite potentials

A Wegner estimate [36] is a assertion about the regularity of the finite volume IDS N_ω^l which may imply the Hölder continuity of the IDS on $\mathbb{R}^d$ or even the existence of its derivative dN/dE, the *density of states* (DOS). Note that in the following result the single site potential may be indefinite, i.e. take values of both signs.

Theorem 2.1 ([34, 35]). *Let $L^p(\mathbb{R}^d) \ni w \geq \kappa\chi_{[0,1]^d}$ with $\kappa > 0$ and $p = 2$ for $d \leq 3$ and $p > d/2$ for $d \geq 4$. Let $\Gamma \subset \mathbb{Z}^d$ be finite, the* convolution vector *$\alpha = (\alpha_k)_{k\in\Gamma}$ satisfy $\alpha^* := \sum_{k\neq 0} |\alpha_k| < |\alpha_0|$, and the single site potential be of* generalized step function *form:*

$$u(x) = \sum_{k\in\Gamma} \alpha_k \, w(x-k). \tag{3}$$

Then there exists for all $E \in \mathbb{R}$ a constant $C = C(E)$ such that

$$\int_\Omega \left[\operatorname{Tr} P_\omega^l([E-\epsilon, E])\right] \, d\mathbb{P}(\omega) \leq C\,\epsilon\, l^d, \quad \forall \epsilon \geq 0. \tag{4}$$

The theorem implies that the DOS exists for a.e. E and is locally uniformly bounded: $dN(E)/dE \leq C(E_1)$ for all $E \leq E_1$. The following result with V. Kostrykin applies to uniform densities.

Proposition 2.2 ([27]). *The assertion of Theorem 2.1 holds true if f is the uniform density on an interval and $\Gamma \subset \{k \in \mathbb{Z}^d | \, k_i \geq 0 \, \forall \, i = 1, \ldots, d\}$.*

First Wegner estimates for indefinite alloy type potentials were derived in [25]. In [16] P. Hislop and F. Klopp combine the techniques from [25] and [9] to prove a Wegner estimate valid for general indefinite single site potentials and for energy intervals at edges of $\sigma(H_\omega)$. They assume the single site potential u not to vanish identically and to be in $C_c \cap l^1(L^p)$ with $p \geq \min(d, 2)$. The density $f \in L_c^\infty$ of the random variable ω_0 (which may be in fact the conditional density w.r.t. $\omega^{\perp 0} := (\omega_k)_{k \neq 0}$) is assumed to be locally absolutely continuous. For any $\beta < 1$ and any energy interval I below the spectrum of the unperturbed operator H_0 they prove

$$\mathbb{P}\{\sigma(H_\omega^l) \cap I \neq \emptyset\} \leq C\,|I|^\beta\, l^d$$

where the constant C depends only on β, d and $dist(I, \sigma(H_0))$. With a sufficiently small global coupling constant λ the same result holds for the operator $H_0 + \lambda V_\omega$ for I in an internal spectral gap of H_0. The results of [16] extend to more general models including certain operators with random magnetic field. The literature on Wegner estimates for multidimensional alloy type models includes [28, 7, 2, 22, 21, 31, 9, 19, 23]. Theorem 2.1 and Proposition 2.2 imply a localization result if the negative part u_- of the single site potential is sufficiently small.

Theorem 2.3 ([34]). *Let H_ω satisfy the assumptions of Theorem 2.1 or Proposition 2.2, let w have cmpact support and E be a boundary point of $\sigma(H_\omega)$. Let furthermore either*

1 V_0 be symmetric w.r.t. refelections along the coordinate axes and $E = \inf \sigma(H_\omega)$, or

2 $\operatorname{supp} f = [\omega_-, \omega_+]$ and for $\tau > d/2$ and $h \geq 0$ sufficiently small let the density f satisfy $\int_{\omega_-}^{h} f \leq h^\tau$ and $\int_{\omega_+ - h}^{\omega_+} f \leq h^\tau$.

Then there exist $\epsilon, r > 0$ such that for $\sum_{\alpha_l < 0} \alpha_l \geq -\epsilon$

$$[E - r, E + r] \cap \sigma_c(H_\omega) = \emptyset \ \text{ and } \ \sigma(H_\omega) \cap [E - r, E + r] \neq \emptyset \qquad (5)$$

The eigenfunctions of H_ω with eigenvalues in $[E-r, E+r]$ decay exponentially.

In [16] this result has been generalized to a larger class of single site potentials using results from [26] and an abstract version of the smallness of u_-. In one dimension the alloy type model has pure point at all energies regardless of the sign properties of u [33]. The proof does not use a Wegner estimate.

3. Sketch of the proof of Theorem 2.1

For simplicity we assume $w = \chi_{[0,1]^d}$. Let $\tilde{\Lambda} := \Lambda \cap \mathbb{Z}^d$, $\mathbb{E}$ denote the expectation w.r.t. $\mathbb{P}$ and $I := [E_1, E_2[$ an energy interval. Thus we have $\mathbb{E}\left[N_\omega^l(E_2) - N_\omega^l(E_1)\right] = l^{-d}\,\mathbb{E}\left[\operatorname{Tr} P_\omega^l(I)\right]$ .We abbreviate $\chi_j := \chi_{[0,1]^d + j}$

and infer from [7] the inequality

$$\mathbb{E}\left[\operatorname{Tr} P_\omega^l(I)\right] \le e^{E_2} C_V \sum_{j\in\tilde{\Lambda}} \left\| \mathbb{E}\left[\chi_j P_\omega^l(I)\chi_j\right]\right\| \tag{6}$$

which corresponds to a partition of unity. Here the constant C_V depends only on V_0, u and f. To estimate $\mathbb{E}[\langle\phi, \chi_j P_\omega^l(I)\chi_j\phi\rangle]$ for any normalized $\phi \in L^2(\Lambda_l)$ we introduce a transformation of coordinates on the probability space Ω. For each cube $\Lambda = \Lambda_l$ denote $\Lambda^+ := \{\lambda - \gamma | \ \lambda \in \tilde{\Lambda}, \gamma \in \Gamma\}$. The operator H_ω^l depends only on the truncated random vector $(\omega_k)_{k\in\Lambda^+} \in \mathbb{R}^{\#\Lambda^+}$. On such vectors acts a block Toeplitz transformation $A_\Lambda := \{\alpha_{j-k}\}_{j,k\in\Lambda^+}$ induced by the convolution vector α. The transformation has an inverse $B_\Lambda = \{b_{k,j}\}_{k,j\in\Lambda^+} = A_\Lambda^{-1}$ which is bounded in the row-sum norm $\|B_\Lambda\| \le \frac{1}{1-\alpha^*}$. We drop now the subscript Λ and denote with $\eta := A\omega$ the vector of the transformed random coordinates. They have the common density

$$k(\eta) = k_\Lambda(\eta) = |\det B|\, F(A^{-1}\eta) \tag{7}$$

where $F(\omega) = \prod_{k\in\Lambda^+} f(\omega_k)$ is the original density of the ω_k. We calculate the potential V_ω written as a function of η (and $x \in \Lambda$):

$$V_\omega(x) = V_{B\eta}(x) = \sum_{j\in\tilde{\Lambda}} \eta_j \chi_j(x).$$

In the new representation of the potential the single site potentials are non-negative, so we can to use a spectral averaging formula [7]

$$\int_{\mathbb{R}} d\eta_j\, k(\eta)\, s(\eta) \le |I| \sup_{\eta_j} |k(\eta)|, \qquad \text{where } s(\eta) := \langle\phi, \chi_j P_{B\eta}^l(I)\chi_j\phi\rangle. \tag{8}$$

Denote $L = \#\Lambda^+$. Fubini, (8) and the fundamental theorem of calculus give

$$\int_{\mathbb{R}^L} d\eta\, k(\eta)\, s(\eta) \le |I| \int_{\mathbb{R}^{L-1}} d\eta^{\perp j} \sup_{\eta_j} |k(\eta)| \le |I| \int_{\mathbb{R}^L} d\eta\, |(\partial_j k)(\eta)|. \tag{9}$$

The last integral equals $|\det A| \int_{\mathbb{R}^L} d\omega\, |(\partial_j k)(A\omega)|$ which is bounded by $\|f'\|_{L^1} \sum_{k\in\Lambda^+} |b_{k,j}|$. The proof of the theorem is finished by the estimate

$$\mathbb{E}\left[\langle\phi, \chi_j P_\omega^l(I)\chi_j\phi\rangle\right] \le |I|\, \|f'\|_{L^1} \|B\|. \tag{10}$$

4. Discussion and Applications

4.1 Localization

The main application of Wegner estimates — apart from establishing the regularity of the IDS — is the proof of localization, i.e. the existence of dense pure point spectrum of H_ω in certain energy regions. For multidimensional Schrödinger operators on $L^d(\mathbb{R}^d)$ the multiscale analysis of J. Fröhlich and T. Spencer [13] provides a method for proving localization by induction over cubes in $\mathbb{R}^d$ of larger and larger side lengths. To start the multiscale induction one has to know certain off-diagonal decay estimates of the kernel of the resolvent of H_ω^l at an initial length scale. For non-negative u there are rather well understood sufficient conditions for this estimates. The Wegner estimate is needed, too, as an a priori estimate for the multiscale analysis, namely to control the induction step when going from one scale to another, larger one. We sketch the main estimates in the multiscale analysis under the assumptions of Theorem 2.3. For the following it is convenient to slightly change the notation and denote by $\Lambda = \Lambda_l(x) = [-l/2, l/2]^d + x$ the cube of side length l centered at x. The characteristic function of $\Lambda_l(x) - \Lambda_{l-2}(x)$, resp. $\Lambda_{l/3}(x)$, is abbreviated as χ^+, resp. χ^-. The cube Λ is called (γ, E)-*good* if the associated resolvent satisfies the following off-diagonal decay estimate in operator norm

$$\|\chi^+(H_\omega^\Lambda - E)^{-1}\chi^-\| \leq e^{-\gamma l}.$$

The initial scale estimate for the multiscale analysis is satisfied if

$$\mathbb{P}\{\forall E \in I : \Lambda_l(x) \text{ is } (\gamma_0, E)\text{-good }\} \geq 1 - l_0^{-\xi_0} \tag{11}$$

holds for $x \in \mathbb{Z}^d, l_0 \in \mathbb{N}$ sufficiently large and $\gamma_0 \geq l_0^{\beta-1}, \beta, \xi_0 > 0$. See [34] how it is derived in the situation of Theorem 2.3. It serves as the induction anchor. The geometric resolvent equation for two cubes $\Lambda \subset \Lambda', \phi \in C_c^1(\Lambda)$ and $z \in \rho(H_\omega^\Lambda) \cap \rho(H_\omega^{\Lambda'})$

$$(H^\Lambda - z)^{-1}\phi = \phi(H^{\Lambda'} - z)^{-1} + (H^\Lambda - z)^{-1}[(\nabla\phi)\nabla + \Delta\phi](H^{\Lambda'} - z)^{-1}$$

and (some weak form of) a Wegner estimate

$$\mathbb{P}\{\sigma(H_\omega^\Lambda) \cap I \neq \emptyset\} \leq C\,|I|^a \text{ vol}(\Lambda)^b$$

with $a \in]0,1], b \in [1, \infty[$ imply for $x, y \in \mathbb{Z}^d, d(x,y) \geq \operatorname{diam} \operatorname{supp} u$

$$\mathbb{P}\{\forall E \in I : \Lambda_l(x) \text{ or } \Lambda_l(x) \text{ is } (\gamma, E)\text{-good }\} \geq 1 - l^{-2\xi}$$

for some $\alpha \in]1, 2[, \gamma, \xi > 0$ and all $l_k, k \in \mathbb{N}$, where $l_{k-1}^\alpha \leq l_k \leq l_{k-1}^\alpha + 6, k \in \mathbb{N}$. This establishes the exponential decay of the resolvent on arbitrary large scales with high probability. Using a priori estimates on the (polynomial) growth of eigenfunctions [3] one concludes that they actually decay exponentially. Details on the multiscale analysis can be found, e.g., in [32] and recent developments in [14, 15].

4.2 Dependent random coupling constants

The proof of Theorem 2.1 uses the transformation of V_ω into an alloy type potential with non-negative single site potential $w \in L^p(\mathbb{R}^d)$ and (negatively) correlated coupling constants $(\eta_k)_{k\in\mathbb{Z}^d}$. Thus we can interpret it as an Wegner estimate for the Schrödinger operator $H_\eta = -\Delta + V_0 + \sum_{k\in\mathbb{Z}^d} \eta_k w(\cdot - k)$. This implies the following Wegner estimate for dependent coupling constants:

Proposition 4.1. *Let $w \geq \kappa\chi_{[0,1]}$, $\kappa > 0$ and for each $l \in \mathbb{N}$ the common density k_l of $\eta^l = (\eta_k)_{k\in\Lambda_l}$ conditioned on $(\eta_k)_{k\notin\Lambda_l}$ be of the form $|\det B_l|\, k_l(\eta) = F(B_l\,\eta^l)$, where $f \in W^{1,1}$ is a probability density, $F_l(\omega) = \prod_{k\in\Lambda_l} f(\omega_k)$ and $B_l := B_l\big((\eta_k)_{k\notin\Lambda_l}\big)$ is a sequence of invertible linear transformations with $\sup\{\|B_l\| \mid (\eta_k)_{k\notin\Lambda_l}\} \leq const\, l^q$. Then for each $E \in \mathbb{R}$ exists a constant $C = C(E)$ such that*

$$\mathbb{E}\left[\operatorname{Tr} P_\eta^l([E-\epsilon, E])\right] \leq C\,\epsilon\, l^{q+d}, \quad \forall \epsilon \geq 0. \tag{12}$$

Wegner estimates for alloy type models with dependent coupling constants have been established in [8] (cf. also [19]). The criteria for their validity are formulated in terms of the conditional densities $h_j(\eta)$ of the random variable η_j w.r.t. the remaining ones $\eta^{\perp j} = (\eta_k)_{k\in\mathbb{Z}^d\setminus j}$. One necessary condition is that the supremum $\|h_j\|_\infty$ is finite. This motivates the comparison of some properties of the common and the conditional densities of the random coupling constants in the next subsection.

4.3 Common and conditional densities

In this subsection we discuss some properties of the common and conditional densities of the random coupling constants. The proofs of Theorem 2.1 and Proposition 2.2 rely on the use of a common density of the type (7). Since one has an explicit formula for the corresponding conditional densities, it is desirable to analyze their properties. Particularly, it is of interest whether the conditional densities are bounded by a constant. If this is the case, the Wegner estimates of [8] and [19] apply by considering the indefinite potential V_ω in its representation $V_{B\eta}$ as an alloy type potential with dependent coupling constants. Lets look first at the common density k_Λ. Its supremum is easily seen to be $|\det B_\Lambda|\,\|f\|_\infty^L$, so it diverges exponentially with the volume of the cube Λ. However, this does not matter since in (9) one takes the integral instead of the supremum. We restrict ourselves now to the one dimensional $d = 1$ case and note that $\Lambda = [0, l[$ and $\Lambda^+ = \{-1, \ldots, l-1\}$. Instead of considering the conditional density $h_j(\eta)$ of η_j w.r.t *all* coupling constants we will study (for two simple examples) the conditional density $\rho_j(\eta) = \rho_j^l(\eta)$ of the variable η_j with respect to the remaining coupling constants $\eta^{\perp j} = (\eta_k)_{k\in\Lambda^+\setminus j}$ in Λ^+.

It is given by $\rho_j(\eta) = \frac{k(\eta)}{g_j(\eta)}$. Here $g_j(\eta) := g_j^l(\eta) = \int k(\eta) d\eta_j$ denotes the marginal density. The question is whether $\sup_j \rho(\eta)$ is finite. If it is finite, does the upper bound depend on l and does the bound diverge as l tends to infinity?

Example 4.2. *Let $u = \chi_{[0,1]} - \chi_{[1,2]}$, i.e. the convolution vector be $(\alpha_0, \alpha_1) = (1, -1)$. Then the assumptions of Theorem 2.1 are (just) not satisfied, since $|\alpha_1| = |\alpha_0|$. However, $A_l = A_\Lambda$ has an inverse B_l with entries $b_{j,k} = \chi_{\{j \geq k\}}$ for all $-1 \leq j, k \leq l-1$. Thus $\|B_l\| = l+1$ and (10) implies for $f \in W^{1,1}$*

$$\mathbb{E}\left[\langle\phi, \chi_j P_\omega^l(I)\chi_j\phi\rangle\right] \leq |I| \, \|f'\|_{L^1}(l+1). \tag{13}$$

This gives a Wegner estimate like (4), but with l^d replaced by $l^{2d} = l^2$. Specialize now to the density $f(x) = -4|x| + 2$ on $[-1/2, 1/2]$ and zero elsewhere. We calculate a lower bound on the supremum by evaluating the conditional density at $\eta = 0$, the unique maximum of $k(\eta)$. For $j \in \{-1, \ldots, l-1\}$ and $l - j$ even we have

$$\sup_\eta \rho_j^l(\eta) \geq \rho_j^l(0) = l - j + 1. \tag{14}$$

This bound diverges as the length l of the interval tends to infinity. Thus the mere invertibility of A is not sufficient to ensure the uniform boundedness of $\sup_\eta \rho_j^l(\eta)$. Remarkably the bound (14) has the same volume growth as the one in (13) involving the common density. If (14) is actually the exact behaviour of $\|\rho_j^l\|_\infty$ then the conditional density may be used to derive (13) and a Wegner estimate.

Example 4.3. *Consider now the density function $f = \chi_{[0,1]}$ and the single site potential $u = \chi_{[0,1]} - \alpha\chi_{[1,2]}$ with $-\alpha = \alpha_1 \in]-1, 0[$. To the corresponding alloy type model the Wegner estimate in Proposition 2.2 applies. We calculate now the supremum of the conditional densities ρ_j. The common density is given by $k(\eta) = \prod_{k=-1}^{l-1} \chi_{[0,1]}(\sum_{\nu=-1}^{k} \alpha^{k-\nu}\eta_\nu)$. For $\eta_{j+1} \in [0,1], \eta_k = 0, \forall k \neq j+1$ we have $k(\eta) = 1$. The marginal density*

$$g_j(\eta) \leq \int d\eta_j \prod_{k=j}^{j+1} \chi_{[0,1]}\left(\sum_{\nu=-1}^{k} \alpha^{k-\nu}\eta_\nu\right)$$

has for $\eta_{j+1} \in [0,1], \eta_k = 0, \forall k \notin \{j, j+1\}$ the upper bound $\int_0^1 \chi_{[0,1]}(\alpha\eta_j + \eta_{j+1}) d\eta_j \leq \alpha^{-1}(1 - \eta_{j+1})$. Particularly, $g_j(\eta) \searrow 0$ for $\eta_{j+1} \nearrow 1$ and thus

$$\|\rho_j\|_\infty = \infty.$$

So proofs of a Wegner estimate which require the conditional density to be bounded cannot be applied to this alloy type potential.

Note that we discussed only mathematical aspects of the use of the common, resp. conditional density. A different question is whether for physical models the dependence between the coupling constants is most naturally expressed in terms of the conditional or common densities, in terms of the correlation coefficients or more general mixed moments.

Acknowledgements It is a pleasure to thank T. Hupfer, W. Kirsch, V. Kostrykin, K. Veselić and S. Warzel for stimulating discussions and helpful comments. This work was partially supported by SFB 237: "Unordnung und große Fluktuationen", the Ruth-und-Gert-Massenberg-Stiftung, both Germany, and the Croatian Ministry of Science.

References

[1] P. Anderson. Absence of diffusion in certain random lattices. *Phys. Rev.*, 109:1492, 1958.

[2] J. M. Barbaroux, J. M. Combes, and P. D. Hislop. Localization near band edges for random Schrödinger operators. *Helv. Phys. Acta*, 70(1-2):16–43, 1997.

[3] Y. M. Berezanskii. Expansion in eigenfunctions of self-adjoint operators. *Am. Math. Soc, Providence (Transl. of Math. Mon., 17)*, 68.

[4] V. Bonch-Bruevich, R. Enderlein, B. Esser, R. Keiper, A. Mirnov, and I. Zyvagin. *Elektronentheorie ungeordneter Halbleiter.* VEB Deutscher Verlag der Wissenschaften, 1984. Russisches Original: Moskau, Nauka, 1981.

[5] R. Carmona and J. Lacroix. *Spectral Theory of Random Schrödinger Operators.* Birkhäuser, Boston, 1990.

[6] J. Combes, P. Hislop, and E. Mourre. Spectral averaging, perturbation of singular spectra and localization. *Trans. Am. Math. Soc.*, 348(12):4883–4894, 1996.

[7] J.-M. Combes and P. Hislop. Localization for some continuous, random Hamiltionians in d-dimensions. *J. Funct. Anal.*, 124:149–180, 1994.

[8] J. M. Combes, P. D. Hislop, and E. Mourre. Correlated Wegner inequalities for random Schrödinger operators. In *Advances in differential equations and mathematical physics (Atlanta, GA, 1997)*, pages 191–203. Amer. Math. Soc., Providence, RI, 1998.

[9] J. M. Combes, P. D. Hislop, and S. Nakamura. The L^p-theory of the spectral shift function, the Wegner estimate, and the integrated density of states for some random Schrödinger operators. *Commun. Math. Phys.*, 70(218):113–130, 2001.

[10] H. L. Cycon, R. G. Froese, W. Kirsch, and B. Simon. *Schrödinger Operators with Application to Quantum Mechanics and Global Geometry.* Text and Monographs in Physics. Springer, Berlin, 1987.

[11] T. Dorlas. *Statistical mechanics.* IOP Publishing Ltd., Bristol, 1999. Fundamentals and model solutions.

[12] A. L. Efros and B. I. Shklovski. *Electronic Properties of Doped Semi-conductors.* Springer, Berlin, 84.

[13] J. Fröhlich and T. Spencer. Absence of diffusion in the Anderson tight binding model for large disorder or low energy. *Commun. Math. Phys.*, 88:151–184, 1983.

[14] F. Germinet and A. Klein. Bootstrap multiscale analysis and localization in random media. *Comm. Math. Phys.*, 222(2):415–448, 2001.

[15] F. Germinet and A. Klein. A characterization of the Anderson metal-insulator transport transition. Preprint, http://www.ma.utexas.edu/mp_arc/, 2001.

[16] P. D. Hislop and F. Klopp. The integrated density of states for some random operators with nonsign definite potentials. http://www.ma.utexas.edu/mp_arc, preprint no. 01-139, 2001.

[17] K. Huang. *Statistical mechanics*. John Wiley & Sons Inc., New York, second edition, 1987.

[18] D. Hundertmark and W. Kirsch. Spectral theory of sparse potentials. In *Stochastic processes, physics and geometry: new interplays, I (Leipzig, 1999)*, pages 213–238. Amer. Math. Soc., Providence, RI, 2000.

[19] T. Hupfer, H. Leschke, P. Müller, and S. Warzel. The absolute continuity of the integrated density of states for magnetic Schrödinger operators with certain unbounded random potentials. *Comm. Math. Phys.*, 221(2):229–254, 2001.

[20] W. Kirsch. Random Schrödinger operators. In H. Holden and A. Jensen, editors, *Schrödinger Operators*, Lecture Notes in Physics, **345**, Berlin, 1989. Springer.

[21] W. Kirsch, P. Stollmann, and G. Stolz. Anderson localization for random Schrödinger operators with long range interactions. *Comm. Math. Phys.*, 195(3):495–507, 1998.

[22] W. Kirsch, P. Stollmann, and G. Stolz. Localization for random perturbations of periodic Schrödinger operators. *Random Oper. Stochastic Equations*, 6(3):241–268, 1998. available at http://www.ma.utexas.edu/mp_arc, preprint no. 96-409.

[23] W. Kirsch and I. Veselić. Wegner estimates for sparse and other generalized alloy type potentials. 2001, to appear in Proc. Indian Acad. Sci., Math. Sci.

[24] F. Klopp. Localization for semiclassical continuous random Schrödinger operators ii: The random displacement model. *Helv. Phys. Acta*, 66:810–841, 1993.

[25] F. Klopp. Localization for some continuous random Schrödinger operators. *Commun. Math. Phys.*, 167:553–569, 1995.

[26] F. Klopp. Internal Lifshits tails for random perturbations of periodic Schrödinger operators. *Duke Math. J.*, 98(2):335–396, 1999.

[27] V. Kostrykin and I. Veselić. Wegner estimate for indefinite Anderson potentials: some recent results and applications. Proceedings Contribution for "Renormalization Group Methods in Mathematical Sciences", July 2001, RIMS, Kyoto, to appear, www.ruhr-uni-bochum.de/mathphys/ivan.

[28] S. Kotani and B. Simon. Localization in general one-dimensional random systems II: continuum Schrödinger operators. *Commun. Math. Phys.*, 112:103–119, 1987.

[29] I. M. Lifshitz, S. A. Gredeskul, and L. A. Pastur. *Introduction to the Theory of Disordered Systems*. Wiley, New York, 1988. Russian original: Nauka, Moscow, 1982.

[30] L. A. Pastur and A. L. Figotin. *Spectra of Random and Almost-Periodic Operators*. Springer Verlag, Berlin, 1992.

[31] P. Stollmann. Wegner estimates and localization for continuum Anderson models with some singular distributions. *Arch. Math. (Basel)*, 75(4):307–311, 2000.

[32] P. Stollmann. *Caught by disorder: A Course on Bound States in Random Media*, volume 20 of *Progress in Mathematical Physics*. Birkhäuser, July 2001.

[33] G. Stolz. Non-monotonic random Schrödinger operators: the Anderson model. *J. Math. Anal. Appl.*, 248(1):173–183, 2000.

[34] I. Veselić. Wegner estimate for some indefinite Anderson-type Schrödinger operators with differentiable densities. preprint, 2000, http://www.ma.utexas.edu/mp_arc/, submitted.

[35] I. Veselić. *Indefinite Probleme bei der Anderson-Lokalisierung*. Ph.D thesis, Ruhr-Universität Bochum, 44780 Bochum, January 2001. http://www-brs.ub.ruhr-uni-bochum.de/netahtml/HSS/Diss/VeselicIvan/.

[36] F. Wegner. Bounds on the DOS in disordered systems. *Z. Phys. B*, 44:9–15, 1981.

ON PRINCIPAL EIGENVALUE OF STATIONARY DIFFUSION PROBLEM WITH NONSYMMETRIC COEFFICIENTS*

Marko Vrdoljak
Department of Mathematics, University of Zagreb,
Bijenička cesta 30, 10000 Zagreb, Croatia
marko@math.hr

Abstract

We consider the eigenvalue problem

$$\begin{cases} -\mathsf{div}\,(A\nabla u) = \lambda\rho u \\ u \in \mathrm{H}^1_0(\Omega) \end{cases}$$

where $\Omega \in \mathbf{R}^d$ is open and bounded, $\rho \in \mathrm{L}^\infty(\Omega)$ and $A \in \mathrm{L}^\infty(\Omega; \mathrm{M}_{d\times d})$ satisfying

$$A(x)\xi\cdot\xi \geq \alpha\xi\cdot\xi\,, \quad \rho(x) \geq c\,, \qquad \xi \in \mathbf{R}^d, \text{ a.e. } x \in \Omega\,,$$

for some $\alpha, c > 0$.

We show that, under appropriate conditions on smoothness of coefficients, the principal eigenvalue depends continuously on coefficients with respect to H-topology for A and L^∞ weak $*$ topology for ρ. An application of this result in optimal shape design problem of optimising the principal eigenvalue is presented.

Moreover, in the same topology for coefficients, we obtain the continuity of corresponding singular values.

Introduction

We consider the eigenvalue problem for the operator

$$\mathcal{A}u = -\mathsf{div}\,(A\nabla u)$$

acting on $\mathrm{H}^1_0(\Omega)$, where $\Omega \subseteq \mathbf{R}^d$ is a bounded domain. Matrix function A will be a measurable function on Ω taking values (almost everywhere) in the set of

* This work is supported in part by the Croatian Ministry of Science and Technology through project 037 015.

matrices described by

$$A\xi \cdot \xi \geq \alpha|\xi|^2$$
$$A\xi \cdot \xi \geq \frac{1}{\beta}|A\xi|^2 , \quad \xi \in \mathbf{R}^d .$$

The set of all such matrix functions will be denoted by $\mathcal{M}(\alpha, \beta; \Omega)$. For the study of spectrum of $\mathcal{A}$ we introduce the linear operator $\mathcal{B} : \mathrm{L}^2(\Omega) \to \mathrm{L}^2(\Omega)$ taking $\mathcal{B}f = u$, where $u \in \mathrm{H}^1_0(\Omega)$ is the solution of the problem $\mathcal{A}u = \rho f$, or in weak formulation

$$(\forall v \in \mathrm{H}^1_0(\Omega)) \int_\Omega A\nabla u \cdot \nabla v \, dx = \int_\Omega \rho f v \, dx .$$

Here, ρ is a L^∞ function, taking values in the segment $[\rho_-, \rho_+]$, with $0 < \rho_- \leq \rho_+$. This solution is unique by the Lax-Milgram lemma and satisfies

$$||u||_{\mathrm{H}^1_0(\Omega)} \leq c||f||_{\mathrm{L}^2(\Omega)} , \tag{1}$$

the constant c depending only on α, ρ_+ and Ω. Since $\mathrm{H}^1_0(\Omega)$ is compactly embedded in $\mathrm{L}^2(\Omega)$, by Rellich's theorem the operator $\mathcal{B}$ is compact. Now a classical result of operator theory states that the spectrum of $\mathcal{B}$ consists of a (complex) sequence of eigenvalues converging to zero.

The positivity of operator $\mathcal{B}$ enables us to obtain more interesting information on spectrum. For that purpose, let us introduce the cone of positive functions in $\mathrm{L}^2(\Omega)$:

$$K = \{u \in \mathrm{L}^2(\Omega) : u \geq 0 \text{ almost everywhere on } \Omega\} .$$

Now, the positivity of $\mathcal{B}$ is simply written as $\mathcal{B}K \subseteq K$.

Theorem 1. **(Krein-Rutman**) *Let K be a closed convex cone in Banach space H which is reproducing (i.e. $K - K = H$), $K \cap (-K) = \{0\}$, and $\mathcal{B} \in \mathcal{L}(H)$ compact, positive on K ($\mathcal{B}(K) \subseteq K$).*

Then the spectral radius $r(\mathcal{B})$ is equal to an eigenvalue (we call it principal) and it possesses an eigenfunction in K.

The proof can be found in [9, 11]. The principal eigenvalue for the operator $\mathcal{B}$ will be denoted by μ^1. The eigenvalue problem

$$\mathcal{B}u = \mu u \tag{2}$$

can be rewritten as

$$\begin{cases} -\mathsf{div}\,(A\nabla u) = \lambda\rho u \\ u \in \mathrm{H}^1_0(\Omega) , \end{cases} \tag{3}$$

with $\lambda = \frac{1}{\mu}$. Consequently, $\lambda^1 = \frac{1}{\mu^1}$ is the least eigenvalue (in absolute value) for operator $\mathcal{A}$.

Many authors studied the case of symmetric coefficients A. With scalar product $\langle u, v\rangle := \int_\Omega \rho uv$ the operator $\mathcal{B}$ becomes symmetric, so its spectrum consists of positive eigenvalues

$$\mu^1 > \mu^2 \geq \mu^3 \geq \cdots$$

tending to zero, while the corresponding eigenfunctions can be chosen such that they form an orthonormal basis for $\mathrm{L}^2(\Omega)$. It was shown in [10] that these eigenvalues depend continuously on coefficients ρ, A with respect to L^∞ weak $*$ topology for ρ and G-topology for A. This result heavily relies on Courant-Fisher characterisation of eigenvalues for (3)

$$\lambda^k = \frac{1}{\mu^k} = \min_{\substack{V \subset \mathrm{H}_0^1(\Omega) \\ \dim V = k}} \max_{v \in V \setminus \{0\}} \frac{\displaystyle\int_\Omega A\nabla v \cdot \nabla v \, dx}{\displaystyle\int_\Omega \rho v^2 \, dx} .$$

The continuity was necessary to get the right relaxation of some optimal shape design problems [1, 2, 8].

The aim of this paper is to examine the continuity property of principal eigenvalue in general (i.e. nonsymmetric) case. Although the main result is much more restrictive, it still allows us to consider some applications in appropriate optimal shape design problems. The second part concerns the singular values of operator $\mathcal{B}$ and their behaviour under the homogenisation process.

1. Homogenisation of eigenvalue problem

Let us now consider the continuity of the principal eigenvalue with respect to coefficients ρ and A. As mentioned in the Introduction there are very precise results under symmetry assumption on matrix function A, concerning G-convergence. The reasonable topology in our case is going to be the H-topology. For completeness we recall its definition and some main results that we shall use. Details can be found in [13, 1].

Definition 1. *A sequence* (A_n) *in* $\mathcal{M}(\alpha, \beta; \Omega)$ *H-converges to* $A_\infty \in \mathcal{M}(\alpha, \beta; \Omega)$ *if for any* $f \in \mathrm{H}^{-1}(\Omega)$ *the sequence* (u_n) *of solutions of*

$$\begin{cases} -\mathsf{div}\,(A_n \nabla u_n) = f \\ u_n \in \mathrm{H}_0^1(\Omega) \end{cases} \tag{4}$$

satisfies

$$\begin{aligned} u_n &\rightharpoonup u_\infty && \text{in } \mathrm{H}_0^1(\Omega)\,, \\ A_n \nabla u_n &\rightharpoonup A_\infty \nabla u_\infty && \text{in } \mathrm{L}^2(\Omega; \mathbf{R}^d). \end{aligned}$$

The last convergence in particular implies that u_∞ is a solution of (4) for $n = \infty$.

This convergence comes from a weak topology on the set $\mathcal{M}(\alpha, \beta; \Omega)$, which is metrisable and compact.

Theorem 2. *Let (A_n) be a sequence in $\mathcal{M}(\alpha, \beta; \Omega)$ that H-converges to A_∞ and (u_n) a sequence in $\mathrm{H}^1_{\mathrm{loc}}(\Omega)$ such that*

$$\begin{array}{rcll} u_n & \rightharpoonup & u_\infty & \textit{in } \mathrm{H}^1_{\mathrm{loc}}(\Omega) \\ \mathsf{div}\,(A_n \nabla u_n) = f_n & \longrightarrow & f_\infty & \textit{in } \mathrm{H}^{-1}_{\mathrm{loc}}(\Omega)\,. \end{array}$$

Then $A_n \nabla u_n \rightharpoonup A_\infty \nabla u_\infty$ in $\mathrm{L}^2_{\mathrm{loc}}(\Omega; \mathbf{R}^d)$.

Let us assume that

$$\rho_n \overset{*}{\rightharpoonup} \rho_\infty \qquad \text{and} \qquad A_n \overset{H}{\rightharpoonup} A_\infty\,.$$

For $n \in \mathbf{N}$ we shall denote by λ^1_n the principal eigenvalue for the eigenvalue problem

$$\left\{ \begin{array}{l} -\mathsf{div}\,(A_n \nabla u) = \lambda \rho_n u \\ u \in \mathrm{H}^1_0(\Omega)\,, \end{array} \right. \tag{5}$$

with corresponding eigenfunction u^1_n. Similarly to the symmetric case, u^1_n is chosen so that

$$\int_\Omega \rho_n (u^1_n)^2\, dx = 1\,. \tag{6}$$

Furthermore, suppose that the sequence (λ^1_n) is bounded.

Then (u^1_n) is a bounded sequence in $\mathrm{H}^1_0(\Omega)$:

$$\alpha \|u^1_n\|_{\mathrm{H}^1_0(\Omega)} \le \int_\Omega A_n \nabla u^1_n \cdot \nabla u^1_n\, dx = \lambda^1_n \int_\Omega \rho_n (u^1_n)^2 = \lambda^1_n\,,$$

so it has a weakly converging subsequence, as well as the sequence of eigenvalues (let us take this subsequence from the beginning):

$$\begin{array}{ll} u^1_n \rightharpoonup u_* & \text{in } \mathrm{H}^1_0(\Omega) \\ \lambda^1_n \longrightarrow \lambda_*\,. & \end{array}$$

Passing to the limit in the equations satisfied by λ^1_n and u^1_n (using Theorem 2) we conclude that

$$-\mathsf{div}\,(A_\infty \nabla u_*) = \lambda_* \rho_\infty u_*\,.$$

Thanks to the fact that $u^1_n \longrightarrow u_*$ in $\mathrm{L}^2(\Omega)$, u_* is nonnegative and not identically equal to zero, since we can pass to the limit in (6):

$$\int_\Omega \rho_\infty (u_*)^2\, dx = 1\,.$$

This implies that λ_* is an eigenvalue and u_* the corresponding eigenfunction for (5) with $n = \infty$.

Unfortunately, we cannot conclude that λ_* is equal to the principal eigenvalue λ^1_∞ in general. The case of smooth coefficients and smooth boundary $\partial\Omega$ is much simpler, since we can use the strong Krein-Rutman theorem that establishes the uniqueness of the positive eigenfunction.

Theorem 3. (Krein-Rutman, strong version) *Let K be a closed, reproducing cone with nonempty interior in Banach space H such that $K \cap (-K) = \{0\}$, and $\mathcal{B} \in \mathcal{L}(H)$ compact, strongly positive on K ($\mathcal{B}(K \setminus \{0\}) \subseteq \mathrm{Int}K$).*

Then $\mathcal{B}$ has unique normalised eigenvector in $K \setminus \{0\}$. Moreover, corresponding (principal) eigenvalue is simple.

In [4] Theorem 3 is applied to problem under consideration (with smoother coefficients, i.e. in Hölder spaces). Combining that results with the previous calculation we obtain the following theorem, while in the next section we give its application to the optimal shape design problem.

Theorem 4. *Let*

$$\begin{aligned} \rho_n &\xrightarrow{\ *\ } \rho_\infty \ \textit{in}\ \mathrm{L}^\infty(\Omega; [\rho_-, \rho_+]) \\ A_n &\xrightarrow{\ H\ } A_\infty \ \textit{in}\ \mathcal{M}(\alpha, \beta; \Omega)\,, \end{aligned}$$

with additional conditions on smoothness of coefficients: for some $0 < \gamma < 1$

$$A_n \in \mathrm{C}^{1+\gamma}(\overline{\Omega})\,, \ \rho_n \in \mathrm{C}^{\gamma}(\overline{\Omega})\,.$$

Furthermore, suppose that the boundary $\partial\Omega$ of Ω is $(d-1)$-dimensional $\mathrm{C}^{2+\sigma}$-manifold for some $\sigma \in \langle 0, 1\rangle$, such that Ω lies locally on one side of $\partial\Omega$.

If we denote by λ^1_n the principal eigenvalue of (5), for $n \in \mathbf{N} \cup \{\infty\}$, then we have

$$\lambda^1_n \longrightarrow \lambda^1_\infty\,.$$

Remark. The boundedness of (λ^1_n) we assumed above follows from the assumptions of Theorem 4, as an application of the maximum principle. This was proved in [6] (Lemma 1.1), for density ρ in (3) equal to 1, but the generalisation to our case is straightforward.

2. Optimal design for principal eigenvalue optimisation

We shall consider an optimal shape design problem of maximising (or minimising) the principal eigenvalue for (3) with coefficients given by

$$A = \chi C + (1 - \chi) D\,, \quad \rho = 1 \text{ (for simpicity)}\,,$$

where χ runs through the set of all measurable characteristic functions on Ω. Here, C and D are constant nonsymmetric matrices. The amount of the first material will also be constrained: $\int_\Omega \chi \, dx \le V$. For given $\chi \in \mathrm{L}^\infty(\Omega; \{0,1\})$ the corresponding principal eigenvalue of (3) will be denoted by $\lambda^1(\chi)$.

Therefore, the problem of optimal shape design can be rewritten as:

$$\begin{aligned} &\lambda^1(\chi) \longrightarrow \max \quad (\text{or } \min) \\ &\chi \in \mathrm{L}^\infty(\Omega; \{0,1\}), \ \int_\Omega \chi \, dx \le V . \end{aligned} \tag{7}$$

Unfortunately, we cannot justify that problem (7) has a solution. This situation is classical for optimal shape design problems. There are also many counterexamples to the existence of optimal designs in literature (see [1]). The first step in dealing with such problems is to choose an appropriate relaxation, and the homogenisation method showed to be the natural approach [12, 5]. However, since this method also has some disadvantages in numerical treatment, the fictitious material method was studied in [3]. Moreover, some modifications were done in [7], which gave similar numerical results as relaxation by the homogenisation method. We shall apply the method presented in [7] to our situation.

We consider the case $D = 0$. Therefore, in the original problem we are looking for a distribution of the given amount of some material over set Ω such that the corresponding principal eigenvalue is maximal or minimal.

The set of admissible controls is

$$\mathcal{H} = \{\rho \in \mathrm{L}^\infty(\mathbf{R}^d) : 0 < \rho_- \le \rho \le 1 \text{ a.e. on } \Omega , \int_{\mathbf{R}^d} \rho \, dx \le V\} .$$

Function ρ is a density function, with value 1 representing the material and small parameter ρ_- the void. The optimisation problem reads:

$$\begin{aligned} &\lambda^1(\rho) \longrightarrow \max \\ &\rho \in \mathcal{H} , \end{aligned} \tag{8}$$

where $\lambda^1(\rho)$ is the principal eigenvalue for

$$\begin{cases} -\mathsf{div}\,((F * \rho)\, C \nabla u) = \lambda u \\ u \in \mathrm{H}^1_0(\Omega) . \end{cases} \tag{9}$$

The function F is a mollifier satisfying (B_R denotes the open ball with radius R and centre at origin, R being a small parameter):

$$\begin{aligned} &F \in \mathrm{C}^2(\mathbf{R}^d) , \ \mathrm{supp} F \subseteq B_R \\ &F \ge 0 , \quad \int_{B_R} F \, dx = 1 . \end{aligned}$$

Theorem 5. *Problem (8) has a solution.*

Proof. Firstly, we notice that the set $\{\lambda^1(\rho), \rho \in \mathcal{H}\}$ is bounded ([6], Lemma 1.1). Let (ρ_n) be a maximising sequence, i.e.

$$\lambda_n^1 := \lambda_n^1(\rho_n) \longrightarrow \lambda^* = \sup_{\rho\in\mathcal{H}} \lambda^1(\rho) < +\infty\,.$$

Passing to a subsequence we have

$$\rho_n^1 \rightharpoonup \rho \quad \text{in } \mathrm{L}^1(\mathbf{R}^d)\,, \tag{10}$$

with $\rho \in \mathcal{H}$. Now, it is obvious that $(F*\rho_n)C \xrightarrow{H} (F*\rho)C$ (strong convergence in any L^p space implies H-convergence), so by Theorem 4 we have $\lambda^* = \lambda^1(\rho)$.

3. Singular values

It is easy to calculate the adjoint of operator $\mathcal{B} : \mathrm{L}^2(\Omega) \to \mathrm{L}^2(\Omega)$. If we take the scalar product in $\mathrm{L}^2(\Omega)$ as $\langle u, v\rangle := \int_\Omega \rho uv$, the adjoint $\mathcal{B}^* : \mathrm{L}^2(\Omega) \to \mathrm{L}^2(\Omega)$ maps $f \in \mathrm{L}^2(\Omega)$ to the solution of the boundary value problem:

$$\begin{cases} -\mathsf{div}\,(A^\tau \nabla u) = \rho f \\ u \in \mathrm{H}_0^1(\Omega)\,. \end{cases}$$

Since $\mathcal{B}$ is compact, linear operator $\mathcal{B}^*\mathcal{B} : \mathrm{L}^2(\Omega) \to \mathrm{L}^2(\Omega)$ is also compact. Moreover, it is positive and self adjoint. Let us denote its positive, self-adjoint square root by $|\mathcal{B}|$. The eigenvalues of $|\mathcal{B}|$ are called the singular values of operator $\mathcal{B}$. They will be denoted by $\sigma^1, \sigma^2, \ldots$, being arranged so that $\sigma^1 \geq \sigma^2 \geq \cdots$ and repeated according to geometric multiplicity. Corresponding eigenfunctions φ^k, $k \in \mathbf{N}$, can be chosen such that they form an orthonormal basis for $\mathrm{L}^2(\Omega)$.

Defining $\psi^k = \frac{1}{\sigma^k}\mathcal{B}\varphi^k$, $k \in \mathbf{N}$ the following formulas are valid

$$\begin{aligned} \mathcal{B}u &= \sum_{k\in\mathbf{N}} \sigma^k \langle u, \varphi^k\rangle \psi^k\,, \quad u \in \mathrm{L}^2(\Omega) \\ \mathcal{B}^*v &= \sum_{k\in\mathbf{N}} \sigma^k \langle v, \psi^k\rangle \varphi^k\,, \quad v \in \mathrm{L}^2(\Omega)\,, \end{aligned}$$

and they are of interest in applications.

A natural question arises: Do the singular values depend continuously on coefficients ρ and A?

In the case of symmetric matrix function A the operator $\mathcal{B}$ is self-adjoint so it coincides with $|\mathcal{B}|$, and singular values and eigenvalues are the same. The answer to the above question is trivial if we consider weak $*$ topology for ρ and G-topology for A [10].

For the nonsymmetric case, let us take two sequences of functions: A_n in $\mathcal{M}(\alpha, \beta; \Omega)$ and ρ_n in $\mathrm{L}^\infty(\Omega; [\rho_-, \rho_+])$, $n \in \mathbf{N} \cup \{\infty\}$. For each n, we define an operator $\mathcal{B}_n$ as before, and denote $\mathcal{B}_n^*\mathcal{B}_n$ by $\mathcal{F}_n : H_n \to H_n$, where H_n is the Hilbert space $\mathrm{L}^2(\Omega)$ with scalar product

$$\langle u, v\rangle_n = \int_\Omega \rho_n uv\, dx\,.$$

The corresponding norm will be denoted by $\|\cdot\|_n$. The sequence of singular values for operator $\mathcal{B}_n$ is

$$\sigma_n^1 \geq \sigma_n^2 \geq \sigma_n^3 \geq \cdots,$$

with the corresponding sequence of singular functions (φ_n^k), $k \in \mathbf{N}$, forming an orthonormal basis for H_n.

Theorem 6. *Let*

$$\begin{aligned} \rho_n &\xrightarrow{\ *\ } \rho_\infty \text{ in } \mathrm{L}^\infty(\Omega; [\rho_-, \rho_+]) \\ A_n &\xrightarrow{\ H\ } A_\infty \text{ in } \mathcal{M}(\alpha, \beta; \Omega)\,. \end{aligned}$$

Then

$$\sigma_n^k \longrightarrow \sigma_\infty^k\,.$$

If the multiplicity of the singular value σ_∞^k is equal to s:

$$\sigma_\infty^{k-1} > \sigma_\infty^k = \sigma_\infty^{k+1} = \cdots = \sigma_\infty^{k+s-1} > \sigma_\infty^{k+s}$$

(with $\sigma_\infty^0 = +\infty$), then there exists a sequence $\overline{\varphi}_n \longrightarrow \varphi_\infty^k$ in $\mathrm{L}^2(\Omega)$, as $n \to \infty$, where $\overline{\varphi}_n$ is a linear combination of singular functions corresponding to singular values $\sigma_n^k, \sigma_n^{k+1}, \ldots, \sigma_n^{k+s-1}$.

Proof. For the proof we follow the approach presented in [10], Chapter 11. The result will be obtained as a consequence of Lemma 11.3 and Theorem 11.5 there, if we verify the following properties of operators $\mathcal{F}_n$.

I. If $u_n, v_n \in \mathrm{L}^2(\Omega)$, $n \in \mathbf{N} \cup \{\infty\}$,

$$\lim_n \|u_n - u_\infty\|_n = 0 \quad \text{and} \quad \lim_n \|v_n - v_\infty\|_n = 0\,,$$

then

$$\lim_n \langle u_n, v_n\rangle_n = \langle u_\infty, v_\infty\rangle_\infty\,.$$

II. The operators $\mathcal{F}_n$, $n \in \mathbf{N} \cup \{\infty\}$, are positive, self-adjoint and the norms $\|\mathcal{F}_n\|_{\mathcal{L}(H_n)}$ are bounded by a constant independent of n. Their eigenvalues are exactly the squares of singular values for operators $\mathcal{B}_n$.

III. If $f_n \in \mathrm{L}^2(\Omega)$, $n \in \mathbf{N} \cup \{\infty\}$ and

$$\lim_n \|f_n - f_\infty\|_n = 0\,,$$

then

$$\lim_n \|\mathcal{F}_n f_n - \mathcal{F}_\infty f_\infty\|_n = 0\,.$$

IV. For any sequence (f_n), $f_n \in \mathrm{L}^2(\Omega)$, $n \in \mathbf{N} \cup \{\infty\}$, such that

$$\sup_n \|f_n\|_n < \infty\,,$$

there exists a subsequence $(f_{n'})$ and a vector $f \in \mathrm{L}^2(\Omega)$ such that

$$\lim_n \|\mathcal{F}_{n'} f_{n'} - \mathcal{F}_\infty f\|_n = 0\,.$$

Statement II follows directly. For the proof of statement I it is sufficient to notice that for any $u_n \in \mathrm{L}^2(\Omega)$, $n \in \mathbf{N} \cup \{\infty\}$ the following holds:

$$\lim_n \|u_n - u_\infty\|_n = 0 \quad \Longleftrightarrow \quad u_n \longrightarrow u_\infty \text{ in } \mathrm{L}^2(\Omega)\,.$$

Let us prove III. For given sequence (f_n) converging in $\mathrm{L}^2(\Omega)$ to f_∞, we have

$$\rho_n f_n \rightharpoonup \rho_\infty f_\infty \text{ in } \mathrm{L}^2(\Omega) \quad (\text{or strongly in } \mathrm{H}^{-1}(\Omega))\,,$$

so Theorem 2 implies $\mathcal{B}_n f_n \rightharpoonup \mathcal{B}_\infty f_\infty$ in $\mathrm{H}^1_0(\Omega)$, or strongly in $\mathrm{L}^2(\Omega)$. Now, an another application of the same theorem (because $A_n^\tau \overset{H}{\rightharpoonup} A_\infty^\tau$, v. [1]) implies

$$\mathcal{F}_n f_n \rightharpoonup \mathcal{F}_\infty f_\infty \text{ in } \mathrm{H}^1_0(\Omega)\,,$$

or strongly in $\mathrm{L}^2(\Omega)$.

The statement IV follows analogously if we take a subsequence $(f_{n'})$ such that $\rho_{n'} f_{n'}$ converges weakly in $\mathrm{L}^2(\Omega)$.

Acknowledgments

I would like to thank Nenad Antonić for many useful discussions and helpful advices during the preparation of this work.

References

[1] G. Allaire: Shape optimization by the homogenisation method, Springer-Verlag, New York, 2002

[2] G. Allaire, S. Aubry, F. Jouve: Eigenfrequency optimization in optimal design, Comp. Meth. App. Mech. Engrg. **190** (2001) 3565–3579

[3] G. Allaire, E. Bonnetier, G. Francfort, F. Jouve: Shape optimization by the homogenization method, Numerische Mathematik **76** (1997) 27–68

[4] H. Amann: Fixed point equations and nonlinear eigenvalue problems in ordered Banach spaces, Siam Review **18** (1976) 620–709

[5] N. Antonić, M. Vrdoljak: Optimal design and hyperbolic problems, Mathematical Communications **4** (1999) 121–129

[6] H. Berestycki, L. Nirenberg, S.R.S. Varadhan: The principal eigenvalue and maximum principle for second-order elliptic operators in general domains, Communications on Pure and Applied Mathematics **47** (1994) 47-92

[7] B. Bourdin: Filters in topology optimization, International Journal for Numerical Methods in Engineering **50** (2001) 2143–2158

[8] S.J. Cox, R. Lipton: Extremal eigenvalue problems for two-phase conductors, Archive for Rational Mechanics and Analysis **136** (1996) 101–117

[9] R. Dautray, J.-L. Lions: Mathematical analysis and numerical methods for science and technology, Vol. III, Springer-Verlag, 1992.

[10] V.V. Jikov, S.M. Kozlov, O.A. Oleinik: Homogenization of differential operators and integral functionals, Springer-Verlag, Berlin, 1994.

[11] M.G. Krein, M.A. Rutman: Linear operators leaving invariant a cone in a Banach space, AMS Translations **10** (1962) 199–325

[12] L. Tartar: An introduction to the homogenization method in optimal design, CIME Summer Course, Troia June 1998.

[13] L. Tartar: Homogenization, compensated compactness, H-measures, unpublished lecture notes

[14] Topics in the mathematical modelling of composite materials, A. Cherkaev, R. Kohn (eds.), Birkhäuser, 1997.

QUALITATIVE ANALYSIS OF SOME SOLUTIONS OF QUASILINEAR SYSTEM OF DIFFERENTIAL EQUATIONS

Božo Vrdoljak
Faculty of Civil Engineering, University of Split
Matice hrvatske 15, 21000 Split, Croatia

Alma Omerspahić
Faculty of Mechanical Engineering Sarajevo, University of Sarajevo
Vilsonovo šetalište 9, 71000 Sarajevo, Bosnia and Herzegovina

Abstract In this paper we study the existence and behaviour of some solutions of the system of quasilinear differential equations. The obtained results contain an answer to the question on approximation as well as stability of solutions whose existence is established. The errors of the approximation are defined by the functions that can be sufficiently small. The qualitative analysis theory of differential equations and the topological retraction method are used [6].

Keywords: quasilinear differential equation, behaviour of solutions, approximation of solutions.

1. Introduction

Let us consider the system of quasilinear differential equations

$$\dot{x} = A(x,t)\,x + F(x,t), \tag{1}$$

where $x\,(t) = (x_1\,(t)\,, \ldots, x_n\,(t))^{\tau}$, $n \geq 2\,; t \in I = \langle a, \infty \rangle$, $a \in \mathbb{R}\,; D \subset \mathbb{R}^n$ is open set, $\Omega = D \times I$, $A(x,t) = (a_{ij}\,(x,t))_{n \times n}$ is the matrix-function with elements $a_{ij} \in C\,(\Omega, \mathbb{R})$, $F(x,t) = (f_1\,(x,t)\,, \ldots, f_n\,(x,t))^{\tau}$ is the vector-function with elements $f_i \in C\,(\Omega, \mathbb{R})$. Moreover, $A(x,t)$ and $F(x,t)$ satisfy sufficient conditions for existence and uniqueness of solution of any Cauchy's problem for system (1) in Ω.

Let

$$\Gamma = \{(x,t) \in \Omega : x = \varphi\,(t)\,,\ t \in I\}\,, \tag{2}$$

where $\varphi(t) = (\varphi_1(t), \ldots, \varphi_n(t))$, $\varphi_i(t) \in C^1(I, \mathbb{R})$, is given function. We shall consider the behaviour of integral curves $(x(t), t), t \in I$, of system (1) with respect to the set

$$\omega = \left\{ (x,t) \in \Omega : \sum_{i=1}^{n} (x_i - \varphi_i(t))^2 < r^2(t) \right\}, \tag{3}$$

where $r \in C^1(I, \mathbb{R}^+)$. The boundary surface of the set ω with respect to the set Ω is

$$W = \left\{ (x,t) \in Cl\omega \cap \Omega : B(x,t) := \sum_{i=1}^{n} (x_i - \varphi_i(t))^2 - r^2(t) = 0 \right\}.$$

Let us denote the tangent vector field to an integral curve $(x(t), t)$, $t \in I$, of (1) by T. The vector ∇B is the external normal on surface W. We have

$$\begin{aligned} T(x,t) &= \left(\sum_{j=1}^{n} a_{1j}x_j + f_1, \ldots, \sum_{j=1}^{n} a_{ij}x_j + f_i, \ldots, \sum_{j=1}^{n} a_{nj}x_j + f_n, 1 \right), \\ \frac{1}{2}\nabla B(x,t) &= \left(x_1 - \varphi_1, \ldots, x_i - \varphi_i, \ldots, x_n - \varphi_n, -\sum_{i=1}^{n} (x_i - \varphi_i)\varphi_i' - rr' \right). \end{aligned}$$

By means of the scalar product $P(x,t) = \left(\frac{1}{2}\nabla B(x,t), T(x,t)\right)$ on W, we shall establish the behaviour of integral curve of (1) with respect to the set ω.

The results of this paper are based on the Lemmas 1 and 2 in [5]. According to these Lemmas the following statement is valid: If the scalar product $P(x,t) < 0$ on W, then every solution of (1), passing through some $(x_0, t_0) \in \omega$, belongs to the set ω (its graph belongs to ω) $\forall t \geq t_0$. If $P(x,t) > 0$ is satisfied on W, then the system (1) has at least one solution that belongs to the set ω ($\forall t \in I$).

We can note that, in all results stated bellow that establish the existence of n-parameter family of solutions belonging to ω (Theorems 1–5, case (a)), the conditions for uniqueness of solutions of (1) can be omitted.

2. The main results

Let

$$\begin{aligned} A_i(x,t) &:= \sum_{j=1(j\neq i)}^{n} \frac{1}{2} |a_{ij}(x,t) + a_{ji}(x,t)|, \quad X_i := x_i - \varphi_i(t), \\ \Phi_i(x,t) &:= \sum_{j=1}^{n} a_{ij}(x,t)\varphi_j(t) + f_i(x,t) - \varphi_i'(t), \quad i = 1, \ldots, n. \end{aligned}$$

Theorem 1. *Let the curve* Γ *be t-axis* ($\varphi(t) = 0$) *and* $g \in C(\Omega)$.

(*a*) *If the conditions*

$$\begin{aligned}\sum_{i=1}^{n} x_i f_i(x,t) &\leqslant g(x,t)\, r^2(t), \\ A_i(x,t) &< -a_{ii}(x,t) + \frac{r'(t)}{r(t)} - g(x,t), \ i = 1,\dots,n\end{aligned}$$

are satisfied on W, *then every solution of* (1), *passing through some* $(x_0, t_0) \in \omega$, *belongs to* ω *for* $t \geq t_0$.

(*b*) *If the conditions*

$$\begin{aligned}\sum_{i=1}^{n} x_i f_i(x,t) &\geqslant g(x,t)\, r^2(t), \\ A_i(x,t) &< a_{ii}(x,t) - \frac{r'(t)}{r(t)} + g(x,t), \ i = 1,\dots,n\end{aligned}$$

are satisfied on W, *then the system* (1) *has at least one solution that belongs to* ω ($\forall t \in I$).

Proof. For the scalar product P we have

$$\begin{aligned}P(x,t) &= \sum_{j=1}^{n} a_{1j}x_j x_1 + f_1 x_1 + \cdots + \sum_{j=1}^{n} a_{nj} x_j x_n + f_n x_n - rr' \\ &= \sum_{i=1}^{n} a_{ii} x_i^2 + \sum_{i=1}^{n-1}\sum_{j=i+1}^{n} (a_{ij} + a_{ji})\, x_i x_j + \sum_{i=1}^{n} f_i x_i - rr'. \quad (4)\end{aligned}$$

Now, according to the conditions of the Theorem, the following estimates for P on W are valid in the cases (a) and (b), respectively:

$$\begin{aligned}P(x,t) &\leq \sum_{i=1}^{n} a_{ii} x_i^2 + \sum_{i=1}^{n-1}\sum_{j=i+1}^{n} \frac{1}{2}\left|a_{ij} + a_{ji}\right| \left(x_i^2 + x_j^2\right) + \sum_{i=1}^{n} f_i x_i - rr' \\ &= \sum_{i=1}^{n} (a_{ii} + A_i)\, x_i^2 + \sum_{i=1}^{n} f_i x_i - rr' \qquad (5) \\ &< \sum_{i=1}^{n} \left(\frac{r'}{r} - g\right) x_i^2 + \sum_{i=1}^{n} f_i x_i - rr' = -gr^2 + \sum_{i=1}^{n} f_i x_i \leqslant 0,\end{aligned}$$

$$\begin{aligned} P(x,t) &\geq \sum_{i=1}^{n} a_{ii}x_i^2 - \sum_{i=1}^{n-1}\sum_{j=i+1}^{n} \frac{1}{2}\left|a_{ij}+a_{ji}\right|\left(x_i^2+x_j^2\right) + \sum_{i=1}^{n} f_i x_i - rr' \\ &= \sum_{i=1}^{n}\left(a_{ii}-A_i\right)x_i^2 + \sum_{i=1}^{n} f_i x_i - rr' \qquad (6) \\ &> \sum_{i=1}^{n}\left(\frac{r'}{r}-g\right)x_i^2 + \sum_{i=1}^{n} f_i x_i - rr' = -gr^2 + \sum_{i=1}^{n} f_i x_i \geqslant 0. \end{aligned}$$

This, according to the Lemmas 1 and 2 in [5], confirms the statements of the Theorem. □

Theorem 2. *Let the curve* Γ *be t-axis.*

(a) If the conditions

$$\sum_{i=1}^{n} f_i(x,t)\,x_i < r(t)\,r'(t)\,,\quad A_i(x,t) \leqslant -a_{ii}(x,t)\,,\ i=1,\ldots,n \qquad (7)$$

are satisfied on W*, then every solution of* (1)*, passing through some* $(x_0,t_0)\in\omega$*, belongs to* ω *for* $t \geq t_0$*.*

(b) If the conditions

$$\sum_{i=1}^{n} f_i(x,t)\,x_i > r(t)\,r'(t)\,,\quad A_i(x,t) \leqslant a_{ii}(x,t)\,,\ i=1,\ldots,n \qquad (8)$$

are satisfied on W*, then the system* (1) *has at least one solution that belongs to* ω $(\forall t \in I)$*.*

Proof. The scalar product P is defined by (4). Further, for the scalar product P, in case of (a) it is valid (5) and according to (7) we have $P(x,t) < 0$ on W. In case of (b) it is valid (6) and according to (8) we have $P(x,t) > 0$ on W. That means that the Theorem 2 holds true. □

Theorem 3. *Let* Γ *is any curve in* Ω *and* $u \in C(\Omega)$.

(a) If the conditions

$$\begin{aligned} \sum_{i=1}^{n}\left|\Phi_i(x,t)\right| &\leqslant u(x,t)\,r(t) + r'(t)\,, \\ A_i(x,t) &< -a_{ii}(x,t) - u(x,t)\,,\ i=1,\ldots,n \end{aligned}$$

are satisfied on W*, then every solution of* (1)*, passing through some* $(x_0,t_0)\in\omega$*, belongs to* ω *for* $t \geq t_0$*.*

(b) If the conditions

$$\begin{aligned}\sum_{i=1}^{n}\left|\Phi_i(x,t)\right| &\leqslant u(x,t)\,r(t)-r'(t),\\ A_i(x,t) &< a_{ii}(x,t)-u(x,t),\ i=1,\dots,n\end{aligned}$$

are satisfied on W, then the system (1) *has at least one solution that belongs to ω ($\forall t\in I$).*

Proof. Here we have:

$$P(x,t)=\sum_{i=1}^{n}a_{ii}X_i^2+\sum_{i=1}^{n-1}\sum_{j=i+1}^{n}\left(a_{ij}+a_{ji}\right)X_iX_j+\sum_{i=1}^{n}X_i\Phi_i-rr'.$$

Using the assumptions of the Theorem, it is sufficient to note that the following estimates for P on W are valid in the cases (a) and (b), respectively:

$$\begin{aligned}P(x,t) &\leqslant \sum_{i=1}^{n}\left(a_{ii}+A_i\right)X_i^2+r\sum_{i=1}^{n}\left|\Phi_i\right|-rr'\\ &< -ur^2+r\left(ur+r'\right)-rr'=0,\\ P(x,t) &\geqslant \sum_{i=1}^{n}\left(a_{ii}-A_i\right)X_i^2-r\sum_{i=1}^{n}\left|\Phi_i\right|-rr'\\ &> ur^2+r\left(-ur+r'\right)-rr'=0.\end{aligned}$$

□

Theorem 4. *Let Γ be any curve in Ω, the matrix A have the properties*

$$\begin{aligned}a_{ii}(x,t) &= a(x,t)+d_{ii}(x,t),\ i=1,\dots,n \quad \textit{and let}\\ A_{ii}(x,t) &:= d_{ii}(x,t),\\ A_{ij}(x,t) &= A_{ji}(x,t):=\frac{1}{2}\left[a_{ij}(x,t)+a_{ji}(x,t)\right],\ i\neq j.\end{aligned}$$

(a) If the conditions

$$\begin{aligned}&(-1)^k\det\begin{pmatrix}A_{11}(x,t) & \cdots & A_{1k}(x,t)\\ \cdots & \cdots & \cdots\\ A_{k1}(x,t) & \cdots & A_{kk}(x,t)\end{pmatrix}>0,\ k=1,\dots,n,\\ &\sum_{i=1}^{n}\left|\Phi_i(x,t)\right|\leqslant -a(x,t)\,r(t)+r'(t)\end{aligned}\tag{9}$$

are satisfied on W, then every solution of (1)*, passing through some $(x_0,t_0)\in\omega$, belongs to ω for $t\geq t_0$.*

(b) If the conditions

$$\det\begin{pmatrix} A_{11}(x,t) & \cdots & A_{1k}(x,t) \\ \cdots & \cdots & \cdots \\ A_{k1}(x,t) & \cdots & A_{kk}(x,t) \end{pmatrix} > 0,\ k = 1,\ldots,n,$$
$$\sum_{i=1}^{n} |\Phi_i(x,t)| \leqslant a(x,t)\, r(t) - r'(t), \tag{10}$$

are satisfied on W, then the system (1) has at least one solution that belongs to ω ($\forall t \in I$).

Proof. For the scalar product P we have:

$$\begin{aligned} P(x,t) &= \sum_{i=1}^{n} a_{ii} X_i^2 + \sum_{i=1}^{n-1}\sum_{j=i+1}^{n} (a_{ij} + a_{ji}) X_i X_j + \sum_{i=1}^{n} X_i \Phi_i - rr' \\ &= ar^2 + \sum_{i=1}^{n} d_{ii} X_i^2 + \sum_{i=1}^{n-1}\sum_{j=i+1}^{n} (a_{ij} + a_{ji}) X_i X_j + \sum_{i=1}^{n} X_i \Phi_i - rr' \\ &= ar^2 + \sum_{i,j=1}^{n} A_{ij} X_i X_j + \sum_{i=1}^{n} X_i \Phi_i - rr'. \end{aligned}$$

According to the conditions (9) and (10) the following estimates for $P(x,t)$ on W are valid in the cases (a) and (b), respectively:

$$\begin{aligned} P(x,t) &< ar^2 + r\sum_{i=1}^{n} |\Phi_i| - rr' \leqslant 0, \\ P(x,t) &> ar^2 - r\sum_{i=1}^{n} |\Phi_i| - rr' \geqslant 0. \end{aligned}$$

According to Lemmas 1 and 2 in [5], the above estimates for $P(x,t)$ imply the statement of the Theorem. □

Now let us consider the systems

$$\dot{x} = C(t)\, x + G(t) + F(x,t), \tag{11}$$
$$\dot{x} = C(t)\, x + G(t), \tag{12}$$

where the matrix-function $C(t) = (c_{ij}(t))_{i,j=1,\ldots,n}$ is real and continuous on interval I, the vector-function $G(t) = (g_1(t),\ldots,g_n(t))^{\tau}$ is real and continuous on interval I. The vector-function $F(x,t)$ has the same properties as in the case of system (1).

Theorem 5. *Let Γ be an integral curve of the system (12) and $u \in C(I)$.*

(a) If the conditions

$$\begin{gathered}\sum_{i=1}^{n} X_i f_i(x,t) < r(t)\, r'(t) \quad \textit{and} \\ \sum_{j=1 (j\neq i)}^{n} |c_{ij}(t) + c_{ji}(t)| \leqslant -2c_{ii}(t),\ i = 1,\ldots,n \quad \textit{or}\end{gathered} \tag{13}$$

$$\begin{gathered}\sum_{i=1}^{n} |f_i(x,t)| < u(t)\, r(t) + r'(t) \quad \textit{and} \\ \sum_{j=1 (j\neq i)}^{n} |c_{ij}(t) + c_{ji}(t)| \leqslant -2c_{ii}(t) - 2u(t),\ i = 1,\ldots,n\end{gathered} \tag{14}$$

are satisfied on W, then every solution of (11), *passing through some* $(x_0, t_0) \in \omega$, *belongs to* ω *for* $t \geq t_0$.

(b) If the conditions

$$\begin{gathered}\sum_{i=1}^{n} X_i f_i(x,t) > r(t)\, r'(t) \quad \textit{and} \\ \sum_{j=1 (j\neq i)}^{n} |c_{ij}(t) + c_{ji}(t)| \leqslant 2c_{ii}(t),\ i = 1,\ldots,n \quad \textit{or}\end{gathered} \tag{15}$$

$$\begin{gathered}\sum_{i=1}^{n} |f_i(x,t)| < u(t)\, r(t) - r'(t) \quad \textit{and} \\ \sum_{j=1 (j\neq i)}^{n} |c_{ij}(t) + c_{ji}(t)| \leqslant 2c_{ii}(t) - 2u(t),\ i = 1,\ldots,n\end{gathered} \tag{16}$$

are satisfied on W, then the system (11) *has at least one solution that belongs to* ω *($\forall t \in I$).*

Proof. For the scalar product $P(x,t)$ we have

$$\begin{aligned} P(x,t) &= \sum_{i=1}^{n} \left[X_i \left(\sum_{j=1}^{n} c_{ij} x_j + g_i + f_i \right) \right] - \sum_{i=1}^{n} X_i \varphi_i' - r r' \\ &= \sum_{i=1}^{n} c_{ii} X_i^2 + \sum_{i=1}^{n-1} \sum_{j=i+1}^{n} (c_{ij} + c_{ji}) X_i X_j \\ &\quad + \sum_{i=1}^{n} \left[X_i \left(\sum_{j=1}^{n} c_{ij} \varphi_j + g_i - \varphi_i' + f_i \right) \right] - r r' \\ &= \sum_{i=1}^{n} c_{ii} X_i^2 + \sum_{i=1}^{n-1} \sum_{j=i+1}^{n} (c_{ij} + c_{ji}) X_i X_j + \sum_{i=1}^{n} X_i f_i - r r'. \end{aligned}$$

The following estimates for $P(x,t)$ on W are valid:

(a) According to the conditions (13) we have

$$\begin{aligned} P(x,t) &\leq \sum_{i=1}^{n} c_{ii} X_i^2 + \sum_{i=1}^{n-1} \sum_{j=i+1}^{n} \frac{1}{2} \left| c_{ij} + c_{ji} \right| \left(X_i^2 + X_j^2 \right) + \sum_{i=1}^{n} X_i f_i - r r' \\ &= \sum_{i=1}^{n} \left[c_{ii} + \sum_{j=1 (j \neq i)}^{n} \frac{1}{2} \left| c_{ij} + c_{ji} \right| \right] X_i^2 + \sum_{i=1}^{n} X_i f_i - r r' < 0, \end{aligned}$$

and according to the conditions (14) we have

$$\begin{aligned} P(x,t) &\leq \sum_{i=1}^{n} \left[c_{ii} + \sum_{j=1 (j \neq i)}^{n} \frac{1}{2} \left| c_{ij} + c_{ji} \right| \right] X_i^2 + \sum_{i=1}^{n} X_i f_i - r r' \\ &\leq \sum_{i=1}^{n} (-u) X_i^2 + \sum_{i=1}^{n} r \left| f_i \right| - r r' = -u r^2 + r \sum_{i=1}^{n} \left| f_i \right| - r r' < 0 . \end{aligned}$$

(b) According to the conditions (15) we have

$$P(x,t) \geq \sum_{i=1}^{n} \left[c_{ii} - \sum_{j=1 (j \neq i)}^{n} \frac{1}{2} \left| c_{ij} + c_{ji} \right| \right] X_i^2 + \sum_{i=1}^{n} X_i f_i - r r' > 0 ,$$

and according to the conditions (16) we have

$$\begin{aligned} P(x,t) &\geq \sum_{i=1}^{n} \left[c_{ii} - \sum_{j=1 (j \neq i)}^{n} \frac{1}{2} \left| c_{ij} + c_{ji} \right| \right] X_i^2 + \sum_{i=1}^{n} X_i f_i - r r' \\ &\geq \sum_{i=1}^{n} u X_i^2 - \sum_{i=1}^{n} r \left| f_i \right| - r r' = u r^2 - r \sum_{i=1}^{n} \left| f_i \right| - r r' > 0 . \end{aligned}$$

According to Lemmas 1 and 2 in [5], the above estimates for $P(x,t)$ imply the statement of the Theorem. □

3. Applications

In this Section two examples (see [2]) and two remarks will be presented to illustrate the possibilities of applying the results obtained in the Section 2.

Example 1. For the model, which is known in plasma physics,

$$\begin{aligned} \dot{x}_1 &= -\nu_1 x_1 + M_1 x_2 x_3, \\ \dot{x}_2 &= -\nu_2 x_2 + M_2 x_1 x_3, \\ \dot{x}_3 &= -\nu_3 x_3 + M_3 x_1 x_2, \end{aligned} \tag{17}$$

where $\nu_i, M_i \in \mathbb{R},\ \nu_i > 0$, the following result holds.

Corollary 1. *Let* $\nu = \min\{\nu_1, \nu_2, \nu_3\}$. *If*

$$|M_1 + M_2 + M_3| \leqslant \frac{\nu}{r} + \frac{r'}{r^2},\ t \in I,$$

then every solution of (17), *passing through some* $(x_0, t_0) \in \omega$, *belongs to* ω *for* $t \geq t_0$, *where*

$$\omega = \left\{(x_1, x_2, x_3, t) \in \mathbb{R}^3 \times I : x_1^2 + x_2^2 + x_3^2 < r^2(t)\right\}. \tag{18}$$

For the function $r(t)$ we can take, for example,

$$r(t) = \alpha e^{-\alpha^2 t},\ 0 < \alpha < \nu,\ t > t_0 \geqslant \frac{1}{\alpha^2} \ln\left(\frac{|M_1 + M_2 + M_3|}{\nu - \alpha}\right).$$

Example 2. Let us consider the Lorenz model, which plays an important role in chemical kinetics, described by the equations:

$$\begin{aligned} \dot{x}_1 &= a(x_2 - x_1), \\ \dot{x}_2 &= bx_1 - x_2 - x_1x_3, \\ \dot{x}_3 &= -cx_3 + x_1x_2, \end{aligned} \tag{19}$$

where a, b, c are given real parameters.

Corollary 2. *Let* $a, c \in \mathbb{R}^+, p = \min\{1, a, c\}$. *If*

$$|a + b| < 2\left(p + \frac{r'}{r}\right),\ t \in I,$$

then every solution of (19), *passing through some* $(x_0, t_0) \in \omega$, *belongs to* ω *for* $t \geq t_0$, *where the set* ω *is defined by* (18) .

If $a \geqslant 1,\ c \geqslant 1,\ |a + b| \leqslant \frac{5}{3}$, for the function $r(t)$ we can take, for example, $r(t) = \alpha e^{-\sqrt{t}},\ \alpha \in \mathbb{R}^+,\ t > 9$.

Remark 1. The obtained results also contain an answer to the question on approximation of solutions $x(t)$ whose existence is established. The errors of the approximation are defined by the function $r(t)$ which can be arbitrarily small $\forall t \in I$.

Remark 2. The obtained results also give the possibility to discuss the stability (instability) of solutions of the system (1) . For example, under the conditions of Theorem 1 (a), every solution of (1) with initial value in ω is r-stable (stable with the function of stability r), if $r(t)$ tends to zero as $t \to \infty$ and $r'(t) < 0$, $t \in I$. However, if we consider the case (b), then established solution in ω is r-instable in case where $r'(t) > 0, t \in I$. The same conclusion holds for other Theorems. See the definitions 1 and 2 in [4].

References

[1] J.DIBLIK, *On Asymptotic behaviour of Solutions of Certain Classes of ordinary Differential Equations,* Journal of Differential Equations 95 (1992), 203-217.

[2] W.H.STEEB, F.WILHELM, *Non-Linear Autonomous Systems of Differential Equations and Carleman Linearization Procedure,* Journal of Mathematical Analysis and Applications 77 (1980), 601-611.

[3] B.VRDOLJAK, *On Classes of solutions of a quasi-linear system of differential equations and certain nonlinear oscillations,* J.Bulyai math. Soc., Budapest, (1988), 510-513.

[4] B.VRDOLJAK, *On behaviour and stability of system of linear differential equations,* Proceedings of the 2nd Congress of Croatian Society of Mechanics, Supetar, 1997, 631-638.

[5] B.VRDOLJAK, *On behaviour of solutions of system of linear differential equations,* Mathematical Communications 2 (1997), 47-57.

[6] T.WAŽEWSKI, *Sur un principe topologique de l'examen de l'allure asymptotique des intégrales des équations différentielles ordinaires,* Ann. Soc. Polon. Math. 20 (1947), 279-313.

HIGH-ORDER ENO AND WENO SCHEMES WITH FLUX GRADIENT AND SOURCE TERM BALANCING

Senka Vuković
University of Rijeka
51000 Rijeka, Vukovarska 58, Croatia
senka.vukovic@ri.hinet.hr

Luka Sopta
University of Rijeka
51000 Rijeka, Vukovarska 58, Croatia
luka.sopta@rijeka.riteh.hr

Abstract We developed a new set of numerical schemes particularly designed for hyperbolic conservation laws with significant source term. These schemes are based on one hand on the essentially non-oscillatory (ENO) and weighted essentially non-oscillatory (WENO) schemes (Harten, Osher, Engquist, Chakravarthy, Shu, Balsara) and on the other hand on the concept of the flux gradient and source term balancing (Bermùdez, Vázquez, Hubbard, LeVeque). In this paper we present results of extended numerical testing of the original ENO and WENO schemes and new schemes on one-dimensional shallow water equations. We perform computations using 2-step and 3-step Runge-Kutta time operator approximation and from lower to higher formal order reconstruction via primitive function for the space operator. On the basis of the obtained numerical results we examine effects of the scheme order increasing in interaction with the introduced improvement by the source term decomposition. We also discuss the time evolution of the numerical error due to variable bed depth in quiescent flow, steady state flow and unsteady flow for the original and for the newly developed schemes.

Keywords: hyperbolic conservation laws, source term, ENO schemes, WENO schemes, C-property, high-order computations

1. Introduction

In this paper we present and analyze two different approaches to the source term – flux gradient balancing [2, 3, 6, 10, 13] that we introduced into the finite difference essentially non-oscillatory (ENO, [4, 5, 12]) and weighted essentially non-oscillatory (WENO, [1, 8, 9, 11, 12]) schemes for the one-dimensional non-homogeneous hyperbolic systems of conservation laws in one-dimension.

Harten, Osher, Engquist, Chakravarthy and Shu [4, 5] developed the ENO schemes and Jiang and Shu [8] extended the idea to the WENO schemes. In the ENO and WENO schemes conservation law system is usually written in the form

$$\partial_t \mathbf{u} = \mathbf{L}(\mathbf{u}, x, t) \tag{1}$$

with $\mathbf{L}$ defined as

$$\mathbf{L}(\mathbf{u}, x, t) = -\partial_x \mathbf{f}(\mathbf{u}) + \mathbf{g}(\mathbf{u}, x, t). \tag{2}$$

Here t is the time, x is the space coordinate, $\mathbf{u}$ is the vector of the conserved variables, $\mathbf{f}$ is the flux and $\mathbf{g}$ is the source term.

For the time integration in (1) a TVD Runge-Kutta-type method is applied [4, 5, 12]. Furthermore, a space discretization with cells $[x_{i-1/2}, x_{i+1/2}]$, $i = 0, \ldots, N$ of uniform width Δx is assumed. Now, if an approximation $\mathbf{u}_i$, $i = 0, \ldots, N$ to the solution $\mathbf{u}$ at the cell center x_i and at any time t is known, the numerical approximations $\mathbf{L}_i$, $i = 0, \ldots, N$ are found as

$$\mathbf{L}_i = -\frac{1}{\Delta x}(\mathbf{f}_{i+1/2} - \mathbf{f}_{i-1/2}) + \mathbf{g}_i. \tag{3}$$

Here $\mathbf{f}_{i+1/2}$, $i = 0, \ldots, N-1$ is the numerical flux at the $(i + 1/2)^{th}$ cell boundary, and $\mathbf{g}_i$, $i = 0, \ldots, N$ is the numerical source term in the ith cell. In order to compute (3) the local characteristic fields are found, i.e. eigenvalues $\lambda^p_{i+1/2}$, left eigenvectors $\mathbf{l}^p_{i+1/2}$ and right eigenvectors $\mathbf{r}^p_{i+1/2}$, $p = 1, \ldots, m$ are evaluated using the approximation $\mathbf{A}_{i+1/2} = \mathbf{A}(\tilde{\mathbf{u}})$ of the local Jacobian matrix of the flux. Here m is the number of conservation laws in the system (1) and $\tilde{\mathbf{u}}$ is an average of the states $\mathbf{u}_i$ and $\mathbf{u}_{i+1}$, for example Roe's or van Leer's. In fact, in this paper we will use only the simple van Leer's average.

Up to now in the ENO and WENO schemes the numerical flux was evaluated using Roe or locally Lax-Friedrichs formulation and ENO or WENO reconstruction via primitive function, while the numerical source term was evaluated pointwise. The ENO and WENO reconstruction of the flux term are high order and shock-capturing but the pointwise evaluation of the source term in the conservation laws with significant right side can give poor numerical results. Particularly, numerical difficulties arise for conservation laws with a geometrical source term, as in conservation laws for shallow water flows, one-dimensional open channel flows, bed-load sediment transport flows, nozzle flows and similar. While in the steady state of any of those flows physically there is a balance between the flux gradient and the external forces i.e. the source term, numerical schemes with pointwise evaluation do not reflect this balance and in the computations steady flows are not conserved. Bermùdez and Vázquez introduced the concept of the exact or approximate conservation property (C-property) and a new approach to the source term modeling in the Q-schemes [3]. This concept requires the scheme to be consistent only with one particular type of steady flow, i.e. the quiescent flow. However, numerical schemes that verify C-property give significantly better results for all the steady and also unsteady flow computations. Here, we propose an application of this concept to the ENO and WENO schemes.

2. ENO and WENO Reconstruction via Primitive Function

ENO and WENO reconstruction [12] solve the following problem: *If values $v_k^{\pm}$, $k = 0, \ldots, N$ of a function $v^{\pm}$ in the cell centers x_k, $k = 0, \ldots, N$ are known, find a numerical approximation for $v^{\pm}$ at $x_{i+1/2}$.*

Both, the $(r+1)^{th}$ order ENO and $(2r+1)^{th}$ order WENO reconstruction on a uniform mesh of cells can be written as

$$v^{\pm}_{i+1/2,r} = \sum_{s=s^{\pm}_{min}}^{s^{\pm}_{max}} \omega_{r,s}(v^{\pm}) \sum_{j=0}^{r} a^{\pm}_{r,s,j} v^{\pm}_{i-r+s+j}. \tag{4}$$

Here the range of indices is bounded with $s^{+}_{min} = 0$, $s^{+}_{max} = r$ or with $s^{-}_{min} = 1$, $s^{-}_{max} = r+1$. The coefficients $a^{\pm}_{r,s,j}$, $j = 0, \ldots, r$, $s = s^{\pm}_{min}, \ldots, s^{\pm}_{max}$, have known values that for example can be found in [12]. Each weight $\omega_{r,s}(v^{\pm})$, $s = s^{\pm}_{min}, \ldots, s^{\pm}_{max}$ is computed from some smoothness indicator that measures smoothness of the function over the sth stencil of points

$$S^{\pm}_{r,s} = \{x_{i-r+s}, \ldots, x_{i+s}\}, \quad s = s^{\pm}_{min}, \ldots, s^{\pm}_{max}. \tag{5}$$

In fact the only difference between the ENO and WENO algorithm is in the evaluation of the weights. In the ENO reconstruction only one stencil is chosen for the reconstruction, while in the WENO reconstruction the weight values are distributed more evenly and then normalized. We can easily see that in both cases

$$\sum_{s=s^{\pm}_{min}}^{s^{\pm}_{max}} \omega_{r,s}(v^{\pm}) \sum_{j=0}^{r} a^{\pm}_{r,s,j} = 1. \tag{6}$$

The last fact regarding ENO and WENO reconstructions that we want to point out is that values of the weights will not change if any constant value is added to the function, i.e.

$$\omega_{r,s}(v^{\pm} + const.) = \omega_{r,s}(v^{\pm}), s = s^{\pm}_{min}, \ldots, s^{\pm}_{max}. \tag{7}$$

3. LLF Formulations

Here we present the LLF version of the ENO and WENO scheme in a form that is not common. The reason for this reformulation is that we wanted to make more obvious the transition from the standard ENO and WENO schemes towards the new balanced algorithms. Simple algebraic manipulations can lead us to the conclusion that the schemes presented in this section are actually the well-known ENOLLF or WENOLLF schemes [4, 5, 12, 1].

The numerical approximation for the flux component in the pth characteristic field in the LLF formulation is given with

$$f^{(p)}_{i+1/2} = \frac{1}{2}((\mathbf{f}_i+\mathbf{f}_{i+1})-\alpha^{(p)}_{i+1/2}(\mathbf{u}_{i+1}-\mathbf{u}_i))\cdot \mathbf{l}^{(p)}_{i+1/2}+P^{(p)}_{i+1/2,+}+P^{(p)}_{i+1/2,-}, \quad (8)$$

where the terms $P^{(p)}_{i+1/2,\pm}$ are polynomial corrections. They are calculated as ENO or WENO reconstructions for functions $v^{\pm}$, which are defined with their cell values

$$v^{\pm}_k = \frac{1}{2}((\mathbf{f}_k \pm \alpha^{(p)}_{i+1/2}\mathbf{u}_k) - (\mathbf{f}_{I^\pm} \pm \alpha^{(p)}_{i+1/2}\mathbf{u}_{I^\pm}))\cdot \mathbf{l}^{(p)}_{i+1/2}, k = 0, \dots, N. \quad (9)$$

Here $I^+ = i$, $I^- = i+1$ and $\alpha^{(p)}_{i+1/2}$ is some numerical approximation for the $\max_{[x_i,x_{i+1}]}|\lambda^{(p)}|$. This reconstruction means that $P^{(p)}_{i+1/2,\pm}$ is equal to the right hand side of the equation (4) with the stencil weights computed from the $v^{\pm}$ values:

$$P^{(p)}_{i+1/2,\pm} = \sum_{s=s^{\pm}_{min}}^{s^{\pm}_{max}} \omega_{r,s}(v^{\pm})\sum_{j=0}^{r} a^{\pm}_{r,s,j}v^{\pm}_{i-r+s+j}. \quad (10)$$

In this formulation only a simple pointwise evaluation is applied to the source term.

3.1 A Semi-Balanced LLF Formulation

We now propose the first modification to the LLF formulation that includes balancing between the source term and the flux gradient [2, 3, 6, 7, 10]. The algorithm for the new schemes is the following.

Our numerical approximation for the flux component in the pth characteristic field is still given with (8), but we use $\alpha^{(p)}_{i+1/2} = |\lambda^{(p)}_{i+1/2}|$ instead of the usual numerical approximations. With this particular choice the first order LLF formulation reduces to the van Leer's Q-scheme of Bermùdez and Vázquez [3]. So this choice for $\alpha^{(p)}_{i+1/2}$ gives us the basis for the construction of a well balanced algorithm.

Then, for the source term p^{th} characteristic field component we apply the decomposition

$$g_i^{(p)} = g_{i+1/2,+}^{(p)} + g_{i-1/2,-}^{(p)} \tag{11}$$

$$\begin{aligned} g_{i+1/2,\pm}^{(p)} &= \frac{1\mp\text{sgn}(\lambda_{i+1/2}^{(p)})}{2\Delta x}\mathbf{G}_{i,i+1}\cdot \mathbf{1}_{i+1/2}^{(p)}\pm \\ &\pm\frac{1}{\Delta x}Q_{i+1/2,+}^{(p)} \pm \frac{1}{\Delta x}Q_{i+1/2,-}^{(p)}. \end{aligned} \tag{12}$$

Here the term $\mathbf{G}_{j,k}$ is conservation law dependant. Formula (11) and the first term in (12) are the same as in the van Leer's Q-scheme of Bermùdez and Vázquez [3]. We add the second and the third term in (12) in order to achieve the balancing for ENO and WENO schemes. So, now the terms $Q_{i+1/2,\pm}^{(p)}$ are polynomial corrections and we compute them using ENO or WENO reconstructions. More precisely, while for $P_{i+1/2,\pm}^{(p)}$ we use functions $v^{\pm}$ defined with (9), for $Q_{i+1/2,\pm}^{(p)}$ we define functions $w^{\pm}$ with cell values

$$w_k^{\pm} = \frac{1}{2}(\mathbf{G}_{I^{\pm},k} \pm \text{sgn}(\lambda_{i+1/2}^{(p)})\beta_{i+1/2}\mathbf{Z}_{I^{\pm},k})\cdot \mathbf{1}_{i+1/2}^{(p)}, k = 0,\ldots,N. \tag{13}$$

The expressions for terms $\beta_{i+1/2}$ and $\mathbf{Z}_{j,k}$ are again conservation law dependent.

We propose to do the ENO or WENO reconstruction as follows. We compute the weights for each stencil using the values of the $v^{\pm}$ functions, we evaluate the polynomial correction $P_{i+1/2,\pm}^{(p)}$ as in (10) and finally we take the polynomial correction $Q_{i+1/2,\pm}^{(p)}$ also to be equal to the right hand side expression in (4), with the same weights but with $w^{\pm}$ values in place of the $v^{\pm}$ values:

$$Q_{i+1/2,\pm}^{(p)} = \sum_{s=s_{min}^{\pm}}^{s_{max}^{\pm}} \omega_{r,s}(v^{\pm})\sum_{j=0}^{r} a_{r,s,j}^{\pm} w_{i-r+s+j}^{\pm}. \tag{14}$$

The motivation for this choice of weights is the following. The numerical approximation $\mathbf{L}_i$, given with (3), contains the difference of the polynomial corrections $P_{i+1/2,\pm}^{(p)} - Q_{i+1/2,\pm}^{(p)}$. So, for the balancing of the algorithm, according with (6), it is fundamental that we use the same weights in (10) and (14). The first obvious idea is to use the same weights for the source term part as those that are used in the original ENO and WENO algorithms for the flux term part. This leads us to (14).

Such an algorithm we will call the semi-balanced and we will use abbreviations ENOSBLLF and WENOSBLLF.

3.2 A Balanced LLF Formulation

Finally, we propose the second balanced modification to the LLF formulation. Actually, the algorithm of this scheme differs from the one of the semi-balanced case only in one detail - we compute the weights for each stencil now using the values of the function $v^{\pm} - w^{\pm}$, and apply all the other relations from the previous subsection, i.e.:

$$P^{(p)}_{i+1/2,\pm} = \sum_{s=s^{\pm}_{min}}^{s^{\pm}_{max}} \omega_{r,s}(v^{\pm} - w^{\pm}) \sum_{j=0}^{r} a^{\pm}_{r,s,j} v^{\pm}_{i-r+s+j}. \tag{15}$$

$$Q^{(p)}_{i+1/2,\pm} = \sum_{s=s^{\pm}_{min}}^{s^{\pm}_{max}} \omega_{r,s}(v^{\pm} - w^{\pm}) \sum_{j=0}^{r} a^{\pm}_{r,s,j} w^{\pm}_{i-r+s+j}. \tag{16}$$

As we already mentioned in 3.1, our numerical approximation $\mathbf{L}_i$, given with (3), contains the difference of the polynomial corrections $P^{(p)}_{i+1/2,\pm} - Q^{(p)}_{i+1/2,\pm}$. This difference, with the choice of weights as we propose in (15) and (16), becomes exactly the ENO or WENO reconstruction of the function $v^{\pm} - w^{\pm}$. Thus, this is the most natural choice of weights for a well balanced algorithm.

We will use abbreviations ENOBLLF and WENOBLLF for the resulting schemes.

4. RF Formulations

4.1 RF Formulation with Pointwise Evaluation of the Source Term

As in the LLF case, we will give a somewhat atypical formulation of the classical ENORF and WENORF schemes. Again, the presented algorithm is identical to the well-known one and our choice of form is aimed to make transition to the balanced variations more obvious.

Typically RF formulation contains the LLF formulation, i.e. if the change in the sign of the eigenvalue occurs an entropy fix is accomplished by the application of the Lax-Friedrichs instead of the Roe formulation. Since we absolved the LLF formulation here we will concentrate only on the case when entropy fix is not needed, i.e. when $\lambda^{(p)}_i \lambda^{(p)}_{i+1} > 0$. In that case the Roe formulation gives

$$f^{(p)}_{i+1/2} = \begin{cases} \mathbf{f}_i \cdot \mathbf{l}^{(p)}_{i+1/2} + P^{(p)}_{i+1/2,+} & \text{if } \lambda^{(p)}_{i+1/2} > 0 \\ \mathbf{f}_{i+1} \cdot \mathbf{l}^{(p)}_{i+1/2} + P^{(p)}_{i+1/2,-} & \text{otherwise} \end{cases} \tag{17}$$

Here again terms $P^{(p)}_{i+1/2,\pm}$ are polynomial corrections. They are computed using ENO or WENO reconstructions for functions $v^{\pm}$ defined with

$$v^{\pm}_k = \frac{1 \mp \operatorname{sgn}(\lambda^{(p)}_{i+1/2})}{2} (\mathbf{f}_k - \mathbf{f}_{I^{\pm}}) \cdot \mathbf{l}^{(p)}_{i+1/2} \tag{18}$$

for $k = 0, \ldots, N$, but again using (10). The source term is of course pointwise evaluated.

4.2 A Semi-Balanced RF Formulation

In this case, we approximate the flux component using (17) and in order to achieve the balancing between the source term and the flux gradient we compute the source term components using (11) and

$$g_{i+1/2,\pm}^{(p)} = \frac{1 \mp \mathrm{sgn}(\lambda_{i+1/2}^{(p)})}{2\Delta x} \mathbf{G}_{i,i+1} \cdot \mathbf{l}_{i+1/2}^{(p)} \pm \frac{1}{\Delta x} Q_{i+1/2,\pm}^{(p)} \tag{19}$$

for $k = 0, \ldots, N$. Motivation for relation (19) is the same as for (12). Furthermore, we compute the polynomial corrections $P_{i+1/2,\pm}^{(p)}$ using (10) for $v^{\pm}$ given with (18) and the polynomial corrections $Q_{i+1/2,\pm}^{(p)}$ using (14) with $w^{\pm}$ given with (13). We give to these schemes abbreviated names ENOSBRF or WENOSBRF.

4.3 A Balanced RF Formulation

As in the LLF case, this formulation is also just a small modification of the semi-balanced one - the weights for each stencil now are computed using the values of function $v^{\pm} - w^{\pm}$, i.e. we use (15) and (16) while everything else is as in the previous subsection. We will refer to these schemes shortly as to the ENOBRF or WENOBRF schemes.

5. The One-dimensional Shallow Water Equations and the C-property

One case of the hyperbolic conservation laws system (1) with

$$\mathbf{u} = \begin{pmatrix} h \\ hv \end{pmatrix}, \mathbf{f} = \begin{pmatrix} hv \\ hv^2 + \frac{1}{2}gh^2 \end{pmatrix}, \mathbf{g} = \begin{pmatrix} 0 \\ -gh\frac{dz}{dx} \end{pmatrix}, \tag{20}$$

are the one-dimensional shallow water equations [2, 3, 6, 7]. In (20) $h = h(x,t)$ is the water depth, $v = v(x,t)$ is the water velocity, g is the acceleration due to gravity and $z = z(x)$ is the bed level. The source term in the second conservation law actually contains also a friction term $-gh\frac{M^2 v|v|}{h^{4/3}}$, with $M = M(x)$ the Manning's friction factor. However, this part of the source term can be evaluated pointwise with no consequence to the numerical accuracy of the computation. So, since this friction term has no significance for our discussion we will omit it.

The key concept in the modeling of the source term in the one-dimensional shallow water equations is the C-property. As Bermùdez and Vázquez defined it

[2, 3], the numerical scheme has the exact C-property if it is exactly compatible with the quiescent steady state solution

$$H = h + z = const. \text{ and } v = 0. \tag{21}$$

We claim that the semi-balanced and the balanced ENO and WENO schemes we proposed have the exact C-property. Thus, we have to prove that for all these schemes $\mathbf{L}_i = 0$, $i = 0, \ldots, N$ (see (3)) in the case of the quiescent flow (21). In order to do this we first have to apply the algorithms from Sections 3 and 4 to the one-dimensional shallow water equations. So we define terms $\mathbf{G}_{j,k}$, $\mathbf{Z}_{j,k}$ and $\beta_{i+1/2}$ as

$$\mathbf{G}_{j,k} = -g\bar{h}_{j,k}\mathbf{Z}_{j,k}, \mathbf{Z}_{j,k} = \begin{pmatrix} 0 \\ -(z_k - z_j) \end{pmatrix}, \beta_{i+1/2} = g\bar{h}_{i,i+1}, \tag{22}$$

where $\bar{h}_{j,k} = \frac{h_j + h_k}{2}$. Also in this application to (1) we assume only simple van Leer's averaging of the states in the local characteristic fields.

Now, let us observe that if in the semi-balanced and balanced schemes we omit the polynomial corrections all our LLF schemes become identical to the van Leer's Q-scheme of Bermùdez and Vázquez, and that scheme has the exact C-property [2, 3]. When in the RF case we omit the polynomial parts only upwinding occurs in the numerical flux and the proof of the C-property is straightforward. So we see that in all our new schemes the crucial part is to prove that polynomial corrections are balanced. In the light of the relations (4), (6) and (7) we are left with the simple task of computing the differences $v_k^{\pm} - w_k^{\pm}$, $k = 0, \ldots, N$.

5.1 The LLF Case

Taking into consideration the definitions of the conservation law dependent terms and the condition of the quiescent flow some computation leads us to

$$v_k^{\pm} - w_k^{\pm} = \frac{((-1)^p g\bar{h}_{I^{\pm},k} \pm g\bar{h}_{i,i+1})(H_k - H_{I^{\pm}})}{4\sqrt{g\bar{h}_{i,i+1}}} = 0. \tag{23}$$

5.2 The RF Case

Since in the case $\lambda_i^{(p)}\lambda_{i+1}^{(p)} \leq 0$ the LLF formulation is used, we only have to compute the differences when the Roe flux formulation is used. Similarly to the previous calculations we find that

$$v_k^{\pm} - w_k^{\pm} = \frac{(-1)^p g\bar{h}_{I^{\pm},k}(H_k - H_{I^{\pm}})}{2\sqrt{g\bar{h}_{i,i+1}}} = 0. \tag{24}$$

We want to point out that with (23) and (24) we proved the exact C-property for all the new schemes that we propose, with LLF or RF formulation, with ENO or WENO reconstruction applied. We particularly emphasize that the proof is valid for both semi-balanced and balanced variations.

6. Test Problems

6.1 The Quiescent Flow Proposed by the Working Group on Dam Break Modeling

In this test problem we use the bed geometry (Figure 1) proposed by the Working group on dam break modeling [6, 13]. The initial condition is the quiescent steady state (21) with water level $H = 10m$ and since no perturbations are introduced this state stays conserved with time. Computational space step and time step are $\Delta x = 2.5m$ and $\Delta t = 0.125s$.

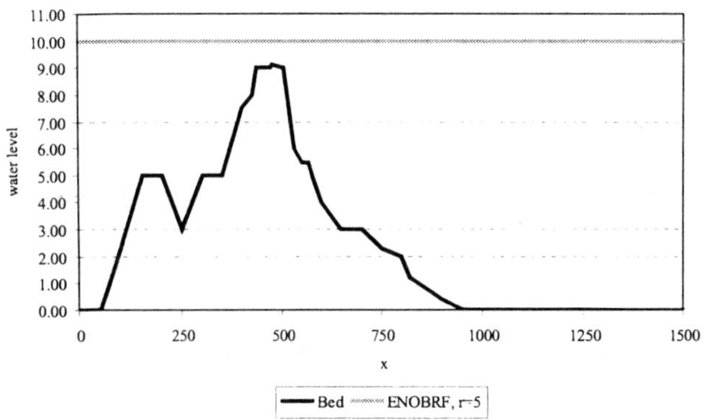

Figure 1. The quiescent flow as proposed by the Working group on dam break modeling (Test problem 6.1)

Table 1. Comparison of ENORF schemes for the quiescent flow as proposed by the Working group on dam break modeling (Test problem 6.1)

Scheme	*Velocity L^∞ Error*	*Water Level L^∞ Error*
ENORF,$r = 1$	$2.87 \cdot 10^{-1}$	$5.29 \cdot 10^{-2}$
ENOSBRF,$r = 1$	$1.55 \cdot 10^{-13}$	$1.49 \cdot 10^{-13}$
ENOBRF,$r = 1$	$1.22 \cdot 10^{-14}$	$1. \cdot 10^{-15}$
ENORF,$r = 5$	$4.99 \cdot 10^{-1}$	$1.66 \cdot 10^{-1}$
ENOSBRF,$r = 5$	$3.34 \cdot 10^{-1}$	$9.80 \cdot 10^{-2}$
ENOBRF,$r = 5$	$2.09 \cdot 10^{-14}$	$1.07 \cdot 10^{-14}$

6.2 The Steady Flow Over a Bump

The bed for this test problem (Figure 2) is given by

$$z(x) = \begin{cases} 0.2 - 0.05(x - 10)^2 \text{ if } 8 \leq x \leq 12 \\ 0 \text{ otherwise} \end{cases} . \tag{25}$$

A subcritical upstream boundary condition $hv(0,t) = 0.18m^3/s$ and a subcritical downstream condition $h(12,t) = 0.406m$ are imposed [6, 13]. This results in a steady transcritical flow with a smooth transition and a hydraulic jump. In the computations space step and time step are $\Delta x = 0.125m$ and $\Delta t = 0.01s$.

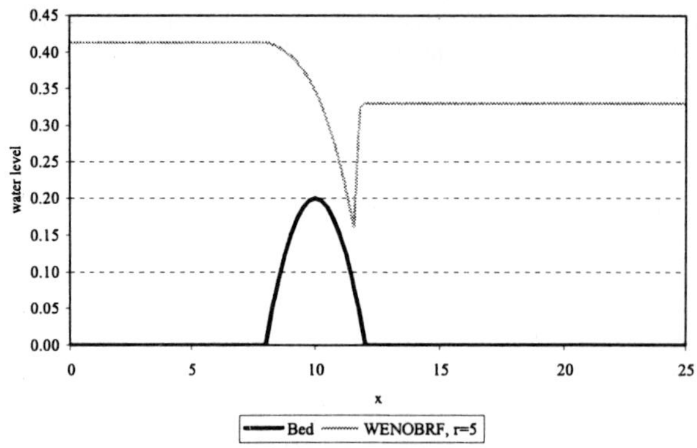

Figure 2. The steady flow with the hydraulic jump (Test problem 6.2)

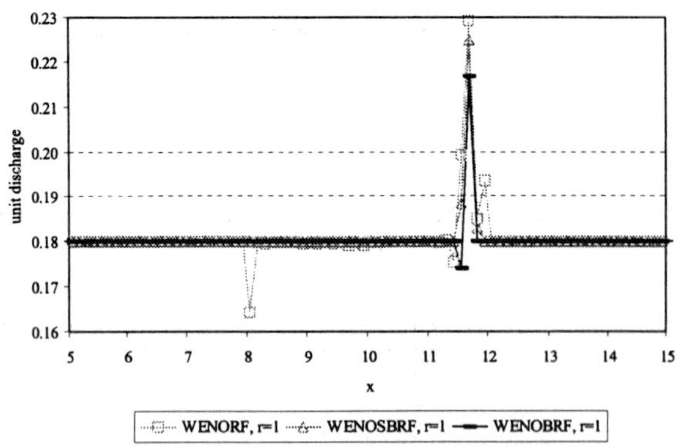

Figure 3. Comparison of WENORF, $r = 1$ schemes for the steady flow with the hydraulic jump (Test problem 6.2), unit discharge hv over the bump

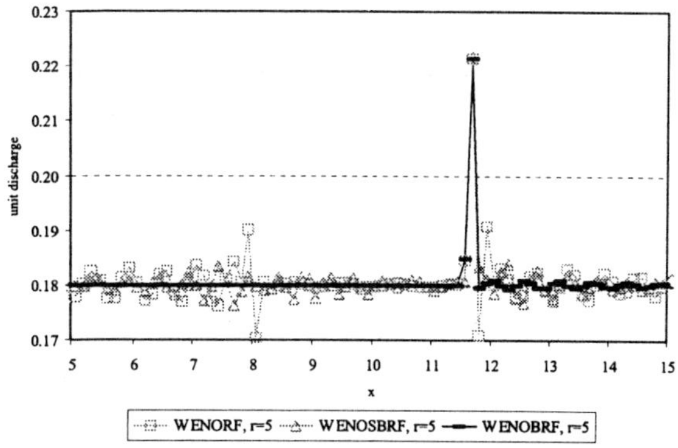

Figure 4. Comparison of WENORF, $r = 5$ schemes for the steady flow with the hydraulic jump (Test problem 6.2), unit discharge hv over the bump

6.3 A Problem Discussed by LeVeque

In this test problem [7] proposed by LeVeque (Figure 5) the bed is given with

$$z(x) = \begin{cases} 0.25(\cos(10\pi(x-0.5))+1) \text{ if } 1.4 \leq x \leq 1.6 \\ 0 \text{ otherwise} \end{cases} \tag{26}$$

and the initial conditions are given with

$$v(x,0) = 0 \text{ and } h(x,0) = \begin{cases} 1.2 - z(x) \text{ if } 1.1 \leq x \leq 1.2 \\ 1 - z(x) \text{ otherwise} \end{cases} . \tag{27}$$

Here no boundary conditions are imposed and the initial condition results in a non-stationary flow that can be described as a two dam break shocks, one travelling upstream and the other travelling downstream and passing over the bump in the bed. Computational space step and time step to be used are $\Delta x = 0.001m$ and $\Delta t = 0.0001s$.

7. Concluding Remarks

For each of the test problems described in Section 6 we performed computations using all the schemes defined in Sections 3 and 4. Even more, each scheme was applied in its low order variation, i.e. using 2-step Runge-Kutta-type time integration and ENO or WENO with $r = 1$ reconstruction and its high order variation, i.e. using 3-step Runge-Kutta-type time integration and ENO or WENO with $r = 5$ reconstruction. From the numerous obtained results we present only those that are significant. In fact, if in a scheme the ENO reconstruction is replaced with the WENO reconstruction the so produced differences in the numerical results are irrelevant. The same is true if the RF formulation

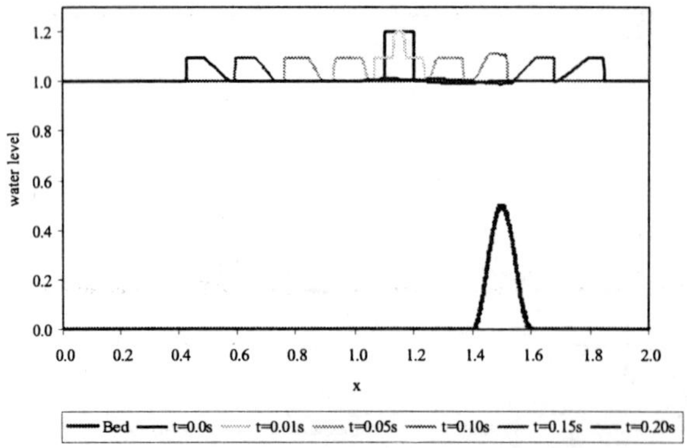

Figure 5. The problem discussed by LeVeque (Test problem 6.3)

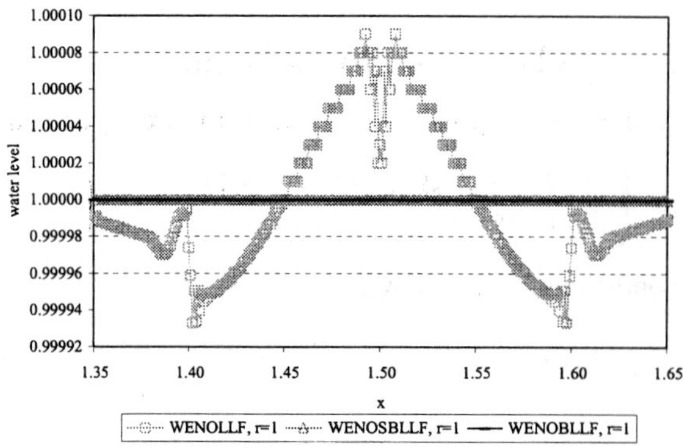

Figure 6. Comparison of order WENOLLF, $r = 1$ schemes for the problem discussed by LeVeque (Test problem 6.3), water level $h + z$ over the bump at $t = 0.04s$

is interchanged with the LLF formulation. However, significant difference in the computations comes from the pointwise evaluation of the source term vs. semi-balancing and balancing and from the low order vs. high order schemes.

First we can conclude that the strategy of the source term pointwise evaluation gives poor numerical results, that the semi-balancing strategy works well in the lower order schemes but has difficulty in higher order schemes, and that the balanced approach gives excellent results in all the cases.

This behavior of the semi-balanced schemes is interesting since we proved that these schemes have the exact C-property. But in these schemes the weights of the stencils are computed only on the base of the flux values, while in the balanced formulations the influence of the source term is included also in the

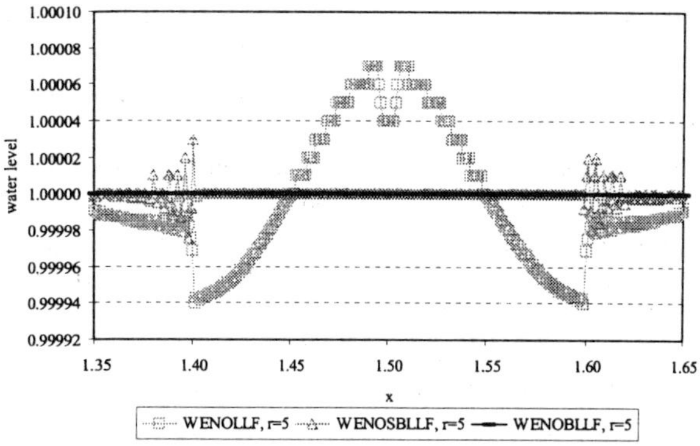

Figure 7. Comparison of order WENOLLF, $r = 5$ schemes for the problem discussed by LeVeque (Test problem 6.3), water level $h + z$ over the bump at $t = 0.04s$

computation of the stencil weights. This difference does not change the analytical C-property schemes but has a deeper physical meaning and as we can see significant numerical consequences in the computer computations.

The second conclusion is that if the bed slope is smooth, as in the test problem 6.3, the numerical error of all the schemes is very small and even the ENO and WENO schemes that do not have the C-property give acceptable results (Figures 6 and 7). This can be explained with excellent properties of the ENO and WENO reconstructions. However, when the bed slope is irregular only balanced schemes give good results. This is very significant in all applications since natural watercourses have irregular bed geometry.

Furthermore, in the test problem 6.2 all the presented schemes have some difficulty with the computation of the discharge in the zone of the hydraulic jump (Figures 3 and 4). But this is in fact a problem common to all known numerical schemes. Particularly this error produces oscillation in a larger area when higher order schemes are applied due to larger stencils, so a cure for this problem must yet be found or at least in these cases lower order schemes should be used.

References

[1] D. S. Balsara and C.-W. Shu, *Monotonicity preserving weighted essentially non-oscillatory schemes with increasingly high order of accuracy*, Journal of Computational Physics 160, 405 (2000), doi: 10.1006/jcph.2000.6443

[2] A. Bermùdez, A. Dervieux, J. A. Désidéri and M. E. Vázquez, *Upwind schemes for the two-dimensional shallow water equations with variable depth using unstructured meshes*, Comput. Methods Appl. Mech. Eng. 155, 49 (1998).

[3] A. Bermùdez and M. E. Vázquez, *Upwind methods for hyperbolic conservation laws with source terms*, Comput. Fluids 23(8), 1049 (1994).

[4] A. Harten and S. Osher, *Uniformly high-order accurate non-oscillatory schemes I*, SIAM Journal on Numerical Analysis 24, 279 (1987).

[5] A. Harten, B. Engquist, S. Osher and S. R. Chakravarthy, *Uniformly high-order accurate non-oscillatory schemes III*, Journal of Computational Physics 71, 231(1987).

[6] M. E. Hubbard and P. Garcia-Navarro, *Flux difference splitting and the balancing of source terms and flux gradients*, Numerical Analysis Report, University of Reading, Department of Mathematics, (1999).

[7] J. Hudson, *Numerical techniques for the shallow water equations*, Numerical Analysis Report, University of Reading, Department of Mathematics, (1999).

[8] G. Jiang and C.-W. Shu, *Efficient implementation of weighted ENO schemes*, Journal of Computational Physics 126, 202 (1996).

[9] G. Jiang and C. Wu, *A high-order WENO finite difference scheme for the equations of ideal magnetohydrodynamics*, Journal of Computational Physics 150, 561 (1999), doi: 10.1006/jcph.1999.6207

[10] R. J. LeVeque, *Balancing source terms and flux gradients in high-resolution Godunov methods: the quasi-steady wave propagation algorithm*, Numerical Analysis Report, Department of Applied Mathematics and Department of Mathematics, University of Washington, (1998).

[11] P. Montarnal and C.-W. Shu, *Real gas computation using an energy relaxation method and high-order WENO schemes*, Journal of Computational Physics 148, 59 (1999), doi: 10.1006/jcph.1998.6105

[12] C. -W. Shu, *Essentially non-oscillatory and weighted essentially non-oscillatory shock-capturing schemes for hyperbolic conservation laws*, NASA/CR-97-206253, ICASE Report No. 97-65, Institute for Comp. App. in Science and Eng., NASA Langley Research Center, (1997).

[13] M. E. Vázquez-Cendón, *Improved Treatment of Source Terms in Upwind Schemes for the Shallow Water Equations in Channels with Irregular Geometry*, Journal of Computational Physics 148, 499 (1999), doi: 10.1006/jcph.1998.6127

Index